Urban Traffic Analysis and Control

Other related titles:

You may also like

- PBTR026 | Li and He | Traffic Information and Control | November 2020
- PBTR018 | Burton and Stevens | Collection and Delivery of Traffic and Travel Information | November 2020

We also publish a wide range of books on the following topics:
Computing and Networks
Control, Robotics and Sensors
Electrical Regulations
Electromagnetics and Radar
Energy Engineering
Healthcare Technologies
History and Management of Technology
IET Codes and Guidance
Materials, Circuits and Devices
Model Forms
Nanomaterials and nanotechnologies
Optics, Photonics and Lasers
Production, Design and Manufacturing
Security
Telecommunications
Transportation

All books are available in print via https://shop.theiet.org or as eBooks via our Digital Library https://digital-library.theiet.org.

IET TRANSPORTATION SERIES 41

Urban Traffic Analysis and Control

The key challenges in the era of ITS

Edited by
Roberta Di Pace and Giulio E. Cantarella

The Institution of Engineering and Technology

About the IET

This book is published by the Institution of Engineering and Technology (The IET).

We inspire, inform and influence the global engineering community to engineer a better world. As a diverse home across engineering and technology, we share knowledge that helps make better sense of the world, to accelerate innovation and solve the global challenges that matter.

The IET is a not-for-profit organisation. The surplus we make from our books is used to support activities and products for the engineering community and promote the positive role of science, engineering and technology in the world. This includes education resources and outreach, scholarships and awards, events and courses, publications, professional development and mentoring, and advocacy to governments.

To discover more about the IET please visit https://www.theiet.org/

About IET books

The IET publishes books across many engineering and technology disciplines. Our authors and editors offer fresh perspectives from universities and industry. Within our subject areas, we have several book series steered by editorial boards made up of leading subject experts.

We peer review each book at the proposal stage to ensure the quality and relevance of our publications.

Get involved

If you are interested in becoming an author, editor, series advisor, or peer reviewer please visit https://www.theiet.org/publishing/publishing-with-iet-books/ or contact author_support@theiet.org.

Discovering our electronic content

All of our books are available online via the IET's Digital Library. Our Digital Library is the home of technical documents, eBooks, conference publications, real-life case studies and journal articles. To find out more, please visit https://digital-library.theiet.org.

In collaboration with the United Nations and the International Publishers Association, the IET is a Signatory member of the SDG Publishers Compact. The Compact aims to accelerate progress to achieve the Sustainable Development Goals (SDGs) by 2030. Signatories aspire to develop sustainable practices and act as champions of the SDGs during the Decade of Action (2020–30), publishing books and journals that will help inform, develop, and inspire action in that direction.

In line with our sustainable goals, our UK printing partner has FSC accreditation, which is reducing our environmental impact to the planet. We use a print-on-demand model to further reduce our carbon footprint.

Published by The Institution of Engineering and Technology, London, United Kingdom

The Institution of Engineering and Technology (the "**Publisher**") is registered as a Charity in England & Wales (no. 211014) and Scotland (no. SC038698).

First published 2024

The Institution of Engineering and Technology
Futures Place,
Kings Way, Stevenage,
Herts, SG1 2UA,
United Kingdom

www.theiet.org

British Library Cataloguing in Publication Data
A catalogue record for this product is available from the British Library

ISBN 978-1-83953-716-5 (hardback)
ISBN 978-1-83953-717-2 (PDF)

Typeset in India by MPS Limited

Cover image credit: Insung Jeon/Getty images

This book is dedicated to my partner, Silvio, the only person who has truly inspired me to "change my mind." I also dedicate it to our children, Claudio Javier and Liliana Audrey, who are my motivation in everything I do in life.

Roberta Di Pace

This book is dedicated to the late Richard E. Allsop, Emeritus Professor of Transport Studies at the University College London, who introduced me and several other scholars to the fine art of optimisation methods in traffic engineering.

Giulio Erberto Cantarella

Contents

List of contributors

Giulio Erberto Cantarella	Department of Civil Engineering, University of Salerno, Italy
Angelo Coppola	Department of Civil, Architectural and Environmental Engineering, University of Naples 'Federico II', Italy
Chiara Colombaroni	Department of Civil, Constructional and Environmental Engineering, Sapienza University of Rome, Italy
Stefano de Luca	Department of Civil Engineering, University of Salerno, Italy
Roberta Di Pace	Department of Civil Engineering, University of Salerno, Italy
Franco Filippi	Department of Civil Engineering, University of Salerno, Italy
Claudio Guarnaccia	Department of Civil Engineering, University of Salerno, Italy
Orlando Giannattasio	Department of Civil, Computer Science and Aeronautical Technologies Engineering, Roma Tre University, Italy
Anna Laura Pala	Department of Computer Engineering, Automatic and Management, Sapienza University of Rome, Italy
Luigi Pariota	Department of Civil, Architectural and Environmental Engineering, University of Naples 'Federico II', Italy
Maria Rella Riccardi	Department of Civil, Architectural and Environmental Engineering, University of Naples 'Federico II', Italy
Antonella Scarano	Department of Civil, Architectural and Environmental Engineering, University of Naples 'Federico II', Italy
Facundo Storani	Department of Civil Engineering, University of Salerno, Italy

The authors would like to thank their friend and colleague Nino Vitetta, with whom they have shared many of the topics covered in this book over the years.

Foreword

More than 40 years ago, I got a Civil Engineering degree in Transportation at the University of Naples discussing a thesis on methods for Traffic Control. Afterwards, I passed the examination to get membership to the Institution of Civil Engineering designing a signalised junction. More than 30 years ago I discussed a lesson on Traffic Control to become associate professor. You may see how much this topic means to me, even though I put it aside for several years, favouring other topics.

Roberta Di Pace 15 years ago – at that time a young post-doc researcher now an associate professor of Traffic Management – pushed me to go back to this topic asking me to become her protégé. She rather quickly surpassed her master with her clever mind and stubborn attitude towards difficulties, features common among people from her homeland who live long due to their positive relationship with an often harsh nature [Her homeland is where the Mediterranean Diet was established long since.]. Since then she has been working hard on Dynamic Traffic Management including Information Systems, as well as several other topics.

Two years ago I proposed to her to write this book as a sort of base camp whilst climbing the mountain of her academic career towards the deserved top. It is a privilege for me to have been invited to participate in this book with some contributions about basic themes.

Giulio Erberto Cantarella

Preface*

Nothing in life is to be feared, it is only to be understood. Now is the time to understand more, so that we may fear less.

Marie Curie

Outline. *In this chapter, the reader may find the contents and the purpose of this book. It delves into the theory of fixed and adaptive urban signal setting design, focusing on both single junctions and network levels. It examines the latest advancements in Intelligent Transportation Systems, especially the influence of connected and cooperative vehicles on traffic analysis and control. Finally, the book presents key strategies aimed at optimising impacts related to safety, fuel consumption, and emissions.*

Since the hunting-gathering era human brains have evolved to be more sensitive to variations in space and/or time of the surrounding environment rather than regularity and uniformity; (mostly unconscious) representations of location over space and evolution over time allowed human beings to survive in challenging conditions. Developing a (mathematical) model of real systems, as common in modern applied sciences, is a more conscious way to follow that ancestral attitude. Even though the future is perfectly determined by the past, according to Beowulf's well-known statement 'Fate will unwind as it must!' (but not to authors' opinion), still, it may not be perfectly forecasted due to lack of enough information about the past, to the uncertainty affecting forecasting methods, ... Thus, however desirable, in several cases, a precise model providing deterministic description and forecasting of system state cannot be developed, and the most general modelling tools include both dynamic and stochastic features together with space characterisation. It should be remarked that any kind of representation or model mentioned above is, as beauty, in the mind of the beholder; therefore dynamics or stochasticity are features of (mathematical) models only, a sort of social constructions agreed by the modeller community, not be confused

*The contents of this book are intended to be consistent with and complementary to those of Cantarella *et al.* (2019, 2024). Accordingly, the following preface is largely based on the prefaces of these books.

Cantarella, G. E., Watling, D., De Luca, S., and Di Pace, R. (2019). *Dynamics and Stochasticity in Transportation Systems: Tools for Transportation Network Modelling*. Elsevier.
Cantarella, G. E., Fiori, C., and Mussone, L. (2024). *Dynamics and Stochasticity in Transportation Systems Part II: Equations and Examples*. Elsevier.

with the object of their applications, such variations in space and/or time in a real system. Along this line of reasoning, observations of the real world are facts, whilst models are opinions about them.

The focus of this book is the use of mathematical modelling methods to assist in the understanding, prediction, policy assessment, and design of transportation systems; but what is a 'transportation system', or most pertinently what do we mean by it for the purposes of this book? First, it contains the infrastructure, the pavements to walk on, the roads on which we cycle, drive or may use a bus or taxi, the train tracks, as well as the fleets of buses, airplanes, and trains that are used to run services and transport goods. Second, it contains the users of the transportation system, namely the people who choose where, when and how to travel, as well as the goods operators and suppliers who decide how and where to transport their goods. Third, it contains the various public and private organisations responsible for planning, operating, pricing, and providing information on the infrastructure. Such a 'system' contains many interacting elements. A traveller may decide to drive to their normal place of work during a busier (congested) period of the day than they would normally, and by doing so contributes additionally to the congestion for that day. This congestion may delay other road users who are using some of the same roads, but perhaps travelling between an entirely different origin and destination to the first traveller. The additional delay experienced may, on the other hand, hold back the second traveller so that traffic is in fact more freely running than it might be on some downstream stretch of road on their intended route. This may cause the responsive traffic signals at an intersection to trigger at a different time, and so influence some other travellers. On the other hand, our second traveller has such a bad experience of travelling that day that they decide to try a different route when they make that trip next time, whereas the first traveller decides that re-adjusting their departure time would be wise in the future. As this happening, a private transport operator decides to introduce a new high-speed train service in the area, which our first traveller then decides to use on some subsequent day, thus alleviating some of the pressure on road capacity. At the same time as all these interactions are ongoing, each minute of every day, a transport planner is deciding on how to adjust to achieve some policy objective, and as a result, introduces on some subsequent day a new high-occupancy vehicle lane for certain hours of the day.

This book focuses on providing models and methods to support the management and control of specific portions (single junctions, arterials, small sub-networks) of the system. Indeed, given some geometrical details of the case study layout and some input data (such as the entry flows etc.), the models and methodologies discussed in the book can be applied for the comparison of different solutions.

The book's catchment users are expected to be closer to transportation engineering thus it is necessary to provide a whole overview about both traffic flow analysis and traffic control without focusing and limiting the book to traffic control. Furthermore, the goal is also to support undergraduate and graduate students who are engaged with these topics, in this sense some of the contributions presented in this book have appeared in a preliminary version in Italian.

P.1 Purpose of this book

Travel and transportation play a central role in the lives of most of the world's population. Transportation provides both a means of trade in moving goods and a way of moving people to engage in employment, education, social, and other activities. If we had observed the same geographical area over a period of past decades, we would likely have seen that the size and structure of employment, production, and residential areas had changed over time, and these changes had in turn changed the requirements and pressures on the transportation system. At the same time, these changes will have made environmental, social, and economic impacts, some positive and some negative, with some winners and some losers. It is natural then to ask whether we can hold a mirror to the past, and use it to see into the future; at least then we may be able to react in a better way to the inevitable changes the mirror shows us, and thereby as a society expend resources more efficiently (in the sense of less negative and more positive impacts). It is only one more step to then realise that the mirror analogy is limited, that unless we believe the future is pre-determined we may influence it by our actions, both as individuals and as organisations. Understanding such influences and their likely consequences then provides a way of not only 'managing' a transportation system more effectively, but also positively engineering it to improve the lives of the people using it.

The current book fits into this wide area of 'transportation planning', particularly the field of traffic flow modelling, and control that directly can support the definition of traffic management policies.

Space and time are two intrinsically important aspects of understanding travellers' needs and what transportation systems can supply. Let us first consider space. The type and density of activities are not distributed evenly across a city or region, and there are fixed geographical features (rivers, mountains, valleys, etc.) that influence the feasibility of different transportation options across an area. Dense, 'vertical' residential areas provide very different challenges to more sparsely distributed ones. There are also complex interactions that play out in the transport infrastructure; a congested road or overcrowded bus may be partially the result of travellers avoiding overloaded facilities, meaning that a good solution will not be understood without considering system-level interactions between the various travel needs of people/organisations and the services and facilities which are provided. Over the last 50 years, the transportation community has developed rather sophisticated ways of representing these kinds of spatial interactions, typically by representing the infrastructure as a network (mathematical 'graph'), and by considering various levels of sophistication in representing the behavioural responses of travellers (e.g., from the perfectly informed traveller to random utility approaches). At the same time, however, it should be mentioned that whilst the individual fields have developed to a high level, it is relatively rare to find a consistent integration of demand modelling and network modelling. There are rather well-developed (if not always consistent) methods, then, for considering 'space'; so what about 'time'? Whilst travel time, as a disincentive in making travel choices, is a central aspect of transportation planning, by the word 'time' we are instead referring here to changes that occur over time ('dynamics'). As there is considerable potential for confusion, let us very early on make a clear distinction: changes on a 'within-day' time-scale are the kind of changes that we would expect to see as we made a journey on a particular day, or if we compared our travel experience with someone travelling by the same route/service but at a different time on that day (there are many other ways to characterise this kind of time, but these examples suffice for now). On the other hand, changes on a 'between-day' time-scale concern, for example, the way in which we might adapt our travel choice next time we make a journey, based on our travel experiences today. Whilst researchers have been aware of both 'within-day' and 'between-day' effects for several decades, it is only relatively recently that a concerted effort has been made to develop tools and methods to explicitly model them. On the within-day scale, this has been achieved by introducing and adapting methods from traffic flow theory for use in network models. On the between-day scale, it has involved bringing in new techniques from both applied mathematics (for deterministic dynamical systems) and probability theory (for stochastic processes).

The subject area of the book will be Traffic Engineering and in particular Traffic Analysis and Control. The methods for traffic analysis, usually derived from Traffic Flow Theory, will be discussed.

Under steady-state conditions, the most used model to describe vehicles flowing along a street (railway, airway, ...) is the so-called fundamental diagram (FD) whilst when steady-state conditions do not hold, within-day dynamics should explicitly be considered through three kinds.

Regarding the methods for traffic flow control, these are based on the description and prediction of traffic flows without any modelling of routing choice behaviour. They include the fixed timing strategies, as well as the variable timing strategies. The latter may be applied offline to support transportation planning or in simple cases when there is no need for adaptive control or online to support real-time traffic management. Online applications require sensors for flow monitoring and within-day dynamic models for flow prediction, or simple data-driven methods.

All these topics are within the scope of this book.

The book is conceived as a research monograph and at the same time intends to be both a textbook and a reference work for transportation academic researchers and upper-level undergraduate and graduate students as well as professionals and consultants.

It will be the first to present the theory concerning the fixed and adaptive urban signal setting design at a single junction and network level. Furthermore, the most recent enhancements about the Intelligent Transportation Systems and the impact of connected and cooperative vehicles on traffic analysis and control will be considered. Finally, the main strategies aiming at the impacts (i.e., safety, consumption, emissions) optimisation will be also included.

P.2 Contribution of this book

Most branches of engineering were founded on physics (and/or chemistry) developed from the late 19th century, and now are well-established. This is the traditional image to common folks of an engineer: a person able to solve practical problems that are well rooted in a specific background, for instance, electronics, hydraulics, etc., through specialised mathematical tools. Good examples within transportation engineering are the analysis and design of components such as vehicles (and their engines), facilities, etc., and traffic engineering, developed by applying a metaphor derived from fluid dynamics, which deals with the behaviour of several vehicles sharing the same facility and the design of traffic control devices, such as traffic lights, ATC, etc.

At the beginning of the 50s of the last century, a new paradigm was introduced, by linking together the contributions of several authors, leading to (abstract) systems engineering, where emphasis is on mathematical representation of a problem rather than its physical background. This is a new type of engineer: a person able to solve a practical problem considering it as a whole through an ever-increasing box of non-specialised mathematical tools.

For what may now be considered a charming synchronicity, John G. Wardrop, in his seminal presentation held on 24 January 1952, and published in June (Wardrop, 1952), founded transportation systems engineering, including both analysis and design. In his paper, he proposed a wide and comprehensive review of traffic engineering, but at the same time, he understood that traffic engineering techniques can be used only to analyse the performance of a single component. He also stated that a transportation system cannot be studied on a single-element basis, but as a whole system indeed (note that he actually used the word 'system').

Hence, starting from a small example, he proposed his now widely known two principles to model travel demand distribution over alternative routes in a transportation network. Then, he stated that these two criteria must be extended to deal with a whole network, where routes are broken into links possibly shared by other routes, even though in the case of a network of roads the theoretical problem becomes very complicated. This way, John G. Wardrop introduced the main elements of any effective model of a transportation system:

- a user behaviour model, which simulates how level-of-service, say journey times, affects user choices, as expressed in his paper by the two criteria (travel demand);
- a performance model, which simulates how user choices, say flows, affect the level of service, say journey times; it is made up of a network model representing topological features and, at a link level, by performance–flow relations derived by applying traffic engineering techniques (transportation supply).

Besides, he greatly stressed *The Value of a Theoretical Approach* as the only effective one, thus stating that models within a specific theory should be developed. These models, now referred to as travel demand assignment to transportation networks, are the basic tools to simulate a transportation system. It is also worth noting that, pointing out that the user behaviour is likely selfish and does not lead towards the most efficient pattern, he stated the need for supply network design.

From the 'seed planted' by J.G. Wardrop the still-growing tree of the modern Traffic and Transportation Theory (TTT) emerged. A general overview of existing problems and tools of TTT is given below to point out the contribution of this book. TTT studies the interactions between the level of service provided by transportation systems and the results of several types of user choice behaviour, which may be regarded in a hierarchical order from bottom to top:

- driving, concerning interactions between users travelling on the same facility and their effects on travel time, ...;
- routing, concerning connections between the origin and destination of the journey, parking location and type, and possibly departing time, ...;
- travelling, concerning transportation mode, time-of-day, destination, frequency, ...;
- mobility, concerning car ownership, driving license acquisition, ...

On top of the above hierarchy, there are the kinds of user behaviour addressed by land-use/transport interaction theories.

Tools of TTT have reached a very advanced and sophisticated level, and large-scale applications are a current practice. Most of these tools are based on explicitly behavioural modelling approaches, which grant clear interpretation of parameters. A taxonomy is given in Table P.1 where for brevity's sake *kinds of choice behaviour other than routing and driving have not been explicitly considered.*

A brief review of these tools is given in the four sub-sections below to introduce the nomenclature used in this book and the contents of the chapters of this book.

P.2.1 Traffic analysis

This sub-section discusses methods for traffic analysis, which addresses the effects of driving choice behaviour, and are usually derived from Traffic Theory.

Under steady-state conditions, the most used model to describe vehicles flowing along a street (railway, airway, ...) is the so-called fundamental diagram (FD) describing the relations among density, flow and (space average) speed. In the stable regime speed is a decreasing function of flow, that can be used to specify travel time functions.

When steady state conditions do not hold, within-day dynamics in a link should explicitly be considered through three kinds of macroscopic models:

Table P.1 Tools of TTT

Modelled behaviour	Problems methods	
	Analysis descriptive – predictive	**Decision prescriptive**
Driving	Traffic analysis (P.2.1)[a]	Traffic control (P.2.3)
Driving and Routing	Transportation systems analysis (P.2.2)	Transportation systems control and design (P.2.4)

[a]Numbers in round parenthesis represent the corresponding sub-section below.

- continuous in space and time;
- discrete in space and continuous in time;
- discrete in space and time.

The full specification of all the above models requires an equation describing the relationship between speed and flow, or between speed and density, derived from the FD, as well as some network equations to develop within-day dynamic assignment models.

All these models are described in Part I of the book.

P.2.2 Transportation systems analysis

This sub-section briefly discusses methods for transportation systems analysis, which can be distinguished into methods for:

- travel demand analysis,
- transportation supply analysis,
- demand-supply interaction analysis, or assignment.

Before applying any of the above methods some preliminary steps should be carried out. The study area is delimited and divided into zones, where a journey starts or ends, and main infrastructures and services are singled out to support journeys between any pair of them. Then, users are distinguished, following a 5W approach, with respect to

WHO: socio-economic characteristics and grouped into categories (or into commodities for freight),
WHY: trip purpose,
WHAT: trip frequency,
WHERE: trip origin and destination (for simplicity's sake we will assume that each journey is defined by a single trip, thus trip-chains or tours are not considered),
WHEN: time of day (morning vs. afternoon peak period, day of week vs. weekend days, winter vs. summer, special events, usual vs. emergency conditions, …).

Once trip origins and destinations have been singled out, itineraries between each pair of origin and destination can be defined, possibly distinguished by category, purpose, …. Then, each itinerary can be broken down into links, each link being a stretch of street, railway, airway, …, with common characteristics. In the most general case, an itinerary is a routing strategy including both pre-trip and en-route choices depending on information available to users.

All these topics are out of the scope of this book.

P.2.3 Traffic control

Most methods for traffic flow control are based on the description and prediction of traffic flows without any modelling of routing choice behaviour.

They include fixed sign strategies, such as priority junctions or roundabouts designed offline, as well as variable sign strategies, such as ramp metering or traffic lights. The latter may be applied offline to support transportation planning or in simple cases when there is no need for adaptive control or online to support real-time traffic management.

Online applications require dynamic data for flow monitoring and within-day dynamic models for traffic flow prediction.

Furthermore, online transportation systems control methods can also be combined with Intelligent Transportation Systems (ITS) such as Advanced Transportation Information Systems (ATIS), which may provide information or indications. In this case, any information or indication provided to users should be consistent with the control strategy and the user reaction leading to closed-loop systems; satisfactory models of such systems are still an open issue. Currently, no method is available to design all features of a transportation system, possibly including an ITS.

All these topics are described in Part II of the book.

P.2.4 Transportation Systems Control and Design

Methods for transportation systems control and design provide optimal features of transportation interventions considering their effects on travel demand, say on user route (and departure time if the case) choice behaviour (quite often under the assumption of constant demand flows). This is usually achieved by considering any model for equilibrium assignment as a constraint embedded within the whole optimisation model underlying the design method.

Methods for transportation systems or better transportation supply control and design may be grouped into:

- Methods for transportation systems control or transportation network capacity design, or for Continuous Network Design, such as network signal setting design with equilibrium constraints, bus frequency design, . . . ;
- Methods for transportation network topology design, or for Discrete Network Design, such as Urban Network Design, including both signal setting design and lane allocation with equilibrium constraints (lane allocation cannot consistently be carried out without including signal setting design too), design of bus lines, ...

According to an arguable but long-since-established tradition, measures regarding restricted area access policy, parking policy, tolls, congestion charge, and the like, are collectively known as Travel Demand Management (TDM) measures. Needless to say, travel demand resulting from the free choices made by users cannot be centrally managed, but only be described. These kinds of interventions should be considered part of transportation supply design within a consistent plan also including the kinds of interventions mentioned above.

All these topics are out of the scope of this book.

P.3 Scope of this book

On deciding on the scope of the book, we decided to make several restrictions, omitting explicit treatment of several important topics, on the grounds that we wish the content of the book to be of a manageable size for the reader to appreciate. This is not, however, intended to suggest a boundary of what is possible with the presented framework; on the contrary, we hope that our work encourages others to cast future modelling advances within the presented conceptual frame.

One of the main limitations of the book is its omission of short-term traffic prediction models. Traditional traffic monitoring systems rely on fixed measurement stations that capture data on flow, occupancy, and sometimes speed. However, the availability of new data sources, such as Floating Car Data (FCD) obtained from GPS-enabled vehicles and mobile devices, opens up fresh opportunities for developing innovative predictive models. As previously mentioned, this approach is not covered in the book.

The book is structured in such a way as to provide the reader with all the tools necessary for understanding therefore this book requires no previous knowledge on the part of the reader. In particular, it focuses on theoretical aspects and does not delve deeply into the practical and numerical elements. The authors therefore reserve the right to publish a second volume in this series, which will instead focus specifically on more operational aspects.

It is organised into four parts and focuses on three main topics: the Traffic Flow Theory, the Traffic Control, the Intelligent Transportation Systems, and the Impacts/Externalities evaluation. The contents of each part/chapter are summarised below.

Part I: Traffic Flow Theory

Chapter 1: It provides a review of common basic definitions and notations on the part topic

Chapter 2: It provides a review of traffic flow steady-state models

Chapter 3: It provides a review of traffic flow dynamic models

Part II: Traffic Control

Chapter 4: It provides a review of common basic definitions and variables on the part topic

Chapter 5: It provides a review of priority junction analysis

Chapter 6: It provides a review of roundabout analysis

Chapter 7: It provides a review of signalised junction analysis

Chapter 8: It provides a review of signalised junction steady-state control (static control) [including Appendix A about the definition of some basic elements]

Chapter 9: It provides a review of signalised junction dynamic control (dynamic control)

Part III: Intelligent Transportation Systems

Chapter 10: It provides a review of advanced traffic management systems

Chapter 11: It provides a review of advanced traveller information systems

Chapter 12: It provides a review of advanced driving assistance systems

Chapter 13: It provides a review of cooperative connected and automated mobility ecosystem

Part IV: Impacts

Chapter 14: It examines externalities in traffic management strategy design

Chapter 15: It provides a review of safety models

Chapter 16: It provides a review of noise models

Chapter 17: It provides a review of emissions and energy consumptions models

Appendix A: It provides a review of control theory basic tools

Postface

Major findings

The book explores the theory behind fixed and adaptive urban signal setting design at both the single junction and network levels. It also addresses the latest advancements in Intelligent Transportation Systems, particularly the effects of connected and cooperative vehicles on traffic analysis and control. Lastly, the book will include key strategies aimed at optimizing impacts such as safety, fuel consumption, and emissions.

Further readings

For completeness, a wide description of most of the tools of TTT may be found in the books by Cascetta (2001, 2009) and Cantarella *et al.* (2019, 2024).

The overview of the historical development of traffic flow theory is provided by Meyer (1979) and Gartner (1983). In more detail, the earliest studies on traffic theory date back to the 1930s with Greenshields, B. D. (1935), who began to identify the representative variables of vehicle outflow. Following this, the first studies on modelling emerged, developed from an in-depth analysis of the relationship between flow, density, and speed (Lighthill and Whitham (1955); Whitham, G. B. (1955); Lighthill, M. J., and Whitham, G. B. (1955); Meyer, M. D. (1956); Wattleworth, J. D. (1956); Greenberg, H. (1959)).

More in general, concerning the traffic analysis and control, other books somehow addressing the same topics are enlisted below:

Ferrara, A., Sacone, S., and Siri, S. (2018). *Freeway Traffic Modelling and Control* (Vol. 585). Berlin: Springer.

Reilly, W. (1997). Highway capacity manual 2000. *TR News*, 193, pp. 22–23.

Tang, K., Boltze, M., Nakamura, H., and Tian, Z. (2019). *Global Practices on Road Traffic Signal Control: Fixed-Time Control at Isolated Intersections*. Cambridge: Elsevier.

Treiber, M., and Kesting, A. (2013). *Traffic Flow Dynamics: Data, Models and Simulation*. Berlin, Heidelberg: Springer.

Concerning the approaches for short-term traffic predictions they can be classified in: explicit and implicit traffic modeling. The explicit approach utilizes mathematical models to represent the interactions among the physical variables that characterize traffic phenomena; some of these models are discussed in the first part of the book. In contrast, the implicit approach derives dynamic relationships directly from time series of observed data and is commonly referred to as a data-driven approach. As previously mentioned, this approach is not covered in the book. The main details of these approaches are discussed in Fusco *et al.* (2016a, 2016b), Isaenko *et al.* (2017).

References

Cantarella, G. E., Fiori, C., and Mussone, L. (2024). *Dynamics and Stochasticity in Transportation Systems Part II: Equations and Examples*. Amsterdam: Elsevier.

Cantarella, G. E., Watling, D., De Luca, S., and Di Pace, R. (2019). *Dynamics and Stochasticity in Transportation Systems: Tools for Transportation Network Modelling*. Amsterdam: Elsevier.

Cascetta E. (2001). *Transportation Systems Engineering: Theory and Methods*. Dordrecht: Kluwer Academic Publisher, 2001.

Cascetta E. (2009). *Transportation Systems Analysis: Models and Applications*. Berlin: Springer.

Ferrara, A., Sacone, S., and Siri, S. (2018). *Freeway Traffic Modelling and Control* (Vol. 585). Berlin: Springer.

Fusco, G., Colombaroni, C., and Isaenko, N. (2016a). Short-term speed predictions exploiting big data on large urban road networks. *Transportation Research Part C: Emerging Technologies*, 73, 183–201.

Fusco, G., Colombaroni, C., and Isaenko, N. (2016b). Comparative analysis of implicit models for real-time short-term traffic predictions. *IET Intelligent Transport Systems*, 10(4), 270–278.

Gartner, N. H. (1983). The historical development of traffic flow theory. *Journal of Transportation Engineering*, 109(2), 159–175.

Greenberg, H. (1959). Traffic flow theory. *Traffic Engineering and Control*, 1(2), 73–79.

Greenshields, B. D. (1935). A study of traffic capacity. *Highway Research Board Proceedings*, 14, 448–477

Isaenko, N., Colombaroni, C., and Fusco, G. (2017). Traffic dynamics estimation by using raw floating car data. In *2017 5th IEEE International Conference on Models and Technologies for Intelligent Transportation Systems (MT-ITS)* (pp. 704–709). Piscataway, NJ: IEEE.

Lighthill, M. J., and Whitham, G. B. (1955a). On kinematic waves. II. A theory of traffic flow on long crowded roads. *Proceedings of the Royal Society A*, 229 (1178), 317–345.

Lighthill, M. J. and Whitham, G. B. (1955b). On kinematic waves. I. Flood movement in long rivers. *Proceedings of the Royal Society A*, 229(1178), 281–316.

Meyer, M. D. (1956). Traffic flow theory. *Highway Research Board Bulletin*, 139, 40–62.

Meyer, M. D. (1979). History of traffic flow theory. *Transportation Research Record*, 722, 7–11.

Reilly, W. (1997). Highway capacity manual 2000. *Tr News*, (193).

Tang, K., Boltze, M., Nakamura, H., and Tian, Z. (2019). *Global Practices on Road Traffic Signal Control: Fixed-Time Control at Isolated Intersections*. Amsterdam: Elsevier.

Treiber, M., and Kesting, A. (2013). Traffic flow dynamics. *Traffic Flow Dynamics: Data, Models and Simulation*, Berlin: Springer, 227, 228.

Wardrop, J. G. (1952). Some theoretical aspects of road traffic research. *Proceedings of the Institution of Civil Engineers*, 1(3), 325–378.

Wattleworth, J. D. (1956). Traffic flow theory and characteristics. *Traffic Engineering*, 26(3), 32–36.

Whitham, G. B. (1955). On kinematic waves: II. A theory of traffic flow on long crowded roads. *Proceedings of the Royal Society A*, 229(1178), 317–345.

Acknowledgement

This project is the result of collaborative work with Prof. Cantarella, to whom I express my gratitude for his trust, support, and concrete contributions to its realisation. Alongside Prof. Cantarella, I also thank my colleague (and friend) Stefano de Luca, with whom I have explored various topics discussed in this book that related to the research activities of the Sustainable Transportation Systems Engineering and Mobility Laboratory (ISTMoS).

Furthermore, many of the subjects developed in the different chapters stem from the research conducted during master's thesis and doctoral studies (including special thanks to Silvio Memoli, my husband, and Facundo Storani, young researcher who has been working with us for several years to whom I wish the best in life).

Of course, I would like to thank all those who contributed to the writing of this book whose names are detailed in the list of contributors. Last but not least, a special thanks to all my colleagues at the ISTMoS laboratory, with particular gratitude to Chiara Fiori and Francesca Bruno. While only some of them contributed directly to this book, each one has played an essential role in my journey.

This book aims to provide tools for the analysis and design of innovative solutions in urban traffic control and to support students and professionals addressing these issues. The content is therefore the result of a cultural project that has led us to introduce traffic analysis and control themes into our master's degree programs. I would like to thank not only our department, where this cultural project is rooted and developed but especially all the students with whom and for whom this project has taken shape.

Over the past decade, I have had the opportunity to engage in various experiences, particularly in research projects such as MOST – the Sustainable Mobility National Research Center, and the NSFC Cooperation with China. Additionally, I have been involved in extracurricular activities, notably serving on the boards of the Italian Society of Transport Professors, the Intelligent Transportation Systems Society Italian Chapter, and the IEEE Women in Engineering Italy Affinity Group. Each of these experiences has contributed to my personal growth and enrichment, fostering the interdisciplinary and cross-cutting perspective that informs the themes explored in this book.

Lastly, I would like to thank the IET production service for their invaluable support and the opportunity they have provided.

Roberta Di Pace

I wish to thank the late Prof. G. Improta, and the recently retired Prof. A. Sforza, who opened me to the vast world of optimisation and greatly supported me during the first 12 years of my academic career just after my graduation.

I wish to thank Prof. A. Vitetta, a good friend of mine since he was a student, with whom I spent much good time in pleasant conversations about what makes life worth living as well as in lively discussions about several issues of Traffic and Transportation Theory.

Giulio E. Cantarella

This book was carried out within the MOST – Sustainable Mobility National Research Centre and received funding from the European Union Next-GenerationEU (PIANO NAZIONALE DI RIPRESA E RESILIENZA (PNRR) – MISSIONE 4 COMPONENTE 2, INVESTIMENTO 1.4 – D.D. 1033 17/06/2022, CN00000023).

Additionally, it received partial funding from: the DIGIT-CCAM project "Progetti di Rilevante Interesse Nazionale" (PRIN 2020 – MUR) and the I CAN BE project "Progetti di Rilevante Interesse Nazionale" (PRIN 2022 – MUR), the local grant n. ORSA214124, ORSA238719, ORSA180377 (University of Salerno).

About the editors

Roberta Di Pace is an associate professor at the University of Salerno, Italy. She is a member of the IEEE; a board member of the Intelligent Transportation System Society – Italian Chapter; and a board member of the IEEE WIE AG Italy. She received the MSc and PhD degrees in Transportation from the University of Naples "Federico II", in 2005 and 2009, respectively. Since 2010 she has been a member of the Laboratory of Transport Systems Engineering and Sustainable Mobility (ISTMoS) within the Department of Civil Engineering at the University of Salerno (DiCiv- UniSA), Italy. She has been an associate professor in Transportation Engineering with the DiCiv- UniSA since 2021. Her research and teaching activities focus on connected cooperative and automated mobility, intelligent transportation systems; advanced traveller information systems; traffic flow modelling and control; network optimisation; transportation networks modelling and simulation; and smart/sustainable mobility. She serves as an Academic Editor for the *Journal of Advanced Transportation, IEEE Transactions on Intelligent Transportation Systems, Mathematical Problems in Engineering*, and *Smart Cities*.

Giulio E. Cantarella is a full professor in Transportation Engineering at the University of Salerno, Italy. On July 26th 1980 he received a Civil Engineering degree in Transportation at the School of Engineering of the University of Naples Federico II, and on November 1st 1994 a full professorship in Transportation Systems. From 1992 to 1994 he was associate professor of Transportation Systems. From 1990 to 1992 he was assistant professor of Operations Research. From 1986 to 1989 he was a fixed-term associate professor of Operations Research. Since November 1st 1999 he has worked at the Department of Civil Engineering of the University of Salerno. He is a member of the Laboratory of Transport Systems Engineering and Sustainable Mobility (ISTMoS) within the Department of Civil Engineering at the University of Salerno (DiCiv-UniSA), Italy. His research and teaching activities focus on traffic assignment, dynamics and stochasticity in transportation systems, network design, traffic analysis and control, and choice behavior. He serves as a Consulting *Editor for Transportmetrica B: Transport Dynamics* and an Associate Editor for the *Journal of Advanced Transportation*.

Part I

Traffic flow analysis

Chapter 1

Basic definitions and notations

Giulio Erberto Cantarella[1]

Make everything as simple as possible, but not simpler.

Albert Einstein

Outline. *This chapter reviews the basic definitions and assumptions about the main traffic observable variables, namely the input data of the modelling stage of the (macroscopic) Traffic Flow Theory. The main relationships among them are introduced and analysed.*

The basic elements of Traffic Flow Theory (TFT) are briefly reviewed below. As said in the introduction a complete discussion of this topic is out of the scope of this book; a recent comprehensive presentation of most of the topics of Traffic Flow Theory is in Treiber and Kesting (2013) (more references at the end of the chapter).

It is worth noting that there is no such thing as a commonly agreed notation, thus each variable below may well be denoted by a different symbol by other authors. In the next chapters, the main existing models of Traffic and Flow Theory will be reviewed and analysed according to definitions and notations introduced in this chapter, distinguishing steady-state and within-day dynamic modelling approaches. Most notations and definitions below, as well as models of TFT presented in the next chapters, are consistent with Di Pace (2019).

Phenomena occurring along a link (*running links*) are discussed in Section 1.1 in accordance with the *uninterrupted flow theory*, and those occurring at a bottleneck (*queuing links*) in Section 1.2 in accordance with the *interrupted flow theory*. In this chapter reference is made to traffic along an urban street or a road, still, most results almost straightforwardly hold for other transportation systems.

1.1 Basic observable variables: running links

Vehicles travelling along the same stretch of road infrastructure, **running links**, may interact with each other and the level of interaction depends on the demand. In particular, if the demand is great enough that the interaction may affect the link performances in terms of mean speed and travel time the congestion phenomenon

[1]Department of Civil Engineering, University of Salerno, Italy

occurs. The main observable variables, or input data, are defined below, and then the relationships between some of them are introduced.

Basic observations over space, a road segment $[x, x+\Delta x]$, and time, an interval $[t, t+\Delta t]$, are the trajectories of each vehicle as described in Figure 1.1:

$x_i(t)$ is the position or the abscissa of (the front of the) vehicle i at time t;
$v_i(x, t)$ is the (longitudinal) speed of vehicle i at time t while traversing point (abscissa) x.

Other observations are the longitudinal and the lateral acceleration, ... Aggregate variables are described below, distinguishing between those at a spot of space and those at an instant of time.

For traffic observed at the instant of time t, for instance, though aerial images, in a road segment $[x, x+\Delta x]$ the following variables can be defined (in logical order) (Figure 1.2):

$n(t; x, x+\Delta x)$ is the number of vehicles at time t between points x and $x+\Delta x$;
$s_i(t)$ is the spacing between vehicles i and $i-1$ at time t; that is the front spacing vehicle to vehicle at time t;
$s(t) = \Sigma_{i=1,\dots,n}\, s_i(t)/n(t; x, x+\Delta x)$ is the mean spacing, among all vehicles between points x and $x+\Delta x$ at time t;
$k(t; x, x+\Delta x) = n(t; x, x+\Delta x)/\Delta x$ is the density between points x and $x+\Delta x$ at time t, measured in vehicles per unit of length;

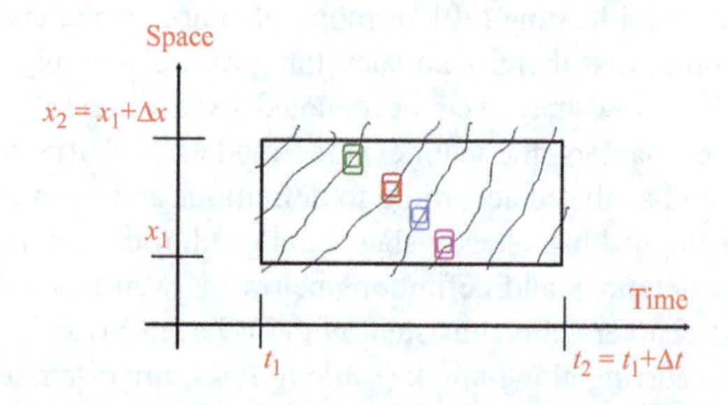

Figure 1.1 Vehicle trajectories over space and time for a running link

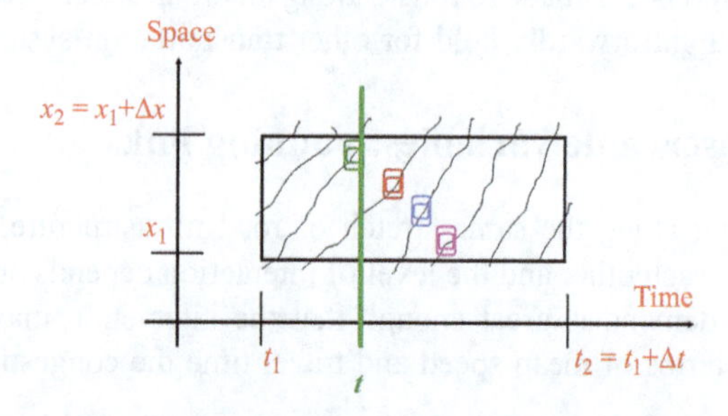

Figure 1.2 Observations at time t for a running link

$v_S(t) = S_{i=1,\ldots,n} v_i \,/ n(t;\ x,\ x+\Delta x)$ is the space mean speed, among all vehicles between points x and $x+\Delta x$ at time t.

For traffic observed at a point with abscissa s, for instance from a gate portal, during the time interval $[t,\ t+\Delta t]$, the following variables can be defined (in logical order) (Figure 1.3):

$m(x;\ t,\ t+\Delta t)$ is the number of vehicles traversing point x during time interval $[t,\ t+\Delta t]$;
$h_i(x)$ is the headway (temporal spacing) between vehicles i and $i-1$ crossing point x;
$h(x) = S_{i=1,\ldots,m} h_i(x) \,/\, m(x;\ t,\ t+\Delta t)$ is the mean headway, among all vehicles crossing point x during time interval $[t,\ t+\Delta t]$;
$f(x;\ t,\ t+\Delta t) = m(x;\ t,\ t+\Delta t)/\Delta t$ is the flow of vehicles crossing point x during the time interval $[t,\ t+\Delta t]$, measured in vehicles per unit of time (also called volume);
$v_T(x) = S_{i=1,\ldots,m} v_i(x) \,/\, m(x;\ t,\ t+\Delta t)$ is the time mean speed, among all vehicles crossing point x during time interval $[t,\ t+\Delta t]$.

Other observable features of a running link are:

f_{MAX} the maximum flow, say the capacity, such that $f \leq f_{MAX}$; capacity is commonly assumed a function of geometrical characteristics of the infrastructure (HCM, 2016);
k_{MAX} the maximum density, often called jam-density k_{iam}, $k \leq k_{MAX}$; maximum density depends on (average) vehicle length and minimum safety distance;
v_o the zero-flow or free-flow speed v_f, say the maximum speed v_{MAX}, such that $v \leq v_o$ maximum speed capacity is commonly assumed a function of geometrical characteristics of the infrastructure (as well as weather and light conditions, ...).

The level of service of a running link may be described by the average travel time, say the time needed to traverse the link (given the time of entry), that can be obtained from the speed values defined above.

Among the aggregate variables introduced above a *flow conservation equation* for a running link holds, it may be formulated as:

$$n(t; x, x + \Delta x) + m(x; t, t + \Delta t) = n(t + \Delta t; x, x + \Delta x) + m(x + \Delta x; t, t + \Delta t) \tag{1.1}$$

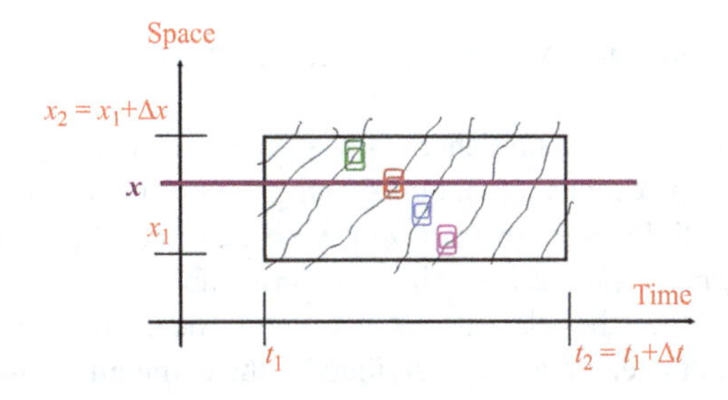

Figure 1.3 Observations at abscissa x for a running link

$$\text{or } \Delta n(x, x + \Delta x; t, t + \Delta t) + \Delta m(x, x + \Delta x; t, t + \Delta t) = 0 \tag{1.2}$$

where

$\Delta m(x, x+\Delta x;\, t, t+\Delta t) = m(x+\Delta x;\, t, t+\Delta t) - m(x;\, t, t+\Delta t)$ is the variation of the number of vehicles between points x and $x+\Delta x$ during Δt;

$\Delta n(x, x+\Delta x;\, t, t+\Delta t) = n(t+\Delta t;\, x, x+\Delta x) - n(t;\, x, x+\Delta x)$ is the variation of the number of vehicles during time interval $[t, t+\Delta t]$ over space Δx.

The flow conservation equation may be generalised to

$$\Delta n(x, x + \Delta x, t, t + \Delta t) + \Delta m(x, x + \Delta x, t, t + \Delta t) = \Delta z(x, x + \Delta x, t, t + \Delta t) \tag{1.3}$$

where

$\Delta z(x, x+\Delta x, t, t+\Delta t)$ be the difference between the number of entering and the number of exiting vehicles during the time interval $[t, t+\Delta t]$, due to entry/exit points (e.g., on/off ramps), between points x and $x+\Delta x$;

Furthermore, let

$\Delta f(x, x+\Delta x, t, t+\Delta t) = \Delta m(x, x+\Delta x, t, t+\Delta t)/\Delta t$ be the variation of the flow over space;

$\Delta k(x, x+\Delta x, t, t+\Delta t) = \Delta n(x, x+\Delta x, t, t+\Delta t)/\Delta s$ be the variation of the density over time;

$\Delta e(x, x+\Delta x, t, t+\Delta t) = \Delta z(y, x+\Delta x, t, t+\Delta t)/\Delta t$ be the (net) entering/exiting flow.

The *general flow conservation* equation (1.3) divided by Δt (omitting independent variables for simplicity's sake) leads to:

$$\Delta n/\Delta t + \Delta f = \Delta e \tag{1.4}$$

Then (1.4) divided by Δs leads to:

$$\Delta k/\Delta t + \Delta f/\Delta x = \Delta e/\Delta x \tag{1.5}$$

Both (1.1) and (1.2) as well as the three equations (1.3), (1.4), and (1.5) are equivalent formulations of the *flow conservation constraint.*

1.2 Basic observable variables: queuing links

Vehicles waiting to cross at a bottleneck, say a flow interruption point (intersections, toll barriers, merging sections, etc.), queue along the upstream stretch of road infrastructure, **queuing links**, may interact with each other and the level of interaction depends on the demand (Figure 1.4). The main observable variables, or input data, are defined below, and then the relationships between some of them are introduced.

Traffic at a bottleneck may be described by the cumulative number of arriving or departing vehicles over time (Figure 1.5). Each time a vehicle joins or leaves the

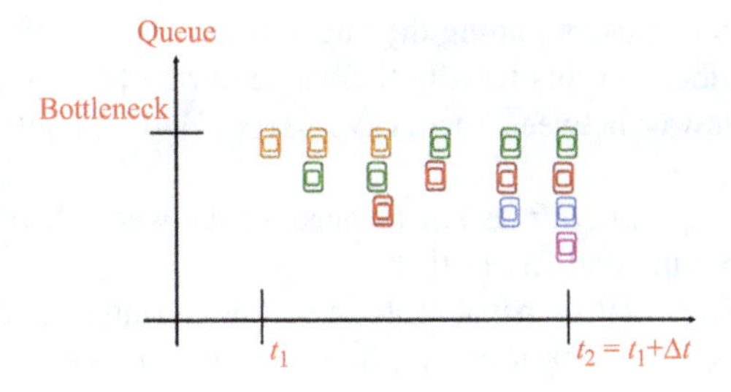

Figure 1.4 Queuing vehicles over time for a queuing link

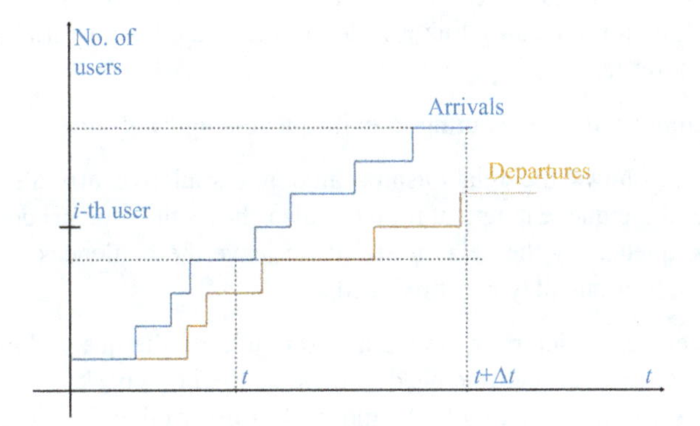

Figure 1.5 Cumulative number of arriving or departing vehicles over time

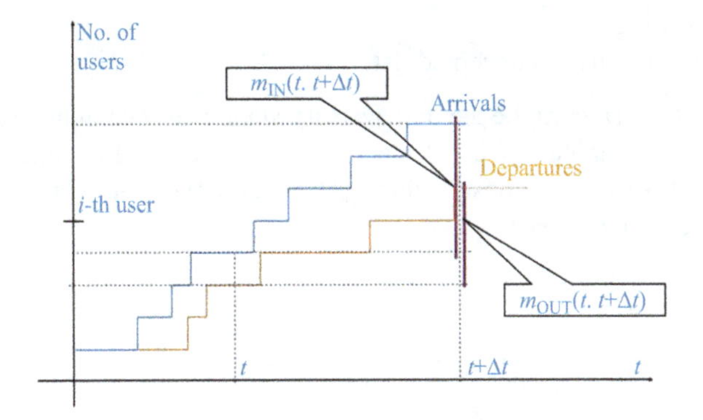

Figure 1.6 Observations for a queuing link

queue a jump can be observed. Of course, the cumulative number of arrivals is always greater than or equal to the cumulative number of departures.

The main variables describing the queuing phenomena over time are (in logical order) (Figure 1.6):

$m_{IN}(t, t+\Delta t)$ number of users joining the queue during $[t, t+\Delta t]$;
$m_{OUT}(t, t+\Delta t)$ number of users leaving the queue during $[t, t+\Delta t]$;
$h_i = t_i - t_{i-1}$ headway between successive users i and $i-1$ joining the queue at times t_i and t_{i-1};
$h(t, t+\Delta t) = S_{i=1,\dots,m}\, h_i/m_{IN}(t, t+\Delta t)$ mean headway between all vehicles joining the queue in the time interval $[t, t+\Delta t]$;
$f_{IN}(t, t+\Delta t) = m_{IN}(t, t+\Delta t)/\Delta t$ arrival (entering) flow during $[t, t+\Delta t]$;
$f_{OUT}(t, t+\Delta t) = m_{OUT}(t, t+\Delta t)/\Delta t$ exiting flow during $[t, t+\Delta t]$;

It is worth noting that the main difference with the basic variables of running arcs is that space $[x, \Delta x]$ is no longer explicitly referred to, but most variables maintain their definitions. Indeed for a running link space is described by the fixed length Δx, whilst for a queuing link it is described by the length variable over time.

Furthermore, let

$n(t)$ be the number of users waiting to exit (queue length) at time t.

Figure 1.7 shows the relationships among cumulative arrivals, cumulative departures, and the queue length at time t. It also shows the time needed for vehicle i to leave the queue, say the delay it affords to leave the bottleneck.

Further definitions may be introduced:

$t_s(t, t+\Delta t)$ average service time among all users joining the queue during the time interval $[t, t+\Delta t]$, say the time need to traverse the bottleneck;
$t_w(t, t+\Delta t)$ average total waiting time among all users joining the queue during the time interval $[t, t+\Delta t]$, say the time spent queuing.

The level of service of a queuing link may be described by the average delay experienced by vehicles that queue to cross the bottleneck, affected by the number of vehicles waiting.

As far as capacity is concerned, let

$f_{MAX}(t, t+\Delta t) = 1/t_s(t, t+\Delta t)$ be the (transversal) *capacity* or maximum exit flow, i.e., the maximum number of users that may be served in the time unit; in the following, it is assumed constant during $[t, t+\Delta t]$ for simplicity's sake (otherwise Δt can be duly redefined);

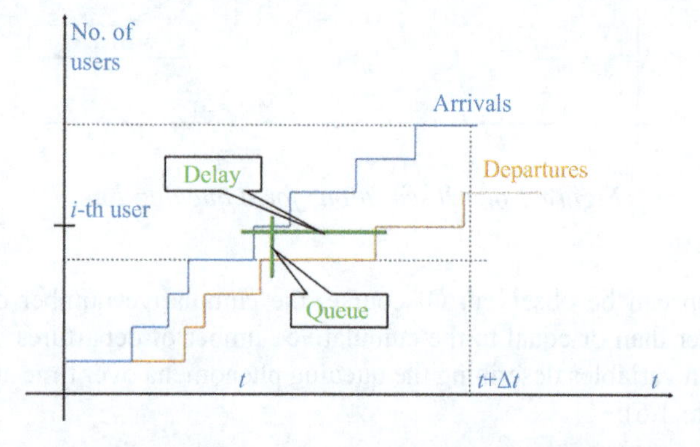

Figure 1.7 Queue length and delay for a queuing link

n_{MAX} is the *storage* (or longitudinal) *capacity* or the maximum number of users that may form the queue before upstream spillback occurs; it is assumed constant over time.

The capacity constraint on exiting flow is expressed by:

$$f_{OUT} \leq f_{MAX} \tag{1.6}$$

The *flow conservation equation* is given by:

$$n(t) + m_{IN}(t, t + \Delta t) = n(t + \Delta t) + m_{OUT}(t, t + \Delta t) \tag{1.7}$$

to be compared with the above Equation (1.1) for running links.

Dividing (1.7) by Δt leads to:

$$\Delta n/\Delta t + [f_{OUT}(t, t + \Delta t) - f_{IN}(t, t + \Delta t)] = 0. \tag{1.8}$$

to be compared with the above Equation (1.4) with $\Delta e = 0$.

1.3 Summary

This chapter introduces basic observable variables of the Traffic Flow Theory distinguishing running and queuing links.

Several technologies are available for data collection, the most effective being images from fixed cameras or moving cameras mounted on drones; these images can be elaborated to provide disaggregate data such as the trajectory of each vehicle in the observation area as well as the speed and acceleration values. These observations are the main observable variables for microscopic analysis. Aggregate observations, such as flow, density, and average speed, ... can be obtained from disaggregate data by simple algebra, they are the main variables for macroscopic analysis.

All the observations described above can be applied to distinguishing traffic per lane (right, middle, left), per type of vehicle (cars, motorcycles, vans, trucks, buses, ...), and other relevant factors. As said above, in this chapter reference has been made to traffic along an urban street or a road, still, most results almost straightforwardly hold for other transportation systems.

References

Some suggestions for further readings are reported below among the huge literature available on these topics.

Daganzo, C. F. (1997). *Fundamentals of Transportation and Traffic Operations.* Bingley: Emerald Group Publishing Limited.

Di Pace, R. (2019). Introduction to the Traffic Flow Theory, appendix 2 in G. E. Cantarella, D. P. Watling, S. de Luca, and R. Di Pace, *Dynamics and Stochasticity in Transportation Systems: Tools for Transportation Network Modelling*. Amsterdam: Elsevier.

Guerrieri, M., and Mauro, R. (2021). *A Concise Introduction to Traffic Engineering: Theoretical Fundamentals and Case Studies*. Cham: Springer.

Kessels, F. (2019). *Traffic Flow Modelling: Introduction to Traffic Flow Theory Through a Genealogy of Models*. Cham: Springer.

Leutzbach, W. (1988). *Introduction to the Theory of Traffic Flow* (Vol. 47). Berlin: Springer.

Transportation Research Board. (2000). *Highway Capacity Manual*. Washington, DC: Transportation Research Board.

Treiber, M., and Kesting, A. (2013). *Traffic Flow Dynamics: Data, Models and Simulation* (Vol. 1). Berlin: Springer-Verlag.

Chapter 2
Steady-state models

Giulio Erberto Cantarella[1]

Slow and steady wins the race.

English proverb

Outline. *This chapter reviews the basic definitions and assumptions about the basic models of the (macroscopic) Traffic Flow Theory under the assumption of steady-state traffic conditions.*

Basic models of (macroscopic) Traffic Flow Theory (TFT) under steady-state conditions are reviewed below. As said in the introduction a complete discussion of this topic is out of the scope of this book; as said in the previous chapter, a recent comprehensive presentation of most of the topics of Traffic Flow Theory is in Treiber and Kesting (2013) (more references at the end of the chapter).

Phenomena occurring along a link (*running links*) are discussed in Section 2.1 in accordance with the *fundamental diagram theory*, and those occurring at a bottleneck (*queuing links*) in Section 2.2 in accordance with the *queueing theory*. In both cases, the real discrete vehicle traffic is assumed described by a mono-dimensional continuous fluid that can be compressed up to a maximum finite value, so that its behaviour is in between a perfect gas, which can be compressed to infinite density, and a perfect liquid, which cannot be compressed at all.

2.1 Stationary models: running links

Modelling approaches of running links under *steady-state conditions* are based on the assumption that the flow during the time analysis interval does not depend on the spatial abscissa, the density and the (space average) speed over the stretch of analysis space do not depend on the instant of time. The main variables are:

$f \geq 0$ the (vehicular) flow (also called volume, sometimes denoted by q);
$k \geq 0$ the (vehicular) density (sometimes denoted by ρ);
$v \geq 0$ the (space average) speed, which may be approximated by the harmonic mean speed.

[1]Department of Civil Engineering, University of Salerno, Italy

Remark. *For simplicity's sake, the same notations are used for both observable variables, described in the previous chapter (1.1), and the modelling variables described in this chapter. Still, the reader must carefully distinguish between their meanings.*

Under steady-state conditions, the following equation holds among the main variables:

$$f = kv \tag{2.1}$$

2.1.1 Running links – deterministic models

Under steady-state conditions, the most commonly used (macroscopic) model to describe vehicles flowing along a street is the so-called Fundamental Diagram (FD) describes the relations among density, flow and (space average) speed, apart from (2.1). When steady state conditions do not hold, within-day dynamic macroscopic models are used, including FD as one of the main equations.

Fundamental Diagram can be dated back to 1935, when the first publication on speed-flow curves 'A Study of Traffic Capacity', presented at the 14th Annual Meeting of the Highway Research Board in 1935 (Greensheilds, 1935), was published,

Under steady-state conditions, a relationship holds between speed and flow:

$$v = v(f) \in [0, vo] \quad 0 \le f \le f_{MAX} \tag{2.2}$$

As said above, two values of speed correspond to each value of flow:

- *stable regime*, high speed and low density, $v(f)$ is monotone decreasing;
- *unstable regime*, low speed and high density, $v(f)$ is monotone increasing.

Combining together (2.1) and (2.2) a monotone decreasing function can easily be defined:

$$v = vv(k) \in [0, vo] \quad 0 \le k \le k_{MAX} \tag{2.3}$$

Similarly, a relationship between flow and density can easily be defined:

$$f = f(k) \in [0, f_{MAX}] \quad 0 \le k \le k_{MAX} \tag{2.4}$$

Main parameters are

f_{MAX} the maximum flow, say the capacity, $f \le f_{MAX}$; capacity is commonly assumed a function of geometrical characteristics of the infrastructure (HCM, 2016);

k_{MAX} the maximum density, often called jam-density k_{iam}, $k \le k_{MAX}$; maximum density depends on (average) vehicle length and minimum safety distance;

v_o be the zero-flow or free-flow speed v_f, say the maximum speed v_{MAX}, $v \le v_o$ maximum speed capacity is commonly assumed a function of geometrical characteristics of the infrastructure (as well as weather and light conditions, ...) (Figure 2.1).

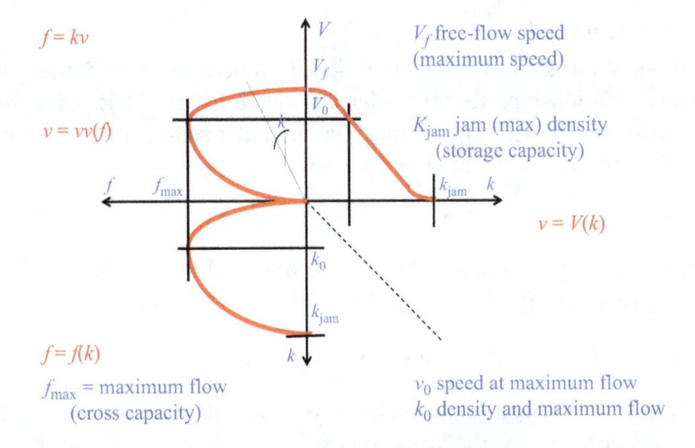

Figure 2.1 Functions of the FD

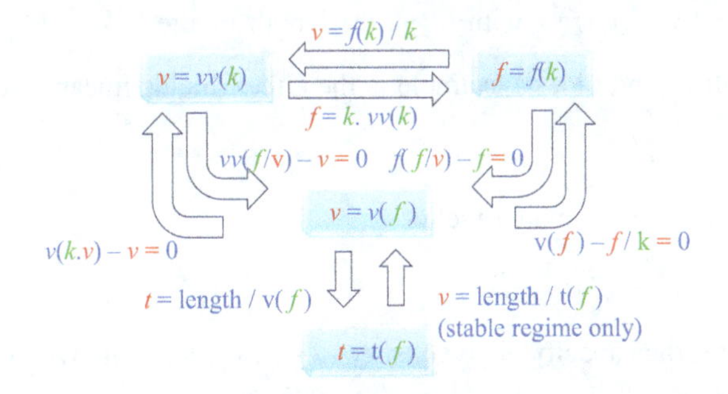

Figure 2.2 Relationships between the functions of FD

Given a speed-flow relationship two further parameters can be defined, let

$v_c = v(f_{MAX})$ be the critical speed, say the speed at maximum flow;
$k_c = f_{MAX} / v_c$ be the critical density, say the density at maximum flow.

Any speed variable is measured in $m \times sec^{-1}$ or $km \times h^{-1}$, flows in vehicles $\times\ sec^{-1}$ or vehicles $\times\ h^{-1}$, and densities in vehicles $\times\ m^{-1}$ or vehicles $\times\ km^{-1}$ (more considerations on this issue below).

Several mathematical formulations have been widely proposed for the fundamental diagram, in recent reviews (Fosu *et al.*, 2020; Romanowska & Jamroz, 2021). Del Castillo (2012) proposes several requirements for a properly defined FD. These models may be based on one function for both stable and unstable regimes or two functions, one for either regime. A few simple examples are given below.

Given any of the functions (2.2), (2.3), or (2.4), the other two can be derived taking into account (2.1). See Figure 2.2, where the travel time – flow function $t = t(f)$ is also reported.

2.1.1.1 Based on $v(k)$

Most FD models are derived from speed density functions (2.3). Starting from the speed-density relationship, the flow-density relationship, (2.4), may be easily derived by using the flow conservation equation under stationary conditions, or the fundamental conservation equation, as:

$$f(k) = v(k)\,k.$$

Then, the flow-speed relationship can be obtained by introducing the inverse speed-density relationship: $k = v^{-1}(v)$, thus

$$f(v) = v\left(k = v^{-1}(v)\right) \text{ with } v^{-1}(v) = vv^{-1}(v).$$

Finally, the flow-speed relationship may be inverted by considering two different relationships, one for the stable regime, $v_{stable}(f) \in [v_c, v_o]$, and the other for the unstable regime, $v_{unstable} \in [0, v_c]$. With reference to relation $v = v_{stable}(f)$ the relationship between the travel time, t, and the flow may be defined:

$$t(f) = \mathrm{L}/v_{stable}(f) \quad \text{with } L \geq 0 \text{ the length of the link}$$

The simplest (and oldest) example is the Greenshields linear model:

$$v(k) = v_0\left(1 - k/k_{jam}\right)$$

Greenshields' linear model yields:

$$f(k) = v_0\left(k - k^2 / k_{jam}\right)$$

In this case, the capacity is given by $f_{MAX} = v_0\, k_{jam}/4$. Moreover, Greenshields' linear model yields: $v^{-1}(v) = k_{jam}(1 - v/v_0)$ thus

$$f(v) = k_{jam}\left(v - v^2/v_0\right)$$

and

$$v_{stable}(f) = (v_0/2)\left(1 + \left(1 - 4f/\left(v_0\, k_{jam}\right)\right)(1/2)\right)$$

$$= (v_0/2)(1 + (1 - f/f_{MAX})(1/2))$$

$$v_{unstable}(f) = (v_0/2)\,(1 - (1 - f/f_{MAX})\,(1/2))$$

The FD based on the Greenshields model (Figure 2.3) is better suited to describe traffic for average values of density, far from 0 and the maximum k_{jam}. Other models are better suited to describe traffic for low or high values of density, they are not shown below for brevity's sake, recent reviews in Fosu *et al.* (2020) and Romanowska and Jamroz (2021).

2.1.1.2 Based on $f(k)$

Other FD models are derived from flow density functions (2.4). Starting from the flow-density relationship, the speed-density relationship, (2.3), may be easily derived by using the flow conservation equation under stationary conditions.

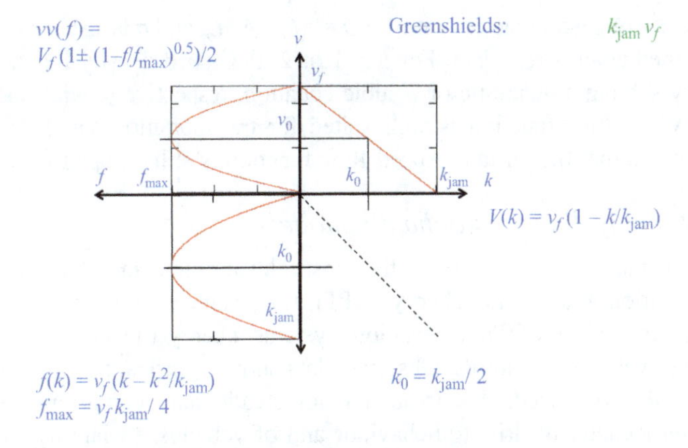

Figure 2.3 FD based on Greenshields speed-density function

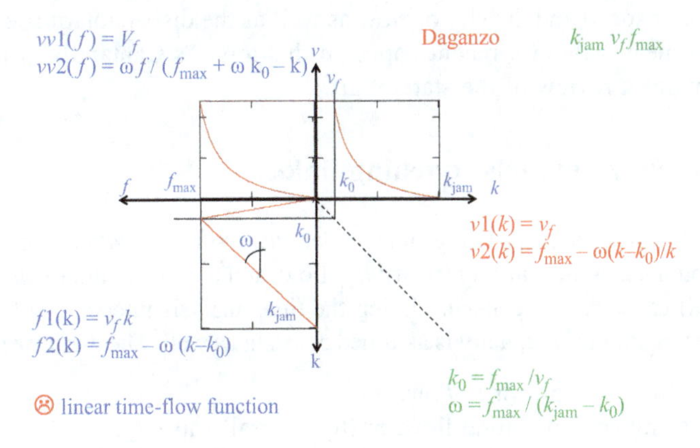

Figure 2.4 FD based on Daganzo flow-density function

The so-called Daganzo (1997) bi-linear speed-flow function is an example of this approach (Figure 2.4). It is often used for dynamic traffic models; on the other hand, it should be noted that it leads to a constant speed-flow function for the stable regime and, thus is not well suited for transportation supply models.

2.1.1.3 Based on $v(f)$

Another approach to FD is based on a travel time – flow function, from which the speed-flow function is easily defined, then the speed-density function can be obtained by solving an equation. For instance, the well-known Bureau of Public Roads (BPR)-like travel time function, often used for transportation supply analysis and demand assignment has been recently proposed (Cantarella *et al.*, 2023) for specifying the BPR-based stable regime speed-flow function:

$$v(f) = v_0/(1 + a\,(f/f_{MAX})b)\ 0 \leq f \leq f_{MAX}$$

where a is the congestion factor, with $1 + a = t(f_{MAX}) / t_o$, and b is a shape coefficient, both calibrated against real data. For $b = 1$ or 2, the speed-density function can be obtained by solving a quadratic or a cubic equation, respectively, with one positive real root. Whilst this function is well suited for transportation supply analysis, it should be noted that the unstable speed-flow function is still an open issue.

2.1.2 Running links – stochastic models

The Fundamental Diagram is one of the most relevant elements of almost any tool of Transportation and Traffic Theory (TTT), for instance for analysing motorway link traffic, as well as of Transportation Systems Theory (TST), for instance for specifying travel time functions. Observed data show significant dispersion around the mean value of speed, due to actual non-steady-state conditions, motorway layout, heterogeneity of driving behaviour and of vehicles, variability of weather and light conditions, etc. Thus, the Stochastic Fundamental Diagram (S-FD) has been recently proposed, allowing us to model both the relationship between the mean value of speed and density or flow as well as the dispersion of speed values around the mean value (A recent paper on this topic is Cantarella *et al.*, 2023, which contains a review of the state of art.).

2.2 Stationary models: queuing links

Modelling approaches of queueing links under *steady-state conditions* are based on the assumptions that the entering (or arrival) flow, and the maximum exiting flow, or (transversal) capacity, are constant during the time analysis interval [0, T], the link storage (or longitudinal) capacity is assumed constant as well. The main variables are:

$f_{IN} \geq 0$ the entering flow, or *in-flow*;
$f_{MAX} \geq 0$ the maximum exiting flow, or (transversal) capacity;
$n_{MAX} \geq 0$ the maximum queue that can be stored without affecting other upstream links, or (longitudinal) capacity;

Further variables that may change over time t are:

$n(t) \geq 0$ the number of users waiting to exit, or *queue length*, at time t;
$f_{OUT}(t) \geq 0$ the exiting flow, or *out-flow*, at time t.

At any time t, the out-flow must respect the following conditions:

$$f_{OUT}(t) \leq f_{MAX} \tag{2.5}$$

$$f_{OUT}(t) = f_{IN} \text{ IF } n(t) = 0 \tag{2.6a}$$

$$f_{OUT}(t) = f_{MAX} \text{ IF } n(t) > 0 \tag{2.6b}$$

If link storage is a major concern, the queue length must respect the following conditions:

$$n(t) \leq n_{MAX} \tag{2.7}$$

Remark. *For simplicity's sake, the same notations are used for both observable variables, described in the Chapter 1, and the modelling variables described in this chapter. Still, the reader must carefully distinguish between their meanings.*

The (average) queuing delay experienced by vehicles waiting to cross an interruption point (intersections, toll barriers, merging sections etc.) is affected by the number of waiting vehicles. This phenomenon may be analysed with models derived from queuing theory, developed to simulate queue evolution over time at a server (administrative counter, bank counter etc.), as shown below with reference to the users of a traffic system.

The whole time t_W needed for a user to leave the system (passing the server) is the sum of the (average) queuing delay accounting for the time spent queuing (pure waiting), and the (average) service time, let

$t_S = 1/f_{MAX}$ be the average service time (time spent at the server);
d_T be the total queuing delay (time spent in the queue) over the time analysis interval [0, T], given by the integral of the queue length over the analysis interval:

$$dT = \int_{[0,T]} n(t)\, dt \tag{2.8}$$

2.2.1 *Queuing links – deterministic models*

In this sub-section, we describe models developed under the assumption that the in-flow and the capacity, or the service time, are represented by deterministic variables constant over time. Two main conditions are distinguished below.

A. Undersaturation, the in-flow is less than capacity, $f_{IN} < f_{MAX}$, or $(f_{MAX} - f_{IN}) > 0$.

Some cases may occur, as described below.

A.1 The queue length at the beginning of the analysis interval is zero, $n(0) = 0$ (Figure 2.5, left):

$$n(t) = 0 \qquad t \in [0,\ T] \tag{2.9a}$$

$$f_{OUT}(t) = f_{IN} \qquad t \in [0,\ T] \tag{2.9b}$$

There is no queue at the end of the analysis interval T: $n(T) = 0$. The total queuing delay is zero ($d_T = 0$), and the entire delay corresponds only to the service time (t_S).

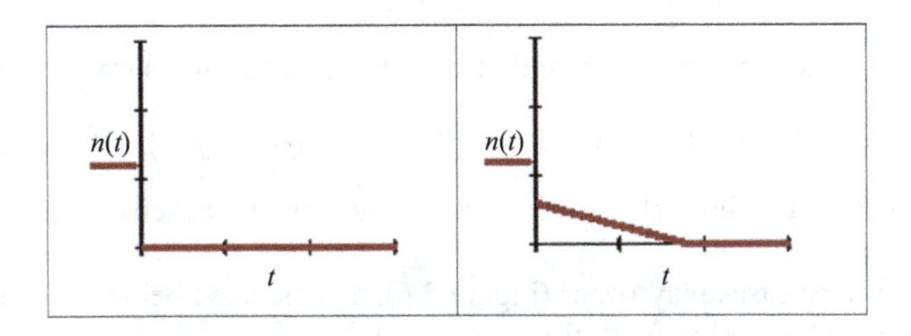

Figure 2.5 Undersaturation: queue over time

A.2 The queue length at the beginning of the analysis interval is strictly greater than zero, $n(0) > 0$. Then the queue length decreases with time and vanishes at time t_0, when the number of queuing users at the beginning of the analysis interval, $n(0)$, plus the number of users who have entered the queue up to time t_0, $f_{IN} \cdot t_0$, is equal to the number of users who have exited the queue up to time t_0, $f_{OUT} \cdot t_0$; since during the interval $[0, t_0]$ the queue is greater than zero, the exiting flow is equal to capacity, $f_{OUT} = f_{MAX}$. Thus:

$$t_0 = n(0) / (f_{MAX} - f_{IN}) \tag{2.10}$$

A.2.1 If the queue vanishes before the end of the analysis time, $t_0 \leq \mathrm{T}$, two successive evolutions over time of the queue occur (Figure 2.5, right).

There is no queue at the end of the analysis interval T: $n(\mathrm{T}) = 0$.

A.2.1a Up to time t_0, the queue length is linearly decreasing with t and the out-flow f_{OUT} is equal to capacity f_{MAX}:

$$n(t) = n(0) - (f_{MAX} - f_{IN}) \cdot t \quad t \in [0, t_0] \tag{2.11a}$$

$$f_{OUT}(t) = f_{MAX} \quad t \in [0, t_0] \tag{2.11b}$$

The total queuing delay is greater than zero:

$$d\mathrm{T} = n(0) \cdot t_0 / 2 = n(0)2 / (2 (f_{MAX} - f_{\mathrm{IN}})) \tag{2.12}$$

A.2.1.b After time t_0, the queue length is zero, $n(0) = 0$, as in case **A.1**:

$$n(t) = 0 \quad t \in [t_0, \mathrm{T}] \tag{2.13a}$$

$$f_{OUT}(t) = f_{IN} \quad t \in [t_0, \mathrm{T}] \tag{2.13b}$$

The total queuing delay is equal to zero: $d_T = 0$.

A.2.2 If the queue does not vanish before the end of the analysis time, $t_0 > \mathrm{T}$, the queue length is linearly decreasing with t and the out-flow f_{OUT} is equal to capacity f_{MAX}, up to time T. Equations (2.11a) and (2.11b) above still hold with $t_0 = \mathrm{T}$.

$$n(t) = n(0) - (f_{MAX} - f_{IN}) \cdot t \quad t \in [0, \mathrm{T}] \tag{2.14a}$$

$$f_{OUT}(t) = f_{MAX} \quad t \in [0, \mathrm{T}] \tag{2.14b}$$

There is a queue at the end of the analysis interval T:

$$n(T) = n(0) - (f_{MAX} - f_{IN}) \cdot T < n(0) \tag{2.15}$$

The total queuing delay accumulated during the analysis interval [0, T] is:

$$dT = (n(0) + n(\mathrm{T})) \cdot \mathrm{T} / 2 = n(0) \cdot \mathrm{T} - (f_{MAX} - f_{IN}) \cdot \mathrm{T}^2 / 2 \tag{2.16}$$

B. Oversaturation, the in-flow is greater than capacity, $f_{IN} > f_{MAX}$, or $(f_{IN} - f_{MAX}) > 0$.

Two further cases may occur (Figure 2.6), as described below. In both cases, the queue length increases with time.

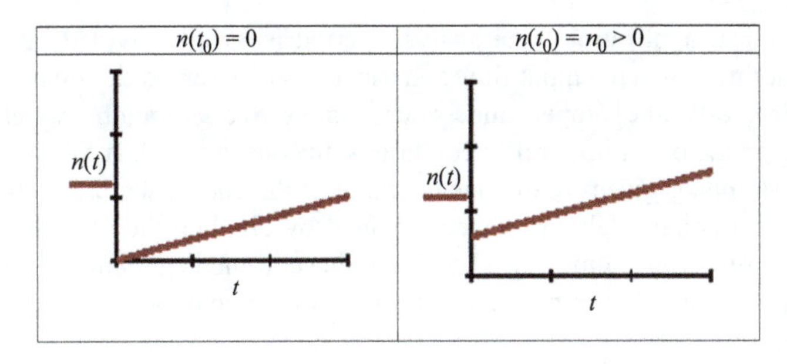

Figure 2.6 Oversaturation: queue over time

B.1 The queue length at the beginning of the period is zero, $n(0) = 0$ (Figure 2.6, left):

$$n(t) = (f_{IN} - f_{MAX}) \cdot t \qquad t \in [0,\ T] \tag{2.17a}$$

$$f_{OUT}\ (t) = f_{MAX} \qquad t \in [0,\ T] \tag{2.17b}$$

The queue length at the end of the interval is given by:

$$(f_{IN} - f_{MAX}) \cdot T \tag{2.18}$$

The total queuing delay accumulated during the analysis interval [0, T] is given by (2.16) with $n(0) = 0$

$$dT = (f_{IN} - f_{MAX}) \cdot T2\ /\ 2 \tag{2.19}$$

B.2 The queue length at the beginning of the period is strictly greater than zero, $n(0) > 0$ (Figure 2.6, right):

$$n(t) = n(0) + (f_{IN} - f_{MAX}) \cdot t \qquad t \in [0,\ T] \tag{2.20a}$$

$$f_{OUT}\ (t) = f_{MAX} \qquad t \in [0,\ T] \tag{2.20b}$$

The queue length at the end of the interval is given by:

$$n(0) + (f_{IN} - f_{MAX}) \cdot T \tag{2.21}$$

The total queuing delay accumulated during the analysis interval [0, T] is given by (2.16) with $n(0) > 0$

$$dT = n(0) \cdot \mathrm{T} + (f_{IN} - f_{MAX}) \cdot \mathrm{T}^2\ /\ 2 \tag{2.22}$$

By comparing the above equations for queue length over time the following general equation for calculating the queue length at time instant t can be formulated (see (1.8) in Chapter 1)

$$n(t) = \mathrm{MAX}\ \{0, (n(0) + (f_{IN} - f_{MAX}) \cdot t)\}. \tag{2.23}$$

where the MAX operator assures that the modelled queue length is not negative.

In practical applications, the analysis period is usually divided into analysis intervals during which the input flow and capacity are assumed stationary. For each of such intervals, the proper undersaturation or oversaturation model is used depending on the prevailing traffic conditions, the queue length at the beginning of each interval being given by the queue length at the end of the previous interval. The (average) unitary delay may be computed by dividing the total delay by the number of users over a time period with no queue at the beginning and the end of the period. An example is reported in Chapter 7 for computing delay at a signalised junction approach.

2.2.2 Queuing links – stochastic models

If flow fluctuations are observed over short time intervals that cannot be modelled through deterministic models, stochastic models are required, as described in this sub-section. The described models are developed under the assumption that the in-flow and the capacity, or the service time, are represented by variables constant over time, the number of users in the system and the time spent in the system are modelled as random variables, with distribution constant over time.

If the system is under-saturated, it can be analysed through (stochastic) queuing theory which includes the particular case of the deterministic models already discussed.

It is particularly necessary to specify the stochastic process describing the sequence of user arrivals, the sequence of service times and the queue discipline.

Main basic results of queueing theory are summarised below for the reader's convenience (several reference books are available, such as Harris *et al.* (2008), see also "Queuing theory" in Wikipedia). Notations are adapted to be consistent within this book. Let

f_{IN}, be the in-flow, already introduced, called the arrival rate λ or the expected value of the arrival flow;

$f_{MAX} = 1/t_s$, be the maximum flow or capacity, already introduced, called the service rate μ (or capacity) of the system, the reciprocal of the expected service time;

$x = f_{IN} / f_{MAX}$, be the traffic intensity ratio or utilisation factor, also called saturation degree, with $x < 1$ for undersaturation, ≥ 1 otherwise;

n be a realisation of the random variable N, the number of users present in the system, including the users queuing N_Q and the user present at the server, if any (the meaning of notation n is thus slightly different);

t_W be the whole time needed for a user to leave the system, sum of the time spent queuing and the service time, already introduced; it is considered a realisation of the random variable TW, the time spent in the system (or overall delay).

The queuing discipline may be:

FIFO First In - First Out (i.e., service in order of arrival);
LIFO Last In - First Out (i.e., the last user is the first served);
SIRO Service In Random Order;
HIFO High In - First Out (i.e., the user with the maximum value of an *indicator* is the first served).

The probability density function (pdf) describing the intervals between two successive arrivals/departures may be:

D = deterministic variable
M = negative exponential random variable
E = Erlang random variable
G = general distribution random variable

The characteristics of a queuing phenomenon can be redefined in the following concise notation (proposed by Kendall):

$$a \,/\, b \,/\, c\,(d, e)$$

where:

a denotes the type of arrival pdf as described above;
b denotes the type of departure pdf as described above;
c is the number of service channels: $\{1, 2, \ldots\}$;
d is the queue storage limit: $\{\infty, n_{max}\}$ or longitudinal capacity;
e denotes the queuing discipline represented as already described above.
Fields d and e, if defined respectively by ∞ (no constraint on maximum queue length) and by *FIFO*, are generally omitted.

In the following we will report the main results for the M/M/1 queue systems, which are commonly used for modelling transportation facilities, such as approaches at a junction, using notations introduced above.

- Probability of n users in the systems:

$$p_n = (1 - f_{IN} / f_{MAX}) \quad (f_{IN} / f_{MAX})^n$$

- Probability of no user in the system:

$$p_0 = (1 - f_{IN} / f_{MAX})$$

- Mean number of users in the systems:

$$E[N] = (f_{IN} / f_{MAX}) / (1 - f_{IN} / f_{MAX}) = x / (1 - x)$$

- Variance of the number of users in the systems:

$$\mathrm{Var}[N] = (f_{IN}/f_{MAX})/(1 - f_{IN}/f_{MAX})^2$$

- Probability that the queue length is less than the (integer) value z

$$p(N < z) = 1 - (f_{IN}/f_{MAX})^z = 1 - x^z$$

- Mean waiting time:

$$E[TW] = (f_{IN}/f_{MAX})/(2f_{MAX}(1 - f_{IN}/f_{MAX}))$$
$$= x/(2f_{MAX}(1 - x))$$

A. Undersaturation, the mean in-flow is less than the mean capacity, $x = f_{IN}/f_{MAX}$, < 1.

The stochastic term of delay may be computed by applying models for the Queuing Theory, briefly outlined above, assuming a queuing system M/M/1; (+∞), FIFO).

The (average) queue length is given by $E[N] = x/(1 - x)$. The maximum queue length n_p with probability p is given by solving equation $1 - x^z = p$ with respect *to* z:

$$np = \ln(1-p)/\ln(x)$$
$$\text{e.g., } n_{p=0.95} \cong -3/\ln x.$$

The (average) unitary delay is given by:

$$x / (2f_{MAX}(1 - x)). \tag{2.24}$$

This delay function has a vertical asymptote at $x = 1$; a tangent approximation close to the saturation, say $x = 0.95$, may be used if a continuous function is needed.

B. Oversaturation, the mean in-flow is greater than the mean capacity, $x = f_{IN}/f_{MAX}$, > 1.

General models are rather complex, thus a heuristic approach is commonly applied, through a transformation of the vertical asymptote to a diagonal one given by the unitary oversaturation delay d_U given by the total delay (2.19) divided by the number of users, $f_{MAX} \cdot T$, exiting the queue during the time interval [0, T]:

$$(f_{IN} - f_{MAX}) \cdot T^2 / (2f_{MAX} \cdot T) = (x - 1) \cdot T / 2 \tag{2.25}$$

An example is reported in Chapter 7 for computing delay at a signalised junction approach.

2.3 Summary

This chapter introduces (macroscopic) steady-state models for running and queuing links. As already stated, most of these models contain parameters to be calibrated against real data.

All the observations described above can be applied to distinguishing traffic per lane (right, middle, left), per type of vehicle (cars, motorcycles, vans, trucks, buses, . . .), As said above, in this chapter reference has been made to traffic along an urban street or a road, still, most results almost straightforwardly hold for other transportation systems.

The FD models have recently been applied to carry out coarse network analyses. In this case, they are called Macroscopic, or better Network, FD models, as introduced by Daganzo and Geroliminis (2008), and Geroliminis and Daganzo (2008); other references on this topic are enlisted below after the main references.

References

Some suggestions for further readings among the huge literature available on these topics are reported below.

Cantarella, G.E., E. Cipriani, A. Gemma, O. Giannattasio, and L. Mannini. (2023). "Multi-vehicle Stochastic Fundamental Diagram Consistent with Transportations Systems Theory." *Proceedings of 4th Symposium on Management of Future Motorway and Urban Traffic Systems (MFTS)*. Dresden.

Cascetta, E. (2009). *Transportation Systems Analysis: Models and Applications*. Berlin: Springer.

del Castillo, J.M. (2012). "Three New Models for the Flow–Density Relationship: Derivation and Testing for Freeway and Urban Data." *Transportmetrica* 8(6):443–465

Di Pace, R. (2019). Introduction to the Traffic Flow Theory, Appendix 2 in G.E. Cantarella, D.P. Watling, S. de Luca, and R. Di Pace. *Dynamics and Stochasticity in Transportation Systems: Tools for Transportation Network Modelling*. Amsterdam: Elsevier.

Fosu, G.O., E. Akweittey, J.M. Opong, and M.E. Otoo. (2020). "Vehicular Traffic Models for Speed-Density-Flow Relationship." *Journal of Mathematical Modeling* 8(3):241–255

Greenshields, B.D. (1935). "A Study of Traffic Capacity." *Highway Research Board Proceedings* 14(1):448–477.

Gross, D., Shortle, J.F., Thompson, J.M., and C.M. Harris. (2008). *Fundamentals of Queueing Theory*. New York: John Wiley & Sons, Inc.

Kessels, F. (2018). *Traffic Flow Modelling: Introduction to Traffic Flow Theory Through a Genealogy of Models*. Cham: Springer.

Romanowska, A., and K. Jamroz. (2021). "Comparison of Traffic Flow Models with Real Traffic Data Based on a Quantitative Assessment." *Applied Science* 11(21):9914.

Simchi-Levi, D., and M.A. Trick. (2013). "Introduction to "Little's Law as Viewed on Its 50th Anniversary." *Operations Research* 59(3):535

Transportation Research Board. (2016). *Highway Capacity Manual 6th Edition: A Guide for Multimodal Mobility Analysis*. Washington, DC: The National Academies Press.

Treiber, M., and A. Kesting. (2013). *Traffic Flow Dynamics: Data, Models and Simulation*. Berlin: Springer.

Further reading on network (macroscopic) fundamental diagram

Daganzo, C.F., and Geroliminis, N., 2008. An analytical approximation for the macroscopic fundamental diagram of urban traffic. *Transp. Res. Part B Methodol.* 42, 771–781.

Geroliminis, N., and Daganzo, C.F., 2008. Existence of urban-scale macroscopic fundamental diagrams: some experimental findings. *Transp. Res. Part B Methodol.* 42, 759–770.

Geroliminis, N., and Sun, J., 2011. Properties of a well-defined macroscopic fundamental diagram for urban traffic. *Transp. Res. Part B Methodol.* 45, 605–617.

Knoop, V., and Hoogendoorn, S., 2013. Empirics of a generalized macroscopic fundamental diagram for urban freeways. *Transp. Res. Rec.* 133–141.

Knoop, V.L., De Jong, D., and Hoogendoorn, S.P., 2014. Influence of road layout on network fundamental diagram. *Transp. Res. Rec.* 2421, 22–30.

Knoop, V.L., Van Erp, P.B.C., Leclercq, L., and Hoogendoorn, S.P., 2018. Empirical MFDs using Google traffic data. In: *IEEE Conf. Intell. Transp. Syst. Proceedings, ITSC 2018*, pp. 3832–3839.

Leclercq, L., Chiabaut, N., and Trinquier, B., 2014. Macroscopic fundamental diagrams: a cross-comparison of estimation methods. *Transp. Res. Part B Methodol.* 62, 1–12.

Tilg, G., Amini, S., and Busch, F., 2020. Evaluation of analytical approximation methods for the macroscopic fundamental diagram. *Transp. Res. Part C Emerg. Technol.* 114, 1–19.

Chapter 3
Dynamic models

Roberta Di Pace[1] *and Chiara Colombaroni*[2]

Traffic was as much an emotional problem as it was a mechanical one.

Tom Vanderbilt

Outline. *This chapter reviews the basic models of the Traffic Flow Theory under the assumption of dynamic traffic conditions.*

The chapter explores non-stationary models, which are categorised based on the level of aggregation of traffic flow variables. User and supply variables are classified as either aggregate or disaggregate.

First, macroscopic models can be identified where user behaviour variables are aggregated (such as arc density or entry flows derived from vehicle positions on the arc) and level of service variables (such as space mean speed and arc performance functions derived from fundamental diagrams). These models can also be classified by time and space. Then an overview of the microscopic models is shown.

The chapter is structured as follows: Section 3.1 introduces non-stationary models, Section 3.2 delves into macroscopic models, and Section 3.3 presents microscopic models.

3.1 Non-steady state models

In accordance with previous chapters the Phenomena occurring along a link (*running links*) in accordance with the *uninterrupted flow theory*, and those occurring at a bottleneck (*queuing links*) are discussed. In this chapter reference is made to traffic along an urban street or a road, still, most results almost straightforwardly hold for other transportation systems.

3.1.1 Running links

In the case of non-stationary models, the main variable describing the uninterrupted flow conditions will be considered as a function of space and time. In this model,

[1]Department of Civil Engineering, University of Salerno, Italy
[2]Department of Civil, Constructional and Environmental Engineering, Sapienza University of Rome, Italy

the macroscopic variables will be represented as

$$f = f(x, t) \tag{3.1}$$

$$v = v(x, t) \tag{3.2}$$

$$k = k(x, t) \tag{3.3}$$

The observed variables, $m(x; t, t + \Delta t)$, the number of vehicles traversing point x during the time interval $[t, t + \Delta t]$ and $n(t; x, x + \Delta x)$, the number of vehicles at time t between points x and $x + \Delta x$, can be averaged (flow with respect to space and density with respect to time) hence density and flow will be consistently defined.

The flow is related to m through the following equation

$$m(x, t, t + \Delta T) = \int_t^{t+\Delta T} f(x, t)dt \tag{3.4}$$

the density is related to n through the following equation

$$n(t, x, x + \Delta x) = \int_y^{y+\Delta x} k(x, t)dx \tag{3.5}$$

The *continuity* (or *conservation*) equation is a partial differential equation for the macroscopic quantities' density, flow, and speed as in the following

$$(\partial f/\partial x) + (\partial k/\partial t) = 0 \tag{3.6}$$

It can describe the density variations in terms of gradients (or differences) of the flow.

The continuity equation may be completed through the general equation constituting the basis of every macroscopic model which is the hydrodynamic equation, which computes the flow as the product of mean speed and density of flow ($f(x;t) = k(x;t)\ v(x;t)$), which can be usually applied in case of stationary conditions; to this aim the time intervals may be fixed with a sufficiently small size to guarantee the flow stationarity condition.

Furthermore, the $f_{IN}(t)$ may be introduced and then the cumulative in-flow may be derived and the $f_{OUT}(t)$ from which the cumulative outflow may be derived; let $f_{IN}(t) = f(x_{1;}t)$ then the cumulative in (*entering*) – flow is given by

$$u(t) = \int_0^t f(x_{1;}z)dz \tag{3.7}$$

and let

$f_{OUT}(t) = f(x_{2;}t)$ then the cumulative out (*exiting*) – flow is given by

$$w(t) = \int_0^t f(x_{2;}z)dz \tag{3.8}$$

With respect to the speed

$$(1/(x_2 - x_1)) \int_{x1}^{x2} v(x_{;}t)dx = v_s(t; x_1, x_2) \tag{3.9}$$

All these variables may be represented in *aggregate* or *disaggregate* ways and non-stationary traffic flow models may be classified on the base of their representation.

In more detail, macroscopic models aggregate users' behaviour variables (e.g., arc density or entry flows, which can be derived from vehicle positions on the arc) as well as level-of-service variables (e.g., space mean speed or arc performance functions, which are obtained from the fundamental diagram). In contrast, microscopic models describe traffic flow dynamics at the level of individual vehicles. Users' behaviour variables are disaggregated (arc density or entry flows can be obtained from the users' position on the arc) as well as the level of service variables (time speed and arc performance functions are derived from the drivers' behaviour models such as the car-following models).

3.1.2 *Queuing links*

Unlike stationary models, in the case of non-stationary models, the input variables are not fixed.

With reference to Figure 3.1(a), the diagram displays the arrivals and departures trajectories, whilst Figure 3.1(b) shows the cumulative value of arrivals and departures. The figures may be analysed with respect to four successive steps and with respect to the time window between t_1 and t_3 during which a capacity variation (reduction) occurs

- at time t_1 the beginning of the queue propagation may be observed
- at time t_2 the link capacity decreases until a minimum value
- at time t_3 the link capacity increases until the f_{IN}
- at time t_5 the queue discharging appears
- at time t_3 the queue length achieves the maximum value

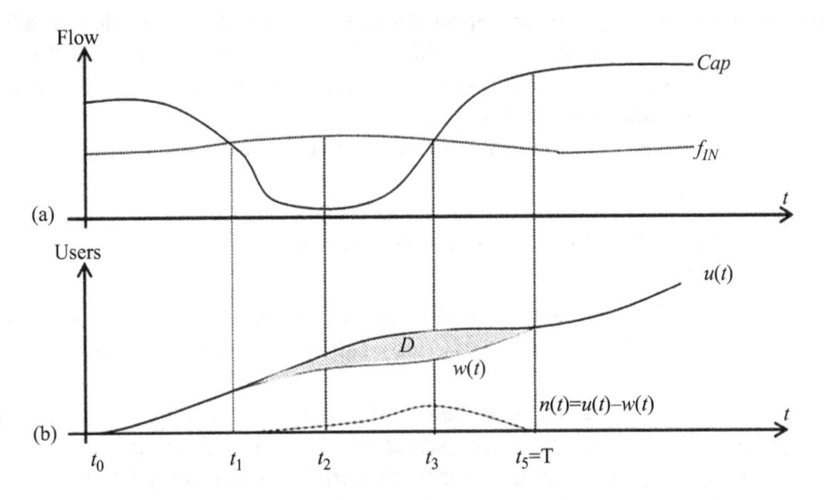

Figure 3.1 (a) Arrivals and departures flows and (b) arrivals and departures cumulative values

and the following equations may be obtained:

$$dn(t)/dt + (f_{OUT}(t) - f_{OUT}(t)) = 0 \tag{3.10}$$

$$n \geq 0$$

$$f_{OUT}(t) \leq Cap(t)$$

or equivalently

$$n(t) = (w(t) - u(t)) \tag{3.11}$$

Then in terms of the general expressions of the average unitary delay, the total delay may be derived.

In particular, with respect to the unitary delay it may be computed as

$$d(t) = n(t)/Cap \tag{3.12}$$

whilst the total delay is given by

$$D(t) = \int_{0T} n(t)dt \tag{3.13}$$

and the average unitary delay during a time interval [0, T] is equal to

$$d(0, T) = D(t)/u(t) \tag{3.14}$$

3.2 Macroscopic models

These models can be classified in accordance with time and space.

Time-continuous and space-continuous macroscopic models are formulated by differential equations in time and space dimensions; this class of models is called *point-based models.* For the *point-based models,* a solution approach in discrete time and space is usually adopted using the finite difference method; this class of models is called *finite difference models.*

Finally, there are the time-discrete and space-continuous macroscopic models which are called *link-based models.*

3.2.1 Continuous-time continuous-space macroscopic models

Two types of models are identified within the class of continuous-time continuous-space models: the *first-order models*, also called *point-based models*, and the *second-order models*.

The main difference between them is that the *first-order models* have only one dynamic equation which is represented by the continuity equation, and it is able to capture only the dynamics of a single variable which is the traffic density, whereas in the second-order models, the acceleration equation is introduced to properly reproduce the traffic inhomogeneity with respect to different vehicles' desired speed.

The considered variables are:

- f the flow measured in vehicles per unit of time;
- k the density measured in vehicles per unit of length;
- v the speed measured in space per unit of time;
- v_k the equilibrium speed as a function of density k.

3.2.1.1 Point-based models

An approach in which the flow is given as a function of density was introduced by Lighthill and Whitham in 1955 and Richards in 1956 thus this class of models is also called LWR models and differs only for the functional form of the fundamental diagram; the corresponding equations are listed below:

$f(x,t) = f(k(x,t))$ which represents the flow – speed relationship
$v(x,t) = v_K(k(x,t))$ which represents the speed – density relationship

and the conservation equation may be rewritten as follows

$$(\partial f/\partial k)(\partial k/\partial x) + (\partial k/\partial t) = 0 \tag{3.15}$$

Furthermore, the previous equation may be rewritten considering the fundamental diagram

$$(v_k + k\partial v_k/\partial k)(\partial k/\partial x) + (\partial k/\partial t) = 0 \tag{3.16}$$

Since the functional form of the fundamental diagram is not specified, these models are generally classified as LWR models, taking inspiration from the initials of their authors.

One of the main limitations of the LWR model is related to the inertial effects; indeed, it assumes that speeds are instantaneously adapted so that accelerations or decelerations of vehicles are unrealistically reproduced too high.

Propagation of density variations
Starting from the conservation equation

$$(\partial f(k(x;t))/\partial x) + (\partial k/\partial t) = 0 \tag{3.17}$$

let the flow in (x, t) depending on the density in the same point (x, t) (that is the drivers adapt their speed instantaneously to any density change)

$$f = f(k(x;t)) \tag{3.18}$$

the conservation equation may be rewritten as in the following

$$(\partial f/\partial k)(\partial k/\partial x) + (\partial k/\partial t) = 0 \tag{3.19}$$

We get a quasi-linear first-order partial differential equation that describes how density varies in the x–t plane; indeed, with $d(k)$ that equals to $(\partial f/\partial k)$, the previous equation may be rewritten as in the following

$$d(k)(\partial k/\partial x) + (\partial k/\partial t) = 0 \tag{3.20}$$

The solution can be found by starting from a given initial solution $k_o = k(x_o,0)$ and moving along a characteristic curve $x = x(t)$ (where density remains constant, $k(x(t),t) = k_0$)

The derivative of $k_o = k(x(t),t)$ with respect to time, t, is given by

$$dk(x(t),t)/dt = (\partial k(x(t),t)/\partial x)(dx/dt) + (\partial k(x(t),t)/\partial t) = 0 \tag{3.21}$$

imposing $k = k_0$

$$(\partial f(k_0)/\partial k)(\partial k/\partial x) + (\partial k/\partial t) = 0 \tag{3.22}$$

and it can be concluded, with respect to the previous equation that

$$(dx/dt) = (df(k_0)/dk) = d \tag{3.23}$$

by integrating, we get that

$$x = dt + x_0 \tag{3.24}$$

thus, the solution of the differential equation is a function φ (differentiable) and the argument of the function is $x - dt$ and the solution equals

$$k = \varphi(x - dt) \tag{3.25}$$

Furthermore, assuming d constant, the characteristic curves are lines whose slope in the time–space plane is $d = df/dk$; these lines describe how a state of density k propagates in the x–t space. Since they represent the propagation of density changes due to the mutual conditioning between vehicles that affects their speed, they are also called kinematic waves.

Some considerations may be made about d which is assumed equal to $\partial f/\partial k$ (see Figure 3.2(a)); indeed, the propagation speed d depends on the density consistently with the steady state flow density relationship (fundamental diagram).

In particular, the density variations may propagate in the driving direction (with positive derivative) in the left part of the diagram (free flow condition, stable flow) and against the driving direction (with negative derivative) in the right part of the diagram (congested condition, unstable flow); if $k = kc$ then $d = 0$, the critical traffic state ($f = fmax$) does not propagate along the traffic stream.

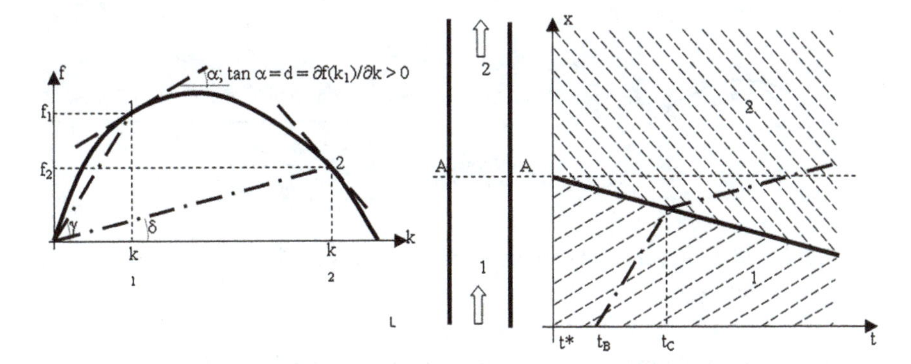

Figure 3.2 (a) Flow–density diagram and (b) waves trajectories in a space–time diagram

Shock waves

Let us consider a road segment (see Figure 3.2(b)) in which at time t^* two different flow conditions may be observed which are respectively stable in the first section (before section A-A) and unstable in the second section (after section A-A); a space–time diagram may be introduced to support the description of the phenomenon. Phenomena in the first section propagate with speed tan (α) and may be represented in the space diagram through parallel segments with an angular coefficient equal to α; the same approach may be applied to section 2. Therefore space–time segments related to section 1 and space–time segments related to section 2, will meet in the wavefront (continuous line) which propagates against the driving direction and whose relative speed equals the slope of the secant connecting the two different states.

A further example may be made if state 2 is represented by point A in Figure 3.3, which is close to the capacity; in this case, the wavefront propagates in driving directions.

This is the case of the density discontinuity which is described through the shock wave propagation; in particular, regarding the previous example, the shock wave speed propagation is computed as $z = (f_2 - f_1)/(k_2 - k_1)$.

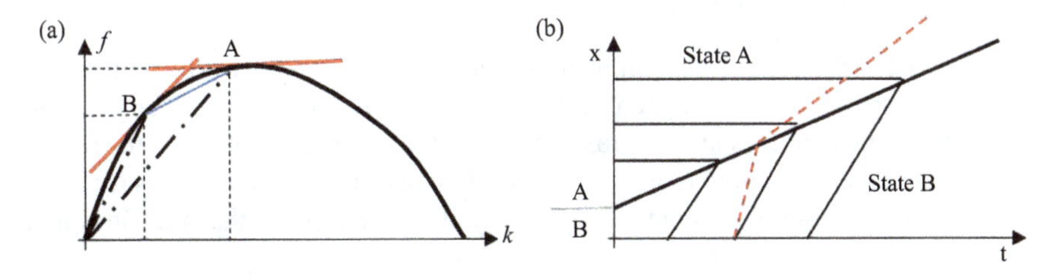

Figure 3.3 (a) Flow–density diagram and (b) waves trajectories in a space–time diagram

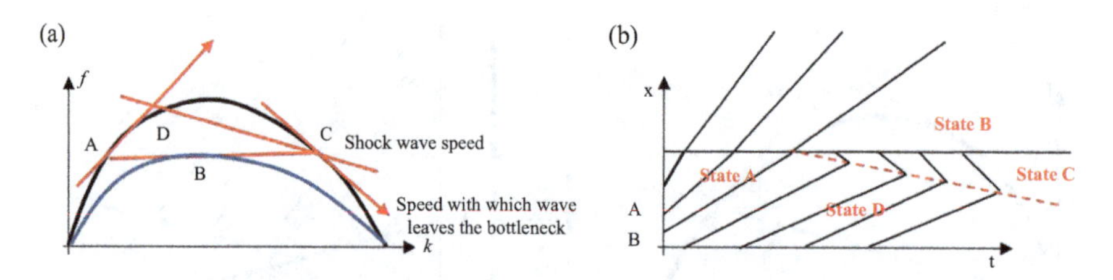

Figure 3.4 (a) Flow–density diagram and (b) waves trajectories in a space–time diagram

Another example may be represented with the capacity variation (this is the case of a bottleneck; see Figure 3.4). Indeed, in this case, it is necessary to consider a different fundamental diagram corresponding to the different features of the facility whose capacity is reduced for example because of a lane closure; in particular, because of the lower capacity, the internal fundamental diagram must be considered. If the flow is lower than the bottleneck capacity, the shock wave propagates stationary in a horizontal direction, when the flow is higher than capacity the flow propagates under capacity restriction thus the shock wave propagates backwards.

3.2.1.2 Second-order models

Two main limitations may be identified in the LWR model: the first one refers to the model assumption of instantaneous adaptation of the vehicle speed, which is certainly an idealisation; the second one is about non-homogeneous traffic; indeed, different desired speeds for different vehicle class may be expected. Therefore, an acceleration equation is introduced, and these models are classified in a second-order model.

The space and time continuous second-order models are formulated through the following equations*

$$(\partial f/\partial k)(\partial k/\partial x) + (\partial k/\partial t) = 0 \tag{3.26}$$

$$f(x,t) = k(x,t)v(x,t) \tag{3.27}$$

and a third additional equation representing the stream acceleration

$$dv/dt = \partial v/\partial t + v\,\partial v/\partial x \tag{3.28}$$

The first-order model is a complete and consistent theoretical scheme that is a very useful tool for analysing road traffic in stationary or near-stationary states. However, it introduces some approximate hypotheses that are unrealistic if one wants to get a more detailed representation of dynamic traffic variations as it can be obtained by microscopic models. Specifically, the assumption that the flow is a

*The expression may be generalised to the case of on-ramps, with in-flows equals to $E^*(y,t)$ thus the conservation equation may be rewritten as $(\partial q/\partial y) + (\partial k/\partial t) = E^*(y,t)$.

function of the density only at the point assumes that vehicles *instantaneously* adapt their speed to density changes. From the above hypothesis, for boundary conditions of the type $dv_1/dt(t)<0$, it follows the non-uniqueness of the solution and therefore the discontinuity of the outflow represented by a shock wave. The solution of the first-order model does not allow for the reproduction phenomena of traffic stream instability, often observed on the highway for sudden increases in density ("stop-and-go" conditions), nor the so-called hysteresis of traffic, often observed in the discharge process of queues on highways.

Higher-order models and additional variable dynamics were introduced in an attempt to overcome the remaining inadequacies. The following paragraph describes one of the best-known of these models, introduced by Payne in 1971 and then extended by Papageorgiu (2010).

Second-order models aim to improve the linear model by introducing additional features that take into account the impossibility for drivers to instantaneously update speed to traffic density. In the context of the analogy with fluid dynamics, these behavioural assumptions result in the introduction of the concepts of relaxation and traffic diffusion, which result in a lag of speed adjustment in space and time, respectively.

The modelling approach adopted is based on a second-order macroscopic dynamic model. The second-order models assume a behaviour for the traffic flow similar to that of a vehicle (follower) which adapts its behaviour, in terms of speed variations, depending on the stimuli it receives from the vehicle in front of it (leader):

$$v(x(t+\tau), t+\tau) = v^e(k(x+\Delta x), t) \tag{3.29}$$

where τ would correspond to the reaction time, $x(t)$ the position of the vehicle at time t, $v(x,t)$ its speed at position x and at time t, Δx the distance from the previous vehicle and v^e is the equilibrium speed expressed as a function of the density k at $(x+\Delta x,t)$. Vehicles no longer adapt their speed instantaneously when they encounter a shock wave, as is assumed in the first-order model, but according to a spatial and temporal lag.

Expanding respectively the first and second members of (3.29) in a Taylor series, we obtain:

$$v(x(t+\tau), t+\tau) = v(x,t) + \tau \cdot v(x,t)(\partial v(x,t))/dx + \tau(\partial v(x,t))/dt \tag{3.30}$$

$$v^e(k(x+\Delta x), t) = v^e(k(x,t)) + \Delta x(\partial k(x,t))/dx(dv^e(k(x,t)))/dk \tag{3.31}$$

After some manipulations, the following equation is obtained:

$$\partial v/dt + (v\partial v/dx) = (((v^e(k) - v))/\tau) - ((c_0{}^2\partial k)/k\partial x) \tag{3.32}$$

$$C : n(v\partial v/dx)$$

$$B : ((v^e(k) - v))/\tau$$

$$A : (c_0{}^2\partial k)/k\partial x)$$

Where $c_0^2 = \xi/\tau$ is a constant and $\xi = -dv^e/dk$ is the decreasing rate of the equilibrium speed when the density increases, assuming a constant rate consisting of a linear relationship between density and speed $v^e(k)$.

Equation (3.32) has many similarities to fluid dynamics, as it expresses the acceleration of the traffic stream by the sum of three terms.

The first term represents the total derivative of the speed, while the second member captures three different aspects of the flow dynamics: a convection term (C) describes how the speed changes depending on the variation of the speed itself in the space, a relaxation term (R) representing how vehicles adapt their speed compared to the equilibrium speed with a certain delay (while in the LWR model the adjustment is instantaneous) and an anticipation term (A) reflecting how vehicles react to downstream traffic conditions.

To simplify the search for the solution of the model by developing a computer program it is convenient to transform the model in a discrete form (in space and time); a road section should be divided into N segments of length l_i, while the temporal discretisation t, with $t = 1,2,\ldots$ is based on a time step (or simulation interval) Δt (Figure 3.5).

The discrete formulation corresponding to the continuous model (3.32) is given by the following equation:

$$\begin{aligned} v_i(t+1) = v_i(t) + (\Delta t/l_i \cdot v_i(t)[v_{i-1}(t) - v_i(t)]) + (\Delta t/\tau[v^e(k_i(t)) - v_i(t)]) \\ - ((v_a \cdot \Delta t[k_{i+1}(t) - k_i(t)])/(\tau l_i[k_i(t) + \kappa])) \end{aligned} \tag{3.33}$$

$$C : (\Delta t/l_i \cdot v_i(t)[v_{i-1}(t) - v_i(t)])$$

$$R : (\Delta t/\tau[v^e(k_i(t)) - v_i(t)])$$

$$A : -((v_a \cdot \Delta t[k_{i+1}(t) - k_i(t)])/(\tau l_i[k_i(t) + \kappa]))$$

where v_a indicates the anticipation constant, $v^e(k_i(t))$ the equilibrium speed deducible from the fundamental diagram, and κ a parameter of the model. As can be seen from the previous equation, the speed in section i at time $(t+1)$ equals the speed in the section itself at time t plus a correction due respectively to:

- Convection (C): vehicles travelling between $i - 1$ and i do not instantly adjust their speed, particularly those that cross-section $i - 1$ at high speed gradually

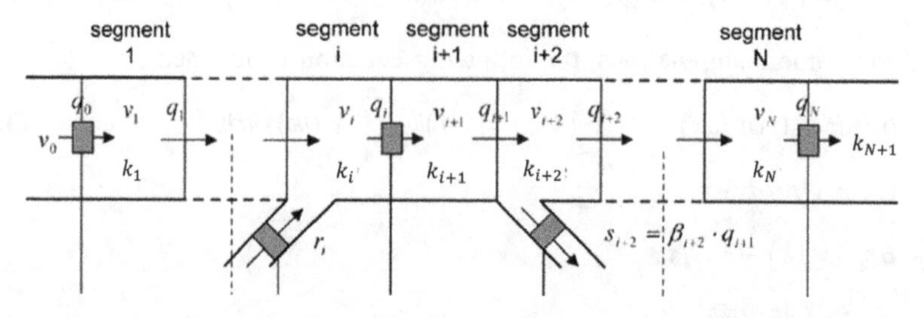

Figure 3.5 Road section schematisation in the discrete formulation

slow down until the drivers reach a speed that they judge comfortable for the successive section.

- Convection is also proportional to the difference in speed between the two sections and therefore the higher this difference, the longer it will take to accelerate or brake and the greater the impact on the average speed in the section; furthermore, C is also proportional to the speed in the section.
- Relaxation (R): it can be deduced that the greater the difference between the current speed and the equilibrium speed, which means a high relaxation of the system with respect to the equilibrium, the higher the acceleration of the traffic stream.
- Anticipation (A): takes into account the fact that vehicles are influenced by downstream conditions, therefore an increase in density in $i+1$ induces vehicles in section i to slow down while a decrease causes them to speed up.

An empirical extension of (3.33) has been proposed in the literature (Papageorgiu *et al.*, 2010) to consider some typical phenomena found in the motion of a current, as well as the flows entering, r_i, and exiting, s_i, at ramps.

Under heavy traffic conditions, a slowdown in flow is observed at the access ramps, which could be traced back to an increase in density downstream of an entrance ramp and on it. In reality, it should be taken into account that vehicles accessing the infrastructure, through lane-changing manoeuvers, induce disturbances to the main flow which translates into a decrease in average speed.

The disturbances just described are considered by adding the following term

$$-(\delta\Delta t/(l_i \cdot n_i))(r_i(t)v_i(t))/(k_i(t)+\kappa)$$

where δ is a model parameter.

The overall structure of the model, appropriately discretised in segments of length l_i, with a temporal discretisation t, with $t = 1,2,\ldots,$ for a motorway section that has access and exit ramps, is provided by the following set of equations:

$$k_i\,(t+1) \;=\; k_i\,(t) + (\Delta t)/(l_i \cdot n_i)\,[q_i - 1\,(t) - q_i\,(t) + r_i\,(t) - s_i\,(t)] \tag{3.34}$$

$$s_i\,(t) = \beta_i\,(t) q(i-1)\,(t) \tag{3.35}$$

$$\begin{aligned} v_i\,(t+1) \;=\; & v_i\,(t) + \;(\Delta t/l_i \cdot v_i\,(t)[v_i - 1(t) - v_i\,(t)]) \\ & + \;(\Delta t/\tau\,[v^e\,(k_i\,(t)) - v_i\,(t)]) \\ & - \;((v_a\,\Delta t[k_i + 1\,(t) - k_i\,(t)])/(\tau l_i\,[k_i\,(t) + \kappa])) \\ & + \;(\delta\Delta t/(l_i \cdot n_i)) \cdot (r_i\,(t) \cdot v_i\,(t))/(k_i\,(t) + \kappa) \end{aligned} \tag{3.36}$$

$$C:\;(\Delta t/l_i \cdot v_i(t)[v_i - 1(t) - v_i\,(t)])$$

$$R:\;(\Delta t/\tau\,[v^e\,(k_i\,(t)) - v_i\,(t)])$$

$$A: -\,((v_a\,\Delta t[k_i + 1\,(t) \;-\; k_i\,(t)])/(\tau l_i\,[k_i\,(t) + \kappa]))$$

$$v^e(k) = v_f \exp\left[-1/a\,(k/k_{cr})\,a\right] \tag{3.37}$$

$$q_i = k_i(t) \cdot v_i(t) \tag{3.38}$$

where (3.34), (3.36), (3.37), and (3.38) respectively represent the conservation equation, the dynamic speed equation, the fundamental diagram, and the state equation; Equation (3.35) instead constitutes a modelling of the exits that occur at a rate β.

The fundamental diagram assumed here is the May model (May, 1990) where the parameters are free flow speed, v_f, and critical density, k_{cr}.

The parameters τ, v_a, δ, κ, v_f, k_{cr}, n_i, and a are assumed to be identical for all segments and they have to be determined through a calibration process.

For each segment, the flow $q_i(t)$ can be calculated from state equation (3.38) and substituted into (3.36). Furthermore, for each instant t, the variables $q_{i-1}(t)$, $v_{i-1}(t)$, $k_{i+1}(t)$, as well as $r_i(t)$ and $\beta_i(t)$, are computed to update the state variables to the values that they assume in the successive instant.

At the end of this process, it is observed that by substituting (3.35) into (3.34), (3.37) into (3.36), and updating the state equation (3.38), the model consists of $2N$ equations, $2N$ state variables, and seven parameters.

Furthermore, to solve the problem, boundary conditions related to the variables below must be defined:

- flow at the origin of the trunk q_0;
- speed at the origin of the trunk v_0;
- density at destination k_{n+1};
- flows on the entrance ramps r_i;
- exit rate from entrance ramps β_i

3.2.2 *Discrete-time discrete-space macroscopic models*

3.2.2.1 First-order models

In this section, the finite difference models as an example of the macroscopic discrete time and discrete space models are discussed. In particular, macroscopic models continuous in time and space dimensions are formulated by differential equations and a solution approach in *discrete time* and *space* is adopted using the finite difference method. In this model, it is assumed that the road segment is divided into cells. The main notations used in the following are enlisted in the previous section in alphabetical order (notations come first) for the reader's convenience (notations come first, then Roman letters, followed by Greek letters). In addition

k is a uniform density in a cell during time interval Δt;
n_i is the number of vehicles on cell i, equals to $k\,\Delta x$;
N_i is the maximum number of vehicles present in cell i;

Cap is the maximum flow rate in cell i;

v is the free flow speed coefficient;
z is the wave speed coefficient;

$y_i(t)$ is the inflow (to cell i) at time t;
$y_{i+1}(t)$ is the outflow (from cell i) at time t;
$\delta = z/v$ with $z \leq v$.

3.2.2.2 Finite difference models

The LWR models need to be solved numerically by *finite-difference* methods. In particular, the road segment is divided into cells of constant length and time in the index k increasing in the downstream direction. Each cell is characterised by uniform density and speed (as a function of the speed–density relationship) and the flow between neighbouring cells which is assumed to be constant during every time interval. In the first step the road is divided into cells (the model is also called Cell Transmission Model; CTM) each one of width Δx. The cell length Δx is the distance a vehicle would travel in a free flow condition in a one-time step; hence it is equal to free flow speed times the length of the time step (also called clock tick), $\Delta x = v\Delta t$. It must be remarked that the relationship between the cell length and the time step corresponds to the *Courant–Friedrich–Lewy* condition for stability of explicit solution methods that is $v\Delta t \leq \Delta x$. In the CTM two quantities are identified: the sending and receiving flows, which are respectively the flow depending on the density upstream and the flow depending on the density downstream.

The Cell Transmission model
Let $N_i(t)$ be the vehicle holding capacity of cell i and k_{jam} the jam density (the maximum number of vehicles which can fit into cell i); thus (Figure 3.6)

$$N_i(t) = k_{jam}\Delta x. \tag{3.39}$$

The conservation equation may be rewritten as

$$n_i + 1(t+1) = n_i(t) + y_i(t) - y_i + 1(t) \tag{3.40}$$

The key quantities of the method can be introduced based on the (trapezoidal) fundamental diagram (see Figure 3.7).

The number of vehicles moving from cell i to cell j is given by

$$\begin{aligned} y_i(t) &= \min\{n_i - 1(t),\ Cap,\ \delta[N_i(t) - n_i]\} \\ &= \min\{k_v,\ Cap,\ z(k_{jam} - k)\} \end{aligned} \tag{3.41}$$

and this is the result of a comparison between the maximum number of vehicles that can be *sent* by cell i directly upstream of the boundary $S_i(t) = \min\{k_v, Cap\}$ and those that can be *received* by the downstream cell i, $R_{i+1}(t) = \min\{Cap, z(k_{jam} - k)\}$.

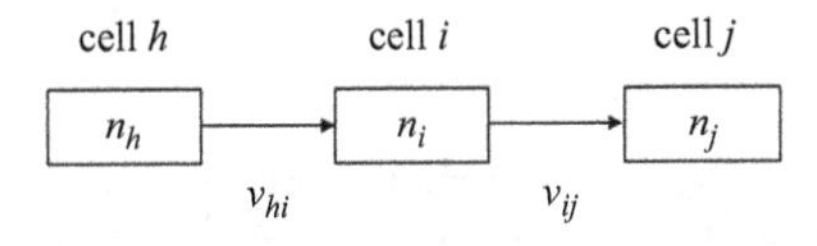

Figure 3.6 Junction represented through cell transmission model

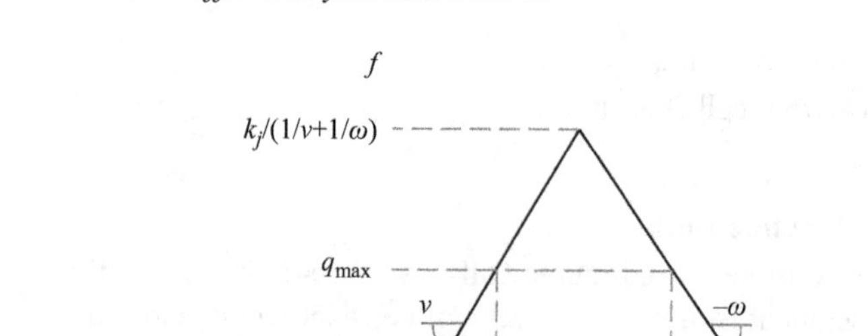

Figure 3.7 Trapezoidal fundamental diagram

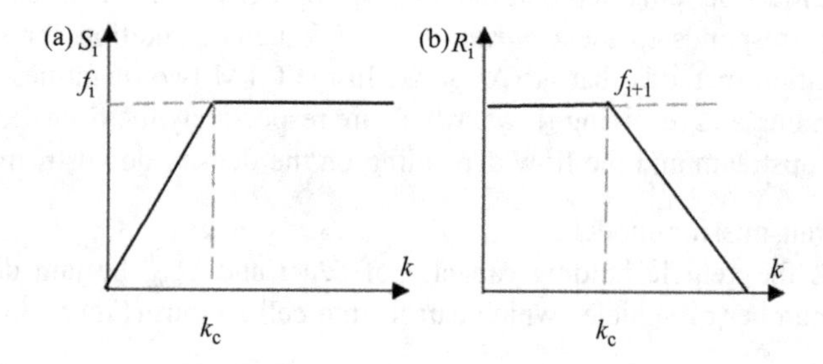

Figure 3.8 The sending (a) and receiving rule of the cell transmission model

Summing up, the flow equation inherently accommodates three traffic conditions:

- in low-level traffic (*uncongested*), the flow is equal to the number of vehicles in cell i at time t, n_i;
- in bottleneck traffic (*flow capacity*), the flow is equal to the saturation flow rate f_i times Δt;
- in oversaturated traffic (*congested case*), the flow is restricted by the jam density and depends on the available space in cell i at time t, $z_i/v[N_i - n_i]$.

This makes it possible to simulate the propagation of blocking back phenomena by considering constraints on the cell outflow equation (*receiving function*).

Hence the flow $y_i(t)$ can be rewritten in accordance with the supply (*sending*)-demand (*receiving*) rule of the cell transmission model (Figure 3.8):

$$y_i(t) = \min\{S_i, R_i + 1\} \tag{3.42}$$

Depending on the density of each cell, the cell flow propagates into the upstream cell (*propagation speed* < 0) or downstream (*propagation speed* > 0). Summing up all steps for the triangular fundamental diagram lead to the supply-demand method constrained as follows: if supply is the limiting factor, information travels upstream; if demand limits the traffic flow, information travels downstream.

3.2.2.3 Second-order models

In Papageorgiou (1990) and Papageorgiou *et al.* (1989), the first discrete second-order traffic model was developed. New terms are added to the PW model, which is discretised in space and time, to reflect the impact of on-ramp and off-ramp flows on the mainstream dynamic behaviour. In addition, it was extended by Kotsialos *et al.* (2002), Papageorgiou *et al.* (1989), and Messmer and Papageorgiou (1990) to consider a highway network using the METANET simulation program, which is commonly referred to as the discrete second-order traffic flow model.

The entering and exiting flows are defined as follows in the METANET model and they are implemented in the METANET model with the relations

$$I_i\,(k) = q(i-1)\,(k) + r_i\,(k) \tag{3.43}$$

$$O_i\,(k) = q_i\,(k) + s_i\,(k) \tag{3.44}$$

Meanwhile, the equation of the speed is:

$$\begin{aligned} v_i(k+1) = {} & v_i(k) + (T/\tau)\,[V(\varrho_i(k)) - v_i(k)] \\ & + (T/L_i)\,v_i(k)[v(i-1)(k) - v_i(k)] \\ & - (\nu\,T[\varrho(i+1)(k) - \varrho_i(k)])/(\tau\,L_i\,[\varrho_i\,(k) + \chi]) \\ & - \delta_{on}T\,\{(v_i(k)\,r_i(k))/(L_i\,[Qi\,(k) + \chi\,])\} \end{aligned} \tag{3.45}$$

where $V(\varrho_i(k))$ is the steady-steady speed–density relation and τ, ν, χ, δ_{on} are model parameters.

The speed equation above mentioned is a discretisation from the PW model considering analogous relaxation, convection, and anticipation terms. Further, it is also considered an additional fourth term, i.e., $-\delta_{on}T\,\{(v_i(k)\,r_i(k))/(L_i[Q_i\,(k)+\chi])\}$, that was implemented in Papageorgiou *et al.* (1989) to model the deceleration of vehicles in the mainline because of the on-ramp entering flow.

In any case, the on-ramp queue dynamics of the flow $r_i(k)$ can be expressed as:

$$r_i(k) = \min\{d_i\,(k) + (l_i(k))/T;\, r_{i\max};\, r_{i\max}(\varrho_{i\max} - \varrho_i(k))/(\varrho_{i\max} - \varrho_{icr})\} \tag{3.46}$$

The METANET model, as it was explained, was extended to consider arbitrary topology networks, where a highway network is represented through a directed graph composed of *freeway links* (with homogeneous geometric characteristics), *origin links* (set the traffic flows from outside into the network), and *nodes* (junctions, bifurcations, on-ramps or off-ramps). When describing traffic behaviour in a highway network, the METANET model can be used in two ways: either in a “non-destination-oriented mode,” where the vehicle’s destination is ignored or in a “destination-oriented mode,” where road users’ choices between alternate paths are explicitly modelled.

3.3 Longitudinal microscopic models

Models able to simulate individual vehicle interactions can be classified as car-following models for longitudinal interactions for vehicles along the same lane and lane-changing models for vehicles travelling along different lanes. In this section, some of the most relevant longitudinal models are discussed. They focus on complex driving behaviour representation.

Concerning the recent enhancement in terms of vehicle automation, traffic flow models must be classified with respect to human-driven vehicles and the automated vehicles (see Chapter 13 for further details about automated vehicles, AVs and connected and automated vehicles, CAVs). With respect to the first class related to human-driven vehicles, there are many well-established mathematical models, such as the Gazis–Herman–Rothery (GHR) model and the Gipps models (1981), among others focussing on the drivers' desired speed, acceleration/deceleration and more in general on the driver's physical actions. Moreover, there are other models considering the psychological factors of the drivers such as the Wiedemann model. Concerning the CAVs models there are many approaches not yet established in the literature. Several studies are based on the enhancement of conventional models (such as the IDM, MIXIC, etc.), others on the modelling of AVs and CAVs through Adaptive Cruise Control and Cooperative Adaptive Cruise Control (see Sections 3.4 and 3.5 for further details). In the following paragraphs, the main conventional and established mathematical models are discussed, and some details of preliminary models for specific ACC representation (non-conventional models) are shown.

3.3.1 Conventional microscopic models

The main notations used are listed below in alphabetical order (notations come first) for the reader's convenience (notations come first, then Roman letters, after Greek letters)

a_n is the desired acceleration of vehicle n at time t
$a_{n, max}$ is the upper bound of the vehicle acceleration
b^*_n is an estimate of the deceleration applied by the preceding vehicle
b_n is the desired deceleration
d'_{n-1} is the leading vehicle deceleration
d_n, is the vehicle acceleration
$d_{n,max}$ is the max deceleration in the braking condition
n is the following vehicle
$n-1$ is the leading vehicle
$s_n(t) = x_n(t) - x_{n-1}(t)$, is the spacing between two successive vehicles, n and $n-1$, front to front
$s_{n, max}$ is the lower bound of the free flow driving condition
$s_{n, min}$ is the upper bound of the braking condition
$s_{n, s}$ is the safety spacing representing the upper bound of the collision condition
s_{n-1} is the effective length

$T = 1/w\ k_{jam}$ is the time shift between two consecutive trajectories with w wave speed and k_{jam} density
u is the speed of vehicles travelling along the highway
v is the vehicle speed
$v'_n(t)$ is the vehicle free flow speed
$v^{\beta}_n(t)$ is the vehicle speed due to the presence of the leading vehicle
$v^{d}_n(t)$ is the desired vehicle speed
v_K is the equilibrium speed as a function of density k
x is the vehicle position
$x_n(t+T)$ is the longitudinal position of vehicle n at time $t + T$
$\Delta v_n\ (t-\tau_n)$ is the speed difference between the subject vehicle and the leading vehicle
Δx is the relative position between the leading and following vehicle
$\delta = 1/k$ is the space shift
λ is the sensitivity parameter which may assume different functional forms
τ_n is the reaction time
L is the average length of the vehicle
η is the random term representing the imperfect driving behaviour equals 0.5 in the case of human-driven vehicles and 0 in the case of CAVs.

In this section, an overview of some main models (the Stimulus–Response models, the Safety, the Optimal Velocity Model, the Intelligent Driver Model, and the Psycho-Physical models) is provided:

3.3.1.1 Stimulus–response models

Gazis–Herman–Rothery (GHR) model

The Gazis–Herman–Rothery (GHR) family of models is probably the most studied model class. The basic relationship between a leader and a follower vehicle is in this case a *stimulus–response* type of function that was first introduced by the General Motors research laboratories (Chandler *et al.*, 1958; Gazis *et al.*, 1961). The framework assumes that each driver responds to a given stimulus in accordance with the following relationship:

$$response = sensitivity \times stimulus \tag{3.47}$$

In general, the following driver's response (the acceleration in the algebraic sense) is strictly influenced by the speed difference between follower and leader, and the space headway.

The first developed model was by Chandler *et al.* (1958); it was a mono-regime model based on a linear expression in which the vehicle acceleration is proportional to the relative speed between follower and leader (stimulus)

$$a_n(t) = \lambda \Delta v_n\ (t - \tau_n) \tag{3.48}$$

If the speed of the leading vehicle is higher than the speed of the following vehicle the acceleration is positive, while is negative if the opposite occurs.

The main limitation of this model was on the independence of the stimulus from the vehicle distance; it follows that, for a given speed difference, the model predicts the same value of acceleration indifferently if the follower is very close or very far from the leading vehicle. Further developments were proposed to increase the model realism; specifically, the sensitivity term was modified in order to be proportional to the speed and inversely proportional to the distance between the vehicles.

The final expression is

$$a_n(t) \ = \alpha \cdot v\beta_n \ (t) \cdot \Delta v_n \ (t - \tau_n)/\Delta x_n \ (t - \tau_n)\gamma \tag{3.49}$$

where $\alpha > 0$, β and γ are model parameters that control the proportionalities.

However, this model is unrealistic in representing human ability. Indeed, if the speed difference is null, it is expected that no reaction occurs (except small random variations). In the case of low-density conditions, however, it is expected that a small adaptation may occur if the spacing is lower than the visibility distance while no reaction occurs if the spacing is longer. A clear drawback of this model is that it does not differentiate between positive and negative speed differences and predicts the same value of either acceleration or deceleration for the same absolute value of speed difference.

To capture differences in driving behaviour Yang and Koutsopoulos (1996) proposed a multi-regime model in which depending on the spacing between vehicles three different driving conditions may occur:

- *emergency*: if the headway is lower than a fixed threshold (h_{lower});
- *free flow driving*: if the headway is higher than a fixed threshold (h_{upper});
- *car-following*: if the headway is between two thresholds above.

$$a_n(t) = \alpha_{acc/dec} \cdot v_n\beta_{acc/dec}(t) \cdot \Delta v_n(t - \tau_n)\lambda_{acc/dec}, /\Delta x_n(t - \tau_n)\gamma_{acc/dec}, \tag{3.50}$$

where $\alpha_{acc/dec}$, $\beta_{acc/dec}$, $\gamma_{acc/dec}$, and $\lambda_{acc/dec}$ are parameters to be calibrated.

Stability in microscopic models

The stability analysis concerns the GHR model. From a theoretical point of view, it is of interest to determine, on the one hand, what are the conditions for system instability (hiccup or collision between vehicles) and, on the other, which law regulates the system at a steady state. The latter aspect is fundamental for studying the system's performance in "normal" conditions, when the vehicles all travel at the same speed, regardless of the behavioural differences between one driver and another, which are considered negligible compared to the macroscopic behaviour of the traffic stream. The conditions that determine the establishment of instability are an equally important aspect since the system performance is extremely degraded. From an engineering point of view, the determination of the stationary state is mainly relevant for the planning and design of the system components; the instability conditions, on the other hand, have relevance mainly for the design of the regulation and control systems.

The following paragraphs are dedicated to the study of the transient, which allows for determining the conditions for the stability of the system, to the study of the stationary state, and to a brief mention of the experiments performed for the calibration of the model.

In physics, an equilibrium system is said to be *stable* if, once perturbed, it returns to its initial position at the end of the perturbation; unstable if, following the disturbance, it moves away indefinitely.

In mathematical terms, a system subjected to perturbations is described by a system of differential equations. The solution of the system is said to be stable if an arbitrarily small variation of the initial values implies, after an interval of anyone, a sufficiently small variation of the solution. The solution of the system is said to be *asymptotically stable* if not only is it stable, but, after an infinite time, it tends to any other solution of the system. In an asymptotically stable system, therefore, all solutions finally converge to a single value.

3.3.1.2 Safety distance

In this class of models, drivers of the following vehicles try to completely preserve the safety distance with respect to the leading vehicle. In particular, the speed is selected by the driver to ensure that the vehicle can be safely stopped in the case that the preceding vehicle should suddenly brake.

The safety distance is computed on the base of the motion equations. Gipps (1981) proposed a multi-regime model in which two driving conditions are identified: the free flow driving and the car-following regime.

Let be

b the constant deceleration,
τ_n the constant reaction time
s_0 the minimum gap

the braking distance of the leading vehicle is given by

$$\Delta x_l = v_l(t)2/2b \tag{3.51}$$

the stopping distance is given by

$$\Delta x_n = v_n(t)\,\tau_n + v_n(t)2/2b \tag{3.52}$$

the distance gap, if the leading vehicle decelerates, should be not smaller than

$$x \geq s_0 + \Delta x_n - \Delta x_l \tag{3.53}$$

and given the last expression, the maximum feasible speed, defined safe speed, v_{safe}, is

$$vsafe = -b\tau_n + [(b_n2(t)\tau_n2 + v_l(t)2 + 2b(s - s_0)](1/2) \tag{3.54}$$

Summing up

$$v_n(t + \tau_n) = \min\{v + a\,\tau_n,\ v_0,\ vsafe\} \tag{3.55}$$

Krauß's model

The two main limitations of the Gipps model concern its unsuitability in the case of unstable traffic flow conditions and the possibility that the model has no solutions due to its analytical formulation. Therefore, the Krauß model, developed by Stephan Krauß in 1997 is a space-continuous model. Krauß model may be considered an alternative approach to that of Gipps, it estimates the speed of the vehicle without deriving it from the acceleration profile of the vehicle. This model is also considered a stochastic version of the Gipps model.

In accordance with the Krauß model, the safe speed is given by the following expression

$$v_{safe} = v_l(t) + (g(t) - v_l(t) \cdot t_r) / \left[(v_l(t) + v_f(t))/2b + t_r \right] \tag{3.56}$$

where

$v_l(t)$ is the speed of the leading vehicle at time t
$g(t) = \Delta x - L$ is the gap between leader and follower at time t
t_r is the drivers' reaction time and
b is the max value of deceleration.

Finally, the desired speed is given as the minimum between the maximum speed, the speed that can be achieved by the vehicle according to its acceleration, and the safe speed as defined above. That is:

$$v_{des} = \min\left[v_{\max}, v(t) + a\Delta t, v_{safe}(t) \right] \tag{3.57}$$

Therefore, the speed and position of the vehicle are computed as in the following

$$v(t + \Delta t) = \max[0, v_{des}(t) - \eta] \tag{3.58}$$

$$x_n(t + \Delta t) = x_n(t) + v(t + \Delta t) \cdot \Delta t \tag{3.59}$$

3.3.1.3 Optimal velocity and desired measures models

Next are the continuous time models, based on first-order differential equations. The two main contributions in the literature concern the *Optimal Velocity Model* (OVM; Bando *et al.*, 1995) and the *Intelligent Driver Model* (IDM, Treiber *et al.*, 2000). In the above class of models, it is supposed that each vehicle has a desired speed depending on the distance between vehicles.

About the OVM, it must be highlighted that the acceleration of the vehicle depends on the desired speed and can be formulated as

$$a_n(t) = (V[\Delta x_n(t)] - v_n(t))/\tau \tag{3.60}$$

where

n is the following vehicle;
$\Delta x_n(t)$ is the spacing between the leading and the following vehicle;
$v_n(t)$ is the speed of the vehicle;
τ is driver sensitivity.

However, one of the main limitations of the model concerns the unrealistic (high) values of maximum acceleration when the drivers' sensitivity is of the same order as the drivers' reaction time, which depends on the difference between the vehicles' speeds (Treiber *et al.*, 2000) that is not considered.

In general, in the IDM formulation, acceleration is a continuous function of speed, distance, and speed difference. Let:

a_0 be the maximum acceleration/deceleration;
$v_n(t)$ be the speed of the following vehicle;
v_0 be the desired speed of the following vehicle;
δ be a model parameter to be calibrated;
Δx_0 be the desired distance, a function of the follower's speed and the speed difference;
$\Delta x_n(t)$ be the gap distance between two vehicles.

The final formulation of acceleration is composed of two terms, the free flow term and the interaction term as detailed in the following:

$$a_n(t) = a_0 \cdot \{1 - [(v_n(t))/v_0\,]\,\delta - [\Delta x_0\,(v_n(t), (v_n(t)))/\Delta x_n(t))]2\} \tag{3.61}$$

This model has no explicit reaction time, and it describes more closely the characteristics of semi-automated driving adaptive cruise control than that of a human driver.

More in general this model can be considered composed of three terms depending on the space headway:

when the distance between the leading and following vehicles is relatively high, the third term becomes negligible, and thus the model acts as a free-flow model, and the vehicle acceleration completely depends on the desired speed;
for closer space headway between vehicles, the following vehicle reduces the free-flow acceleration depending on the third term.

3.3.1.4 Psycho-physical or action-point models

The basics of the *psycho-physical* models are the introduction of "perceptual thresholds" used to define threshold values for actions, representing the driver's perception and reaction to it.

These thresholds or *action points* introduced by Michaels (1963) are expressed as a function of speed difference and spacing between two successive vehicles in a car following the regime. In general, thresholds can alert drivers or provide more freedom depending on the spacing whether it is small or large. The key point is the introduction of the driver's perception of vehicle distance by the effect of different relative speed perceptions due to the visual angle threshold.

Wiedemann's model

The perception thresholds car-following models assume that drivers can react to changes in spacing or relative velocity only when some thresholds are reached. These thresholds represent different perception levels, corresponding to different driving regimes, that characterised different driver's behaviour.

Multi-regime models are more realistic but have some drawbacks: introduce discontinuities in the mathematical formulation and require cumbersome solution methods. In the last years, great advances in computer science opened the opportunity to develop even more complex models to implement sophisticated micro-simulation models.

Wiedemann and Reiter (1992) formalised a model that classifies such regimes and introduced different mathematical relationships to describe the corresponding drivers' behaviour to apply them in a simulation model. In more detail: The threshold values differentiate the driving regime into four parts:

- Uninfluenced free-flow driving (very far vehicles)
- Closing process-approaching slower vehicle (consciously influenced by a slower front vehicle)
- (Car-) Following process (unconscious influence)
- Emergency braking (avoid collision)

Wiedemann's model represents human perceptions and reactions through a set of thresholds of the desired distances that delimit these four driving regimes. The types of driver's reactions are different depending on the two cases of either approaching or leaving vehicles, respectively represented on the right and the left part of the plane. On each part, the different regimes are identified by thresholds defined by specific mathematical functions that express the desired distance.

The feasible set of solutions is limited on both sides by a lower bound corresponding to the distance between stopped vehicles and an upper bound corresponding to the visibility limit.

The desired distance between standing vehicles AX is the lower bound that is given by the sum of the length L of the vehicle and the desired distance between the bumpers of the two vehicles, which is defined by the combination of the calibration parameters α and β, the latter of which multiplies a normally distributed driver-dependent variable ε_S that considers the safety needs of the driver:

$$AX = L + \alpha + \beta\varepsilon S \tag{3.62}$$

Behind this horizontal line traffic states are unfeasible and would correspond to a collision.

Let us now consider the case of approaching vehicles, represented on the right part of the plane.

From the bottom (low distance), the desired minimum distance for approaching vehicles is limited by a lower bound (represented by the line ABX in Figure 3.9) that denotes the perceptual threshold for the minimum distance that requires emergency braking to avoid a possible collision:

$$ABX(t) = AX + BX(t) \tag{3.63}$$

$$BX(t) = (\gamma + \delta\varepsilon S)\sqrt{(v_n(t))} \tag{3.64}$$

where AX is the distance between stopped vehicles defined above and BX is a speed-depending term; v is the time-dependent speed of the slowest vehicle, ε a normally distributed driver-dependent parameter, and g and d are calibration

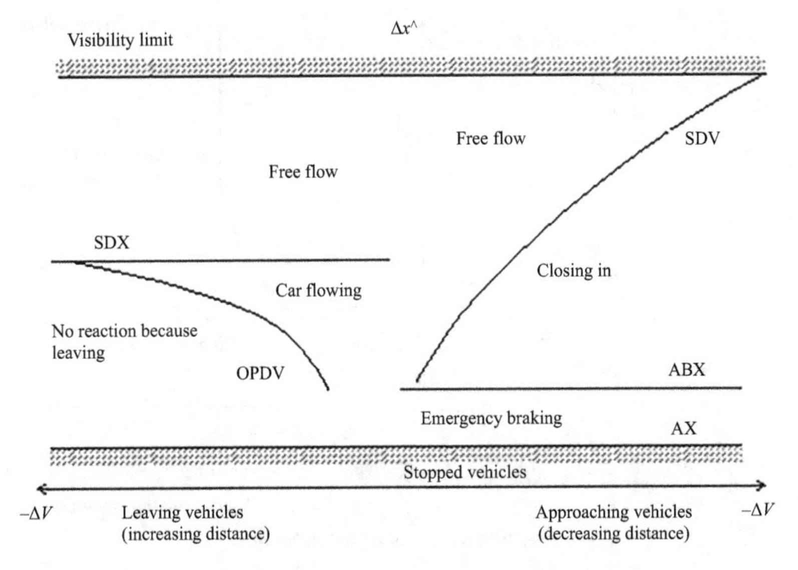

Figure 3.9 Different traffic regimes in the car-following model for perception thresholds (simplified figure from Wiedemann and Reiter, 1992)

parameters. The threshold *ABX*, because of the *BX* term, is an increasing less-than-linear function of the vehicle speed.

At the top, the desired distance for approaching vehicles is limited by an upper bound, (represented by the curve *SDV* in Figure 3.10) that denotes the point where the driver notices that he is approaching a slower vehicle and selects the desired spacing. This perceptual threshold of speed variations is not constant but increases with the relative distance Δx:

$$SDV(t) = (\Delta x(t) - AX)/(\eta + \zeta \cdot (\varepsilon_1 + \varepsilon_E)) \tag{3.65}$$

where $\eta + \zeta$ are calibration parameters and ε_E is a normally distributed random variable that considers the drivers' estimation ability. The left part of the plane describes the case of leaving vehicles, whose relative distances are increasing.

Apart from the lower bound that delimits the unfeasible region corresponding to standing vehicles, in the semi-plane that describes the case of leaving vehicles, there are large portions in which the following driver has no reaction because the distance from the leader or the relative speed is too high, and even increasing. The existence of conscious reactions by a driver following a faster vehicle is limited by two thresholds, corresponding to the perception capability of growing distances.

The upper threshold (represented by the line *SDX* in Figure 3.10) identifies the perceptual capability of a driver to recognise spacing differences when he is leaving the following process because the leading vehicle is becoming too far. The mathematical expression is very similar to the function *ABX*, with the remarkable difference in considering the speed of the following vehicle instead of the leader, other than the inclusion of an additional random term ε that is independent of the driver

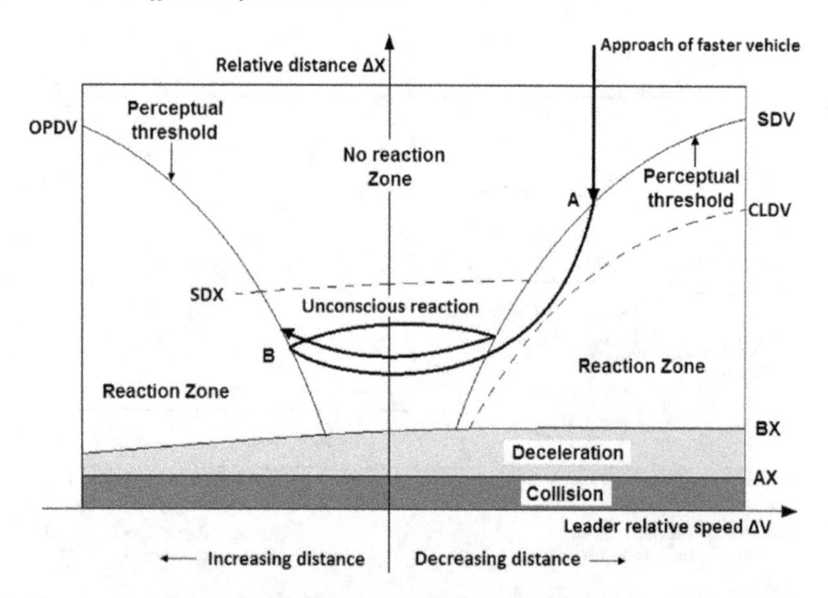

Figure 3.10 Car-following model of Wiedemann, thresholds and one vehicle trajectory vehicle that approaches and then follows a leading vehicle in $\Delta x - \Delta v$ plane (Source: Wiedemann, 1974)

characteristics and the corresponding calibration parameter λ:

$$SDX(t) = AX + \lambda\cdot(\varepsilon_E + \varepsilon)BX(t) \tag{3.66}$$

$$BX(t) = (\gamma + \delta\varepsilon S)\sqrt{(v_n + 1\ (t))} \tag{3.67}$$

The lower threshold (represented by the line *OPDV* in Figure 3.10) identifies the perceptual capability of a driver to recognise small speed differences at short but increasing distances.

$$OPDV(t) = \varepsilon_3\cdot\xi\cdot(\varepsilon_E + \varepsilon)\cdot SDV(t) \tag{3.68}$$

being ξ a calibration parameter.

After having defined the different thresholds, we can now describe the drivers' behaviour in the traffic regimes bounded by such thresholds.

Free-Flow regime. Traffic is in a free flow regime when no conditioning is perceived by the driver who tends to reach or maintain his desired free speed by applying an acceleration of the vehicle.

$$dv_n + 1/dt = \varphi\cdot v_{\max} - v_{\max}/(v_d + \vartheta\cdot(v_{\max} - v_d))\cdot v_n + 1 \tag{3.69}$$

which is a function of the current speed v_{n+1}, the desired speed v_d, and the maximum speed $v_{\max}$, and φ and ϑ are calibration parameters.

In the case of approaching vehicles (decreasing distances), the free flow regime is above the *SDV* curve; in the case of leaving vehicles (increasing distances), the free flow regime may occur even for small distances, if the leading vehicle is significantly faster (that is, below the lower threshold *OPDV*), other than for big distances, when the driver cannot perceive differences in spacing (that is, above the upper threshold *SDX*).

Car-following regime. In the portion of the $\Delta x - \Delta v$ plane comprised between the thresholds *SDX* and *OPDV*, the combination of spacing and relative speed is small enough that they are perceived by the following driver. The driver follows the leader at quite the same speed, and he does not consciously react to the movements of the leader but tries to keep acceleration low. This is represented by the lowest value of acceleration and deceleration $a_{\min}$

$$dv_n + 1/dt = a_{\min}\,(\varepsilon_3 + \varepsilon) \tag{3.70}$$

That multiplies the random component composed by a driver-dependent term ε_3, which considers the driver's ability to control acceleration and a driver-independent normalised random term ε.

Wiedemann assumed that unconscious reactions occur for small differences in speed and distance modelled in the region between the thresholds OPDV, ABX, or SDX. He modelled this unconscious car-following behaviour by assuming an oscillating process and keeping the acceleration constant until one of the thresholds OPDV, ABX, or SDX that limit the unconscious car-following region is reached.

Closing-in regime. Traffic states between *ABX* and *SBV* correspond to a traffic regime consisting of a closing-in process in which the following driver realises that he is temporarily faster than the leader. After a short delay due to the physical reaction, he begins to decelerate to reduce his speed up to that of the leader and keep a distance no lower than *ABX*. The deceleration of the follower exceeds that of the leader by a quantity depending on the relative speed Δv and the remaining distance $(ABX - \Delta x)$ from the minimum desired spacing *ABX*:

$$dv_n + 1/dt = 1/2{\cdot}[\Delta v_2/(ABX - \Delta x)] + dv_n/dt + (\mu \cdot (\varepsilon_2 + \varepsilon))/\rho_{n+1} \tag{3.71}$$

The last random term considers the human error in estimating the distance from the preceding vehicle and introduces a factor ρ_{n+1} that models the learning process of the driver during the conscious approach: the longer a driver is accelerating, the better he will estimate the motion of the leader vehicle.

Emergency braking. Under the *ABX* threshold of minimum desired distance, the traffic regime consists of emergency braking, when vehicles are so close that the follower brakes at a faster rate than normal; the expression is similar to the previous one, but since the spacing has exceeded the ABX threshold, the intensity of the deceleration includes an additional term that reduces the maximum deceleration $b_{\max}$ depending on how much the desired distance has been exceeded:

$$dv_n + 1/dt = 1/2{\cdot}[\Delta v_2/(ABX - \Delta x) + dv_n/dt] + [(ABX - \Delta x)/BX\; b_{\max}] \tag{3.72}$$

Stopped vehicles. This no-motion condition is the lower boundary of the emergency braking regime. No traffic state is feasible under this threshold.

Figure 3.10 exemplifies a hypothetical vehicle trajectory in the $\Delta v - \Delta x$ plane that starts from a very far distance from the vehicle in front then approaches it and finally follows it from behind. The diagram can be analysed proceeding along the trajectory from the top to the bottom. Assuming the relative distance with the front vehicle is decreasing, the following driver moves along these traffic states:

- Free-flow traffic regime: front-to-rear distance is bigger than the sight distance (green region in the figure) and there is *no reaction* by the vehicle;
- Closing-in regime: the vehicle perceives a vehicle ahead (orange region in the figure) and starts decelerating as a *reaction*;
- Unconscious car-following regime: for small Δv and Δx, the driver tries to keep the stable state: the relative speed Δv and the spacing Δx oscillate around the equilibrium in the unconscious zone.

3.3.2 Non-conventional microscopic models

3.3.2.1 MIXIC model

In the context of V2V communication (see Section 3.5), the MICroscopic Model for Simulation of Intelligent Cruise Control (MIXIC) developed by van Arem *et al.* (1997; 2006) assumes that the relative speed of the following vehicle with respect to the leading vehicle is set to zero whilst the space gap equals the desired speed. The acceleration profile is described through two components:

(i) the controlling component, which delivers reference values,
(ii) the vehicle model component, which converts the reference values into realised values.

In particular, let

v_{int} be the intended speed,
v be the current speed,
k be the speed error factor constant,
a_p be the leading vehicle's acceleration,
v_p be the leading vehicle's speed,
r be the current gap to the leading vehicle,
r_{safe} be the safe following distance,
r_{system}, be the following distance based on the system time setting
r_{min} be the minimum following distance, computed as a function of the deceleration capabilities of the following vehicle and the leading vehicle
r_{ref} be the reference gap to the leading vehicle, computed as the max(r_{safe}, r_{system}, r_{min})
k_a, k_v, and k_d are the constant factors
$a_{ref,\Delta v}$ is defined as the acceleration based on the speed difference (intended speed and current speed) of the following vehicle and it is computed through

$$a_{ref}, \Delta v = k \cdot (v_{\text{int}} - v) \tag{3.73}$$

$a_{ref,d}$ defined as the acceleration based on the speed and gap differences between the following and leading vehicles and it is computed through

$$a_{ref,d} = k_a \cdot a_p + k_v \cdot \left(v_p - v\right) + k_d \cdot \left(r - r_{ref}\right) \tag{3.74}$$

the final acceleration reference for the vehicle control is the minimum of both acceleration references.

3.3.2.2 ACC and CACC models

More recently Shladover *et al.* (2012) developed a control algorithm similar to MIXIC to estimate the speed of an ACC-equipped vehicle in the next time steps. The method aims to compute the acceleration of the vehicle on the basis of two control modes respectively based on the speed and the gap. Concerning the first one, it aims to keep the speed of the following vehicle close to the speed limit, whilst in the latter case, of gap control mode, it aims to keep the desired gap between the two vehicles. The final and decisive acceleration is computed as the minimum between the car-following and the free-flow acceleration.

Let

k_g and k_s be the gap and speed control constants,
s be the current space gap,
t_d be the desired time headway,
v_d is the desired speed,
s_d (= $t_d v$) is the desired distance between two vehicles, the car-following acceleration is computed as in the following:

$$a = k_g \cdot (s - s_d) + k_s \cdot (v_d - v) \tag{3.75}$$

Finally, let

a be the maximum acceleration
b be the maximum deceleration

the free-flow acceleration is only controlled by the speed and it is computed as:

$$a_f = \max\left(\min(k_s(v_d - v), a, b\right) \tag{3.76}$$

3.4 Summary

This chapter introduces dynamic models for running and queuing links, in particular with reference to the macroscopic and microscopic approaches. As already stated, most of these models contain parameters to be calibrated against real data.

It must be highlighted that other approaches are also discussed in the literature such as mesoscopic modelling (Di Gangi *et al.*, 2016). Furthermore, the literature has also focused on some simplified microscopic models (Storani *et al.*, 2022) or some hybrid approaches that can be used for large-scale applications (multi-scale models; Sadid and Antoniou, 2023).

Finally, concerning the microscopic models focusing on car-following driver's behaviour representation, in this chapter an overview of the mathematical models is provided, moreover simplified data-driven models with a limited number of parameters should also be considered.

References

Some suggestions for further readings among the huge literature available on these topics are reported below.

Chandler R.E., Herman R., and Montroll E.W. (1958) Traffic dynamics: Studies in car following. *Operations Research* 6: 165–184

Daganzo, C.F. (1995) Requiem for second-order fluid approximations of traffic flow. *Transportation Research B* 29(4): 277–286.

Di Gangi, M., Cantarella, G. E., Di Pace, R., and Memoli, S. (2016) Network traffic control based on a mesoscopic dynamic flow model. *Transportation Research Part C: Emerging Technologies* 66: 3–26.

Gazis D.C., Herman R., and Rothery R.W. (1961) Nonlinear follow the leader models of traffic flow. *Operations Research* 9: 545–567.

Gipps P.G. (1981) A behavioural car-following model for computer simulation. *Transportation Research B* 15: 105–111.

Kotsialos A., Papageorgiou M., Diakaki C., Pavlis Y., and Middelham F. (2002) Traffic flow modeling of large-scale motorway networks using the macroscopic modeling tool METANET. *IEEE Transactions on Intelligent Transportation Systems* 3: 282–292

Krauß, S. (1997) Towards a unified view of microscopic traffic flow theories. *IFAC Proceedings Volumes* 30(8): 901–905.

Krauß, S. (1998) Microscopic modeling of traffic flow: Investigation of collision free vehicle dynamics. Ph.D. Thesis, University of Cologne, Germany.

Krauß, S., Nagel, K. and Wagner P. (1999) The mechanism of flow breakdown in traffic flow models. *Proceedings of the 14th International Symposium of Transportation and Traffic Theory* (abbreviated presentations)

Lighthill, M.J., and Whitham, G.B. (1955) On kinematic waves II. A theory of traffic flow on long crowded roads. *Proceedings of the Royal Society of London. Series A. Mathematical and Physical Sciences* 229(1178): 317–345.

May, A.D. (1990) *Traffic flow fundamentals*. Englewood Cliffs, NJ: Prentice-Hall

Messmer, A., and Papageorgiou, M. (1990) METANET: A macroscopic simulation program for motorway networks. *Traffic Engineering Control* 31: 466–470

Michaels R.M. (1963) Perceptual factors in car following *Proceedings of the Second International Symposium on the Theory of Road Traffic Flow*. Paris: OECD, pp. 44–59.

Papageorgiou M., Blosseville J.-M., and Hadj-Salem, H. (1989) Macroscopic modelling of traffic flow on the Boulevard Périphérique in Paris. *Transportation Research Part B* 23: 29–47.

Papageorgiou, M. (1990) Modelling and real-time control of traffic flow on the Southern part of Boulevard Périphérique in Paris: part I: modelling. *Transsporatation Research Part A* 24: 345–359.

Papageorgiou, M., Papamichail, I., Messmer, A., and Wang, Y. (2010) Traffic simulation with METANET. In *Fundamentals of traffic simulation* (pp. 399–430). Springer, New York.

Payne, H.J. (1971) Models of freeway traffic and control. *Mathematical Models of Public Systems*. Simulation Councils. Inc., Vista, CA.

Payne, H.J. (1979) FREFLO: A macroscopic simulation model of freeway traffic. *Transportation Research Record* 722, 68–77.

Payne, H.J. (1979) Research directions in computer control of urban traffic systems, in W.S. Levine, E. Lieberman, and J.J. Fearnsides (eds.), American Society of Civil Engineers, New York.

Richards P.I. (1956) Shockwaves on the highway. *Operations Research* 4: 42–51.

Ross, P. (1988) Traffic dynamics. *Transportation Research B* 22(6): 421–435.

Sadid, H., and Antoniou, C. (2023) Modelling and simulation of (connected) autonomous vehicles longitudinal driving behavior: A state-of-the-art. *IET Intelligent Transport Systems* 17(6): 1051–1071.

Shladover, S.E., Su, D., and Lu, X.-Y. (2012) Impacts of cooperative adaptive cruise control on freeway traffic flow. *Transportation Research Record: Journal of. Transportation Research Board* 2324(1): 63–70.

Storani, F., Di Pace, R., and De Luca, S. (2022) A hybrid traffic flow model for traffic management with human-driven and connected vehicles. *Transportmetrica B: Transport Dynamics* 10(1): 1151–1183.

Treiber, M., and Kesting, A. (2013) Traffic flow dynamics. *Traffic Flow Dynamics: Data, Models and Simulation*, Springer, Berlin.

van Arem, B., de Vos, A.P., and Vanderschuren, M. (1997) *The microscopic traffic simulation model MIXIC 1.3*, no. REPORT INRO-VVG 1997.

van Arem, B., van Driel, C.J.G., and Visser, R. (2006) The impact of cooperative adaptive cruise control on traffic-flow characteristics. *IEEE Transactions on Intelligent Transport Systems* 7(4): 429–436.

Whitham, G.B. (1974) *Linear and Nonlinear Waves* Wiley, New York.

Wiedemann R. (1974) Simulation des Strassenverkehrsflusses. *Schriftenreihe des Institutes für Verkehrswesen der Universität* Karlsruhe.

Wiedemann, R., and Reiter, U. (1992) Microscopic traffic simulation: The simulation system MISSION, background and actual state. *Project ICARUS (V1052) Final Report*, 2: 1–53.

Yang Q.I., and Koutsopoulos H.N. (1996) A microscopic traffic simulator for evaluation of dynamic traffic management systems. *Transportation Research Part C* 4(3): 113–129.

Part II

Traffic control

Chapter 4
Common definitions

Giulio Erberto Cantarella[1]

> You look at where you're going and where you are and it never makes sense, but then you look back at where you've been and a pattern seems to emerge.
>
> Zen and the art of motorcycle maintenance – R.M. Pirsig

Outline. *This chapter reviews the basic definitions and assumptions about the road junction analysis and control.*

Definitions for Junction Analysis and Control common to all kinds of junctions are briefly reviewed below. As said in the introduction a complete discussion of this topic is out of the scope of this book; a recent comprehensive presentation of most of the topics of Junction Analysis and Control is in Roess *et al.* (2010) see also HCM (2022) (other references at the end of the chapter). In the next chapters, the main existing methods for junction Analysis and Control will be reviewed and analysed according to the definitions introduced in this chapter.

A *junction* (or intersection) is a road transportation system made up of several road infrastructures entering/exiting a common area. Users – vehicles or pedestrians – willing to pass through the common area may conflict with others. Therefore, user movements are to be controlled by signals (according to rules that are generally country-specific) mainly for safety reasons. To this end, some formal definitions are reviewed below.

4.1 Manoeuvres and conflict points

A *manoeuvre* is made up of all users moving from the same entry to go toward the same exit. Points where two manoeuvres may conflict are called *conflict points*; they are classified as (Figure 4.1):

[1]Department of Civil Engineering, University of Salerno, Italy

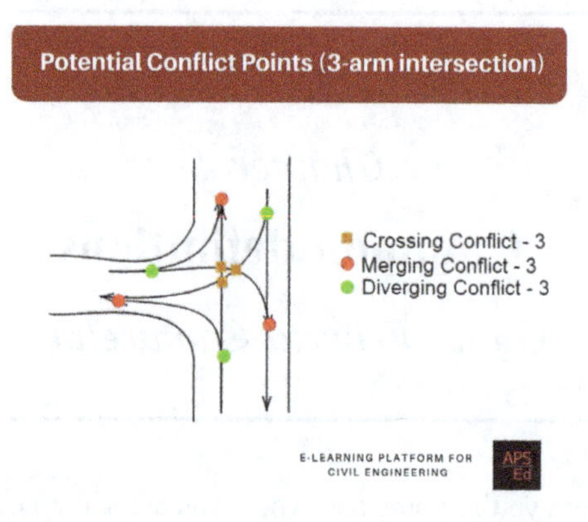

Figure 4.1 Types of conflict points (Source: https://www.apsed.in/conflict-points-intersection)

- Major conflict:
  - Crossing
- Minor conflicts:
  - Merging
  - Diverging
  - Weaving, a merging followed by a diverging

4.2 Control types

Conflicting manoeuvres should be not allowed to pass through the junction at the same time. Control rules, described to users by signals, try to eliminate conflict points whether possible or to downgrade from major to minor types,

At this aim two conflicting manoeuvres may be separated:

Over time, through

- signals fixed over time, as in a *priority junction*; a *roundabout* may be considered a (circular) sequence of priority junctions,
- signal variables over time, as in a *signalised junction*;

Over space

- horizontally, for instance, re-design the layout to change a crossing conflict into a weaving one, controlled as a priority junction,
- vertically, for instance, re-design the layout to change a crossing conflict into a converging one, controlled as a priority junction; this kind of junction is very expensive and has a large impact on the surroundings thus is mostly used for motorway entering/exiting ramps.

4.3 Junction design

The design of road junction may be broken down into three main steps:

1. definition of the geometric characteristics, mainly the width of each entry, say the part of street available for entering the junction, a part of the space used for exiting the

2. junction, parking, ...;
3. definition of the junction layout, say the division of each entry into lanes, at least 2.5 metres;
4. choice of the type of control.

The choice between different types of control requires a careful analysis of the effects such as the level of service provided, the level of safety, the maintenance costs, the realisation costs, the environmental impacts, ... The following chapters present capacity, delay and queue indicators to analyse the level of service provided by priority junctions (see Chapter 5), roundabouts (see Chapter 6), single and networks signalised junctions static (see Chapters 7 and 8), and dynamic (see Chapter 9).

4.4 Approaches, streams, and incompatibilities

Once the junction layout has been defined, a set of (one or more) lanes always receiving the same signal is called an *approach*, each approach is described by the number of lanes and their width, and each approach is associated with a *stop line* where users wait to pass through the junction. There can be specialised approaches for pedestrians, buses, emergency vehicles, ...

The set of (one or more) manoeuvres moving from the same approach is called a *stream*, thus there is a biunivocal correspondence between approaches and streams; indeed, some authors used those words as synonyms. Several variables are associated with each stream, as shown in Chapter 1. Two main flow variables, already introduced in Chapters 1 and 2, are recalled below:

f is the *arrival flow*, say the number of users arriving at the corresponding approach in a time unit (hour, minute, second); if a stream contains different types of users, such as cars, lorries, buses, ..., duly equivalence factors are introduced so that the arrival flow, as well as any other flow, is measured in pcu (passenger car unit) per time unit; anyhow a pedestrian flow is measured in persons per time unit; it can be observed (see Chapter 7) or computed through models for transportation systems analysis;

f_{MAX} is the (average) *maximum flow* or *capacity* that can move from the approach to pass through the junction, measured in pcu or persons; it can be computed from the geometric characteristics of the approach, such as width, slope, ... it can be observed (see Chapter 7) or computed from geometric characteristics of the street.

Two streams are called *compatible* if their manoeuvres may pass through the junction at the same time, that is there is no conflict point between any pair of them. Otherwise, they are called *incompatible*. The main goal of junction control is for the user to safely pass through the junction, thus avoiding two incompatible streams that have the right to pass at the same time.

Compatibilities may be described by a symmetric square binary matrix with as many rows and columns as streams and an entry equal to 1 if the streams are compatible, equal to 0 otherwise. A *compatibility clique* is a set of streams that may have the right to pass at the same time. A *maximal compatibility clique* is a clique to which no more streams can be added. More details and examples are in the following chapters.

4.5 Summary

This chapter introduces common definitions of Junction Analysis and Control. Assuming steady-state conditions, the following chapters in Part II present indicators to analyse the level of service provided by priority junctions (Chapter 5), roundabouts (Chapter 6), single and network signalised junctions – static (Chapters 7 and 8), and dynamic (Chapter 9).

References

Some suggestions for further readings are reported below among the huge literature available on these topics, see also the reference list at the end of each of the chapters.

HCM (2022). *Highway Capacity Manual 7th Edition: A Guide for Multimodal Mobility Analysis*. National Academies of Sciences Engineering and Medicine, Transportation Research Board.

Roess R. P., Prassas E. S., and McShane W. R. (2010). *Traffic Engineering*. Pearson College Div.

Chapter 5
Priority rules analysis

Orlando Giannattasio[1]

Make everything as simple as possible, but not simpler.

Albert Einstein

Outline. *This chapter introduces basic definitions and notations common to the priority rules. Some basic elements concerning the capacity and delay estimation are provided and discussed.*

5.1 Data

The necessary data for the illustrated methodology are like those of any capacity analysis, i.e. those aimed at providing a detailed description of the geometry, type of control, and flows affecting the junction.

The data relating to the geometry include the number of lanes, their use, canalisations, accumulation lanes where present, and the presence of flared secondary abutments. In fact, if two or more manoeuvres share the same lane, the capacity will decrease and vice versa, if each manoeuvre has its own specific lane available, the capacity will increase since at the same time two vehicles will be able to use the same time gap available in the priority stream.

Traffic volumes must be distinguished based on individual manoeuvres and therefore manoeuvre x will be associated with a value q_x of vehicle flow affected by the specific manoeuvre x. The flows must be specified for each manoeuvre. For a more realistic analysis of the problem the analysis is conducted for a peak period of 15 min, the hourly flows must therefore be brought back to the peak value with the help of a peak hour factor (or PHF). If the data is provided per quarter of an hour, PHF is assumed equal to 1.

The presence of traffic-lighted junctions upstream of the junction on the mainstream causes a non-random flow pattern and modifies the capacity of the secondary streams. This effect can be considered negligible if the traffic-lighted junction is more than 400 m away.

[1]Department of Civil, Computer Science and Aeronautical Technologies Engineering, Roma Tre University, Italy

The capacity is also reduced due to the presence of pedestrians crossing the junction, thus producing a delay in the manoeuvres of vehicles coming from both the non-priority stream and the priority stream.

5.1.1 Priority of the stream

In the application of the methodology, the priorities for precedence are identified as shown in Figure 5.1:

Grade 1: crossing the junction in the priority stream (manoeuvres 2, 5); turn right from the priority stream (manoeuvres 3, 6);
Grade 2: left turn from the priority stream (manoeuvres 1, 4); turn right into the priority stream (manoeuvres 9, 12);
Grade 3: crossing the junction in the non-priority stream (manoeuvres 8, 11);
Grade 4: Left turn from non-priority stream (manoeuvres 7, 10).

5.1.2 Conflict between the streams

From the above, it can be deduced that depending on the manoeuvre considered, the possible points of conflict change, directly connected to the nature of the manoeuvre itself. One of the most important parameters that we will need to calculate the capacity is the vehicle flow in conflict with the manoeuvre x for which we want to calculate the capacity and which we will indicate with $q_{c,x}$. Of all the manoeuvres, the most difficult is that of Grade 4 as it is the last in order of priority and is the one that proposes the most complex conflict point scenario. In fact, all the vehicular flows coming from the priority stream and the vehicular flows of the non-priority stream coming from the opposite direction make the right turn and cross the stream.

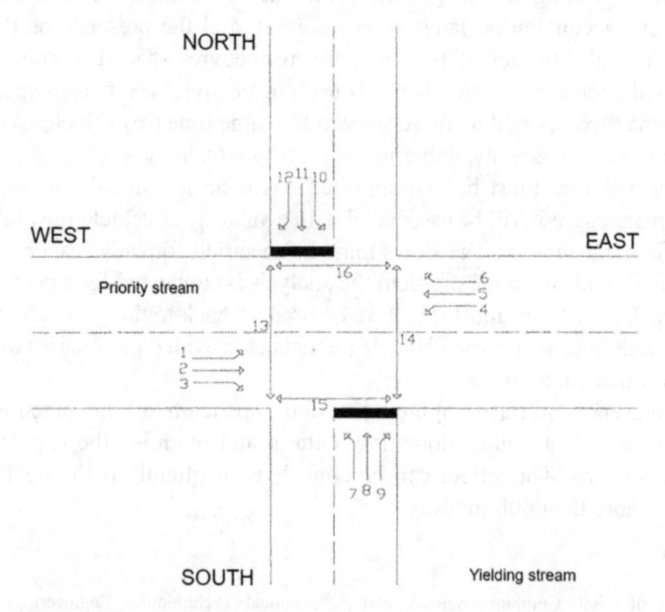

Figure 5.1 Manoeuvres at a junction without traffic lights

5.1.3 Critical gap and follow-up time

The critical gap t_c was defined as the minimum temporal distance between two consecutive vehicles of the priority stream that allows a user of the secondary stream to cross the junction.

The time gaps smaller than the critical one will be considered inadequate, while larger gaps will be deemed adequate by drivers of the non-priority stream. The critical gap is not directly observable and furthermore, it changes over time and from user to user. The critical gap is an intrinsic characteristic of each user.

The follow-up time t_f was defined as the time interval between the departures of two subsequent vehicles belonging to the secondary stream during the same time gap as the mainstream. The follow-up time can be directly observed.

With reference to cars, basic values for the times t_c and t_f obtained in US experiences are indicated in Table 5.1.

The critical gap is calculated for each secondary manoeuvre using the following equation:

$$t_{c,x} = t_{c,base} + t_{c,HV} \cdot P_{HV} + t_{c,G} \cdot G - t_{c,T} - t_{3,LT}$$

where

$t_{c,x}$ critical gap for manoeuvre x (measured in [s]);
$t_{c,base}$ base critical gap (as shown in Table 5.1);
$t_{c,HV}$ correction factor for heavy vehicles (for two-lane priority stream = 1);
P_{HV} percentage of heavy vehicles on no priority stream;
$t_{c,G}$ correction factor for the slope (for Figure 5.1 the values are 0.1 for manoeuvres 9 and 12; 0.2 for manoeuvres 7, 8, 10 and 11);
G slope expressed in % value;
$t_{c,T}$ correction factor for each part of a two-stage gap-acceptance process (1.0 for the first or second stage; 0.0 if there is only one stage);
$t_{3,LT}$ correction factor for the geometry of the junction (0.7 for the left turn manoeuvre from the non-priority stream in T-junctions; 0.0 in other cases).

Follow-up time is calculated for each secondary manoeuvre using the following equation:

$$t_{f,x} = t_{f,base} + t_{f,HV} \cdot P_{HV}$$

Table 5.1 Critical gaps and baseline follow-up times

Manoeuvers	$t_{c,base}$ **Two lane main stream**	$t_{c,base}$ **Four lane main stream**	$T_{f,base}$
Left turn from main	4,1	4,1	2,2
Right turn from secondary	6,2	6,9	3,3
Crossing from secondary	6,5	6,5	4
Left turn from secondary	7,1	7,5	3,5

where

$t_{f,x}$ follow-up time for manoeuvre x (measured in [s]);
$t_{f,base}$ follow-up time (reported in Table 5.1);
$t_{f,HV}$ correction factor for heavy vehicles (for two-lane priority stream = 0.9; for four-lane priority stream = 1);
P_{HV} percentage of heavy vehicles of a non-priority stream.

5.1.4 Potential capacity

The potential capacity associated with the single manoeuvre calculated using the gap acceptance model considered is evaluated by applying the following equation:

$$c_{p,x} = q_{c,x}\cdot\left(\left(e\hat{}\left(-q_{c,x}\cdot t_{c,x}\right)/3600\right)/\left(1-e\hat{}\left(\left(-q_{c,x}\cdot t_{f,x}\right)/3600\right)\right)\right)$$

where

$c_{p,x}$ potential capacity for manoeuvre x (measured in [veh/h]);
$q_{c,x}$ vehicle flow in conflict with manoeuvre x (measured in [veh/h]);
$t_{c,x}$ critical gap associated with manoeuvre x;
$t_{f,x}$ follow-up time associated with manoeuvre x.

For the above relationship for the calculation of the potential capacity $c_{p,x}$ to be valid, the following hypotheses must be verified:

- there are no situations of regurgitation from junctions close to the one considered;
- a separate lane is provided for each manoeuvre of the secondary stream;
- arrivals in the priority stream are not influenced by a traffic light upstream;
- no other Grade 2, 3 and 4 manoeuvres prevent the manoeuvre in question.

5.1.5 Effective capacity

To obtain the effective capacity $c_{m,x}$ of a generic manoeuvre x it is necessary to consider the various factors that can constitute a disturbance to the manoeuvre itself. It is possible to distinguish disturbance factors caused by vehicles, which vary depending on the degree of the manoeuvre x for which the capacity is being calculated, as well as factors caused by interference with pedestrians and the geometry of the junction.

5.1.6 Disturbances due to inferences between vehicles

In the case of Grade 1 manoeuvres, there is no impediment by other vehicles and therefore the effective capacity of the manoeuvre coincides with the capacity of the priority current. This means that vehicles involved in Grade 1 manoeuvres should not experience any delays. This consideration is not completely accurate and therefore any small delays are considered by introducing corrections. The capacity associated with the Grade 1 manoeuvre is the capacity of the priority current considered as undisturbed, i.e., in the absence of the non-priority current being introduced. The basic capacity is estimated at 1,700 veh/h for each direction. On extended stretches, it is estimated that the total for the two directions is 3,200 veh/h. The base value is calculated assuming the following conditions:

- Width of the travel lanes not less than 3.60 m;
- Width of the platforms not less than 1.80 m;
- Overtaking is allowed along the entire length of the route;
- Traffic flows made up of cars only;
- No impediment to transit traffic (presence of checkpoints);
- Flat terrain;
- Flow is equally distributed in both directions.

If the infrastructure on which the considered priority stream flows does not comply with these conditions, it is necessary to introduce corrective coefficients which allow the basic capacity to be corrected and the actual capacity to be obtained.

Users who must carry out Grade 2 manoeuvres are obliged to give priority only to vehicles which must carry out Grade 1 manoeuvres. Since there are no other impediments, for these manoeuvres it is assumed, indicating with j the generic Grade 2 manoeuvre:

$$c_{m,j} = c_{p,j}$$

Users who must perform Grade 3 manoeuvres are obliged to give priority to users who must perform Grade 1 and Grade 2 manoeuvres. Therefore, not all the acceptable gaps identified by users who must perform Grade 3 manoeuvres are available, given the right of precedence of higher-level manoeuvres. The importance of this impediment factor depends on the probability that the vehicles of the priority stream that want to make the left turn are waiting for an adequate gap at the same time as the vehicles that intend to make a Grade 3 manoeuvre. The greater the value of this probability, the greater the capacity-reducing effect of all Grade 3 manoeuvres.

It is, therefore, necessary to calculate the probability that the Grade 2 manoeuvres in conflict with the Grade 3 ones operate in a so-called queue-free state, i.e., the probability that there are no vehicles that want to carry out a Grade 2 manoeuvre waiting to carry it out. It is given by:

$$p_{0,j} = 1 - \left(v_j / c_{m,j}\right)$$

where $j = 1, 4$ (mainstream left-turning Grade 2 manoeuvres).

The effective capacity of Grade 3 manoeuvres is therefore calculated with the aid of a correction factor that considers the impediments caused by higher-grade manoeuvres. The correction factor f_k is given by:

$$f_k = \prod_j p_{0,j}$$

with k Grade 3 manoeuvres.

Therefore, the effective capacity of Grade 3 manoeuvres is given by:

$$c_{m,k} = c_{p,k} \cdot f_k$$

Grade 4 manoeuvres (present only in 4-leg junctions) are influenced by Grade 1, Grade 2, and Grade 3 manoeuvres. Also in this case, the estimate of the probability that each of the vehicles wishing to carry out Grade 1, 2 or 3 operate in queue-free state conditions. However, since these individual probabilities (i.e., individually for Grade 1, 2 or 3 manoeuvres) are not independent of each other, we cannot calculate the total probability as a product probability since we would obtain an overestimate of the impediment effect on the Grade 4 manoeuvre. For this reason, we use a specific relationship:

$$p' = 0.65p'' - p''/(p'' + 3) + 0.6\, radq(p'')$$

with:

p' corrective factor which considers the manoeuvres of turning left from the priority current (Grade 2) and crossing the non-priority current (Grade 3);
$p'' = p_{0,j} \cdot p_{0,k}$;
$p_{0,j}$ probability of queue-free state for the left turn manoeuvre from the priority stream (Grade 2);
$p_{0,k}$ probability of queue-free state for the manoeuvre to cross the non-priority stream (Grade 3).

The correction factor for calculating the effective capacity of the Grade 4 manoeuvre is given by:

$$f_l = p' \cdot p_{0,j}$$

where

l left turning manoeuvres of Grade 4 secondary currents (as in Figure 5.1 manoeuvres 7 and 10);
j Grade 2 right-turn manoeuvres of conflicting secondary currents (as in Figure 5.1 manoeuvres 9 and 12).

Therefore, the effective capacity of Grade 4 manoeuvres is given by:

$$c_{m,l} = c_{p,l} \cdot f_l$$

5.1.7 Disturbances due to interference with pedestrians

The flows of secondary streams must take pedestrian flows into account. The hierarchies that are established between the different vehicular and pedestrian manoeuvres are shown in Table 5.2.

A corrective factor that considers the interference caused by pedestrian manoeuvres is given by:

$$f_{pb} = \left(v_x \cdot \left(w/S_p\right)\right)/3600$$

Table 5.2 Relative hierarchies between vehicular and pedestrian manoeuvres

Vehicle manoeuvre	Pedestrian flow with priority	Disturb factor for pedestrian $p_{p,x}$
v_1	v_{16}	$p_{p,16}$
v_4	v_{15}	$P_{p,15}$
v_7	v_{15}, v_{13}	$P_{p,15}\ P_{p,13}$
v_8	v_{15}, v_{16}	$P_{p,15}\ P_{p,16}$
v_9	v_{15}, v_{14}	$P_{p,15}\ P_{p,14}$
v_{10}	v_{16}, v_{14}	$P_{p,16}\ P_{p,14}$
v_{11}	v_{15}, v_{16}	$P_{p,15}\ P_{p,16}$
v_{12}	v_{16}, v_{13}	$P_{p,16}\ P_{p,13}$

where

f_{pb} pedestrian blocking factor, a fraction of the time a lane of an apron is blocked during 1 h;
v_x number of pedestrian groups where x represents manoeuvres 13, 14, 15 or 18 (as shown in Figure 5.1);
w_j lane width (measured in [m]);
S_p pedestrian speed assumed equal to 1.2 m/s.

The correction factor for calculating the effective capacity is given by:

$$p_{p,x} = 1 - f_{pb}$$

If the number of pedestrians present is significant, the correction factor $p_{p,x}$ is included in the equation for f_k and for f_l. The equation for f_k then becomes:

$$f_k = \prod_j (p_{0,j}) \cdot p_{p,x}$$

While the equation for f_l becomes:

$$f_l = p' \cdot p_{0,j} \cdot p_{p,x}$$

where $p_{p,x}$ takes the value $p_{p,13}\ p_{p,15}$ for stream 7 and $p_{p,14}\ p_{p,16}$ for stream 10 (Figure 5.1).

5.1.8 Capacity changes due to junction geometry

In the case of secondary abutments, the presence of different manoeuvres on the same lane prevents vehicles from coming together at the stop line. Shared lane capacity can be calculated as:

$$c_{sh} = \left(\sum_y v_y\right) / \left(\sum_y (v_y / c_{m,y})\right)$$

where

c_{sh} represents shared lane capacity (measured in [veh/h]);
v_y manoeuvre flow y in the lane (measured in [veh/h]);
$c_{m,y}$ effective capacity of manoeuvre y in the lane (measured in [veh/h]).

In the described methodology, it is assumed that for the main streams the accumulation lane for left turns exists. If it does not exist, crossing manoeuvres (and sometimes right turns) can be delayed by vehicles waiting for an acceptable gap to be able to carry out the manoeuvre. To take this possibility into account, the factors $p^*_{0.1}$ and $p^*_{0.4}$ can be calculated considering the probability that there is no queue in the respective shared lanes of the main road:

$$p^*_{0,j} = 1 - ((1 - p_{0,j})/(1 - (v_{i1}/s_{i1} + v_{i2}/s_{i2}))$$

where

$p_{0,j}$ probability of queue-free state for manoeuvre j in the presence of an accumulation lane for left turns on the main road;
j 1, 4 (left turning manoeuvres of the main stream);
i_1 2, 5 (main current crossing manoeuvres);
i_2 3, 6 (right turning manoeuvres of the main stream);
s_{i1} saturation flow for main stream crossing manoeuvres (measured in [veh/h]);
s_{i2} saturation flow for main stream right turning manoeuvres (measured in [veh/h]);
v_{i1} main stream flow (measured in [veh/h]);
v_{i2} mainstream right turn flow (measured in [veh/h]).

The use of $p^*_{0,1}$ and $p^*_{0,4}$ instead of $p_{0,1}$ and $p_{0,4}$ allows us to consider the potential presence of queues on the main road with a shared left turn lane.

5.1.9 Two-stage gap acceptance

If there is a traffic island of considerable width between the two carriageways of the main road, the approach is divided into two stages, as shown in Figure 5.2.

The conflicting flows are evaluated by considering for the first stage the flows of the priority current coming from the left and, for the second stage, those of the priority current coming from the right.

To calculate the effect due to the two-stage gap acceptance process, proceed as follows: first, the capacity of the manoeuvre is calculated by assuming a single-stage gap acceptance process for the entire junction. Subsequently, the capacity values for the first and second stages c_I and c_{II} are calculated using the appropriate values of critical gap and follow-up time taken from Table 5.1. These capacity values are calculated considering the conflict flows related to each stage using the equation for $c_{p,x}$.

The capacity for the generic manoeuvre that considers the two-stage gap acceptance process is therefore determined by initially calculating a correction factor and an intermediate variable y using the following equations:

$$a = 1 - 0.32 \cdot e\hat{}(-1.3\, radq(m))\, for\ m > 0$$

$$y = (c_I - c_{m,x})/(c_{II} - v_L - c_{m,x})$$

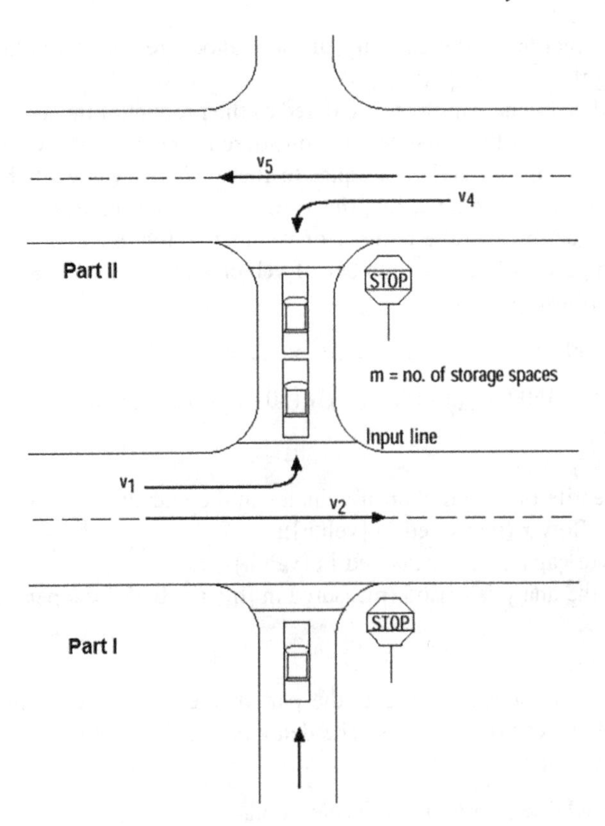

Figure 5.2 Diagram of a junction with manoeuvres involved in the two-stage gap acceptance procedure

where

m number of vehicles that can be accumulated in the centreline;
c_{I} capacity for stage I of the process (measured in [veh/h]);
c_{II} capacity for stage II of the process (measured in [veh/h]);
v_L main stream left turn flow (V1 or V4) (measured in [veh/h]);
$c_{m,x}$ manoeuvre capacity considering the total conflict flow for both stages of the process (measured in [veh/h]).

The total capacity c_T of the junction referring to the generic manoeuvre considering the two-stage gap acceptance process is therefore calculated using the equations:

$$c_T = a/(y\wedge(m+1)-1)\cdot[y\cdot(y\wedge m-1)\cdot(c_{II}-v_L)+(y-1)\cdot c_{m,x}]\ for\ y\neq 1$$

$$c_T = (a/(m+1))\cdot[m\cdot(c_{II}-v_L)+c_{m,x})]\ for\ y=1$$

5.1.10 Queue estimation

Both theoretical studies and field observations have shown that the probability distribution of the queue length for a secondary manoeuvre at a non-signalised junction is a function of the capacity of the manoeuvre and the volume of traffic served during the analysis period.

The average queue length is calculated as the product of the average delay per vehicle and the flow of the manoeuvre considered. The expected value of the total delay (measured in [veh*h/h]) is equal to the expected value of the number of vehicles in the average queue, i.e., the value of the total hourly delay is numerically the same as the average number of vehicles in the queue.

The 95th percentile of the number of vehicles in the queue can be calculated using the relationship:

$$Q_{95} = 900 \cdot T[v_x/c_{m,x} - 1 + radq((v_x/c_{m,x} - 1)\hat{}2 \\ + ((3600/c_{m,x}) \cdot (v_x/c_{m,x}))/150 \cdot T] \cdot (c_{m,x}/3600)$$

where

Q_{95} 95° percentile of the number of vehicles in the queue;
v_x manoeuvre flow x (measured in [veh/h]);
$c_{m,x}$ manoeuvre capacity x (measured in [veh/h]);
T duration of the analysis period (measured in [h]; $T = 0.25$ for a period of 15 min).

5.1.11 Delay

The delay due to control represents the part of the delay due exclusively to the presence of flow control measures. The delay due to the control is made up of the following components:

- time wasted due to the initial deceleration,
- time in the queue,
- stop at the precedence signal,
- time wasted due to the final acceleration.

The calculation of the delay can be done using the following relationship:

$$d = 3600/c_{m,x} + 900 \cdot T \cdot [v_x/c_{m,x} - 1 + radq((v_x/c_{m,x}) - 1)\hat{}2 \\ +((3600/c_{m,x})(v_x/c_{m,x}))/450 \cdot T)] + 5$$

where

d delay due to control (measured in [s/veh]);
v_x manoeuvre flow x (measured in [veh/h]);
$c_{m,x}$ manoeuvre capacity x (measured in [veh/h]);
T duration of the analysis period (measured in [h]; $T = 0.25$ for a period of 15 min).

The analytical model for calculating the control delay assumes that demand is less than capacity for the analysis period. If the degree of saturation is greater than 0.9, the length of the analysis period affects the calculation of the average delay due

Table 5.3 Criteria for determining the level of service for junctions with priority currents

Level of service	Average delay due to control [s/veh]
A	0–10
B	>10–15
C	>15–25
D	>25–35
E	>35–50
F	>50

to the control. In most cases, an analysis period of 15 min is recommended. If demand exceeds capacity in the period considered, the length of the period will need to be increased to include the over-saturation period.

5.1.12 Level of service

The service level for a junction with priority currents is not defined for the junction itself but is determined based on the value of the delay due to the control and is defined for each of the secondary current manoeuvres. The criteria for determining the service level are shown in Table 5.3.

5.2 Conflict between the streams

This chapter describes the operations necessary to conduct the analysis of a junction without priority currents.

5.2.1 Data

The necessary data to develop the procedure concern the geometry, extent and composition of the flows, and the study period to be considered. As far as geometry is concerned, it is necessary to define the number and configuration of the lanes for each abutment. As regards the flows affecting the junction, it is necessary to know for each junction the extent of the flows v for each manoeuvre and the percentage of heavy vehicles. The analysis is conducted considering the hour as the reference period, therefore, to report the findings to hourly values it is necessary, in general, to define a PHF peak factor. If the data relates to time, $PHF = 1$.

The flow values v are converted to the equivalent hourly values V using the peak hour factor:

$$V = v/PHF$$

and are associated with each lane of each settlement either as a function of observations or through estimation.

5.2.2 Geometric group definition

For each abutment, it is necessary to determine the geometric group which is defined according to the number of lanes of the opposite and conflicting abutments as indicated in Table 5.4.

5.2.3 Conflict between the streams

The saturation time spacing for each lane for conflict level i manoeuvres h^s_i is determined, starting from a basic value, and considering the turns and the vehicle composition, by means of the relation:

$$h_i^s = h_i + h_{dx}^{corr} \cdot p_{dx} + h_{sx}^{corr} \cdot p_{sx} + h_{vp}^{corr} \cdot p_{vp}$$

where

h_i is the base value derived from the values indicated in Table 5.5;
h^{corr}_{dx}, h^{corr}_{sx}, h^{corr}_{vp} are the correction factors for right turn manoeuvres, left turn manoeuvres, and the presence of heavy vehicles respectively;
p_{dx}, p_{sx}, p_{vp} are the percentages of right turn manoeuvres, left turn manoeuvres, and heavy vehicles respectively present on the certificate considered.

The values of the correction coefficients used to consider turning manoeuvres and heavy vehicles are indicated in Table 5.6.

5.2.4 Definition of the initial value of the start split times

The initial value of the departure split times is set equal to $h_d = 4$ s.

Table 5.4 Criteria for defining the geometric group

Geometric group	**Junction configuration**	**Number of lanes**		
		Considered stand	**Opposite stand**	**Stand in conflict**
1	4 brackets or *T*	1	I	1
2	4 brackets or *T*	1	I	2
3a/4/a	4 brackets or *T*	1	2	1
3b	*T*	1	2	2
4b	4 brackets	1	2	2
5	4 brackets or *T*	2	1 or 2	1 or 2
6	4 brackets or *T*	3	3	3

Table 5.5 Basic saturation temporal spacing values as a function of conflict level and geometric group

Level of conflict	Number of vehicles	Group 1	Group 2	Group 3a	Group 3b	Group 4a	Group 4b	Group 5	Group 6
1	0	3.9	3.9	4.0	4.3	4.0	4.5	4.5	4.5
2	1	4.7	4.7	4.8	5.1	4.8	5.3	5.0	6.0
	2							6.2	6.8
	≥3								7.4
3	1	5.8	5.8	5.9	6.2	5.9	6.4	6.4	6.6
	2							7.2	7.3
	≥3								7.8
4	2	7.0	7.0	7.1	7.4	7.1	7.6	7.6	8.1
	3							7.8	8.7
	4							9.0	9.6
	≥5								12.3
5	3	9.6	9.6	6.7	10.0	9.7	10.2	9.7	10.0
	4							9.7	11.1
	5							10.0	11.4
	≥6							11.5	13.3

Table 5.6 Correction factors as a function of the geometric group

Factor	Group 1	Group 2	Group 3a	Group 3b	Group 4a	Group 4b	Group 5	Group 6
h^{corr}_{sx}	0.2	0.2	0.2	0.2	0.2	0.2	0.5	0.5
h^{corrd}_{x}	–0.6	–0.6	–0.6	–0.6	–0.6	–0.6	–0.7	–0.7
h^{corr}_{vp}	1.7	1.7	1.7	1.7	1.7	1.7	1.7	1.7

5.2.5 Degree of saturation

For each settlement, the calculation of the degree of saturation x is carried out given by the product of the flow and the temporal spacing of the departures:

$$x = (V \cdot h_d)/3600$$

5.2.6 Current value of start split times

The current value of the departure split times for each lane, a function of the traffic flows on opposite and conflicting positions, is calculated by considering the probability that for each lane there is a vehicle waiting on the stop line.

The probability is a combined probability based on every case of vehicles present and not present in each lane.

The expected value of the distribution is given by:

$$h_d = \Sigma_i p[C_i] \cdot h_i^s$$

where

$p[C_i]$ is the probability that conflict level i occurs;
h^s_i is the saturation time gap for level i.

The probability $p[C_i]\cdot$ for each level of conflict is calculated according to the degree of saturation of the attestation considered; each case is described schematically in Table 5.7.

Therefore, with reference to Table 5.7, the probability for each degree of conflict $P(C_i)$ can be calculated according to the following equations:

$$P(C_1) = (1 - x_O) \cdot (1 - x_{CS}) \cdot (1 - x_{CD})$$

$$P(C_2) = (x_O) \cdot (1 - x_{CS}) \cdot (1 - x_{CD})$$

$$P(C_3) = (1 - x_O) \cdot (x_{CS}) \cdot (1 - x_{CD}) + (1 - x_O) \cdot (1 - x_{CS}) \cdot (x_{CD})$$

$$P(C_4) = (x_O) \cdot (1 - x_{CS}) \cdot (x_{CD}) + (x_O) \cdot (x_{CS}) \cdot (1 - x_{CD}) + (1 - x_O) \cdot (x_{CS}) \cdot (x_{CD})$$

$$P(C_5) = (x_O) \cdot (x_{CS}) \cdot (x_{CD})$$

If there are multiple lanes, it must be considered that, compared to the case developed for a single lane, at each conflict level there can be multiple vehicles for the same manoeuvre. Consequently, separate saturation time spacings are calculated for each vehicle that obstructs the considered vehicle for each conflict level. In the case of a junction with two lanes in each direction, the possible combinations between the number of vehicles in each abutment for each conflict level are shown in Table 5.8. These cases can, in turn, be divided according to the lane occupied for each station, so that we go from the 27 cases listed in Table 5.8 to 64 cases. A similar reasoning can be developed for junctions with three lanes for each abutment.

Table 5.7 Conflict levels and probabilities of each case

Level of conflict	Stand considered	Opposite stand	Conflict with left stand	Conflict with right stand	Probability
1	✓	–	–	–	$(1 - x_O)(1 - x_{cs})(1 - x_{CD})$
2	✓	✓	–	–	$(x_O)(1 - x_{cs})(1 - x_{CD})$
3	✓	–	✓	–	$(1 - x_O)(x_{cs})(1 - x_{CD})$
3	✓	–	–	✓	$(1 - x_O)(1 - x_{cs})(x_{CD})$
4	✓	✓	–	✓	$(x_O)(1 - x_{CS})(x_{CD})$
4	✓	✓	✓	–	$(x_O)(x_{cs})(1 - x_{CD})$
4	✓	–	✓	✓	$(1 - x_O)(x_{cs})(x_{CD})$
5	✓	✓	✓	✓	$(x_O)(x_{cs})(x_{CD})$

Table 5.8 Conflict levels and probabilities of each case

Level of conflict	Number of vehicles in conflict	Considered stand	Opposite stand	Conflict with left stand	Conflict with right stand
1	0	1	0	0	0
2	1	1	1	0	0
2	2	1	2	0	0
3	1	1	0	1	0
		1	0	0	1
3	2	1	0	2	0
		1	0	0	2
4	2	1	1	0	1
		1	1	1	0
		1	0	1	1
4	3	1	2	1	0
		1	1	2	0
		1	0	1	2
		1	0	2	1
		1	2	0	1
		1	1	0	2
4	4	1	2	2	0
		1	2	0	2
		1	0	2	2
5	3	1	1	1	1
5	4	1	1	2	1
		1	2	1	1
		1	1	1	2
5	5	1	2	2	1
		1	2	1	2
		1	1	2	2
5	6	1	2	2	2

5.2.7 *Convergence of start splits*

If the expected value of the temporal spacing of departures varies, for each attestation, by a quantity greater than a previously set limit (e.g. 0.01 second) it is necessary to recalculate the values of the degree of saturation otherwise the service time s can be calculated as $s = hd - m$ where m is the start-up time and is considered equal to 2 s for sites falling into geometric groups 1 to 4 and 2.3 seconds for those falling into groups 5 and 6.

5.2.8 *Capacity*

The capacity for each claim is not a value that can be calculated directly. Given the flow for each station, it is possible to calculate the average waiting value for each vehicle and, depending on this spacing, the fraction of time in which the resources of a given station can be exploited can be calculated.

Table 5.9 Criteria for determining the level of service for junctions without priority currents

Level of service	Average delay [s/veh]
A	0–5
B	>5–10
C	>10–20
D	>20–30
E	>30–40
F	>45

Therefore, two different definitions of capacity can be considered:

- The maximum flow value for a claim keeping the values on the other claims unchanged;
- The maximum flow value for each settlement keeping the proportion between the flows present in the different settlements' constant.

5.2.9 Delays

The average delay for each vehicle is evaluated for each lane and each stand using the relationship:

$$d = s + 900T[(x - 1)radq((x - 1)\hat{}2((s \cdot x)/400T))]$$

where

d average delay for each vehicle (measured in [s/veh]);
s service time (measured in [s]);
x degree of saturation of the lane or of the stand;
T duration of the analysis period (measured in [h]; $T = 0.25$ for a period of 15 min).

The junction delay is given by the weighted average of the delays for each stand.

5.2.10 Level of service

The service level for a junction without priority currents is not defined for the junction itself but is determined based on the value of the delay. The criteria for determining the service level are shown in Table 5.9.

5.3 Summary

This chapter provides an overview of fundamental definitions and notations associated with priority rules. It discusses basic elements related to capacity and delay estimation, offering insights into their application and significance.

This chapter deals with the priority rules with a main focus on the methods for the analysis, considering that they can be included as sub-models in larger models for Transportation Systems Analysis and Design.

References

Some suggestions for further readings are reported below among the huge literature available on these topics.

Di Gangi, M., and Mussone, L. (2010). Linee guida per la progettazione e verifica funzionale delle intersezioni non semaforizzate (pp. 1–201). Maggioli.

Esposito, T., and Mauro, R. (2003). *Fondamenti di infrastrutture viarie: La geometria stradale.*

Esposito, T., and Mauro, R. (2003). *Fondamenti di infrastrutture viarie: La progettazione funzionale delle strade.*

Kimber, R. M. (1980). *The traffic capacity of roundabouts. TRRL Laboratory Report* 942. Transport and Road Research Laboratory, Crowthorne, United Kingdom, 63.

Kimber, R. M., and Hollis, E. M. (1977). Flow/delay relationships for major/minor priority junctions. *Traffic Engineering & Control*, 18(Analytic).

Kimber, R. M., and Hollis, E. M. (1978). Peak-period traffic delays at road junctions and other bottlenecks. *Traffic Engineering & Control*, 19(N10).

Kimber, R. M., and Hollis, E. M. (1979). Traffic queues and delays at road junctions (No. LR909 Monograph).

Krogscheepers, C., Robinson, B., and Rodegerdts, L. (2001). Roundabout operations: a summary of FHWA's – '*Roundabouts: An Informational Guide*'. *20th South African Transport Conference*, 2001.

Wu, N. (1994). An approximation for the distribution of queue lengths at unsignalized intersections. In *Proceedings of the second international symposium on highway capacity* (Vol. 2, pp. 717–736). Victoria, Australia.

[illegible] with the [illegible] rules [illegible] the method [illegible] analysis considering that they can be considered as sub-models [illegible] system analysis [illegible].

References

[illegible]

Di Gangi, M. and Russo, F. (2010) [illegible]

[illegible] T. and [illegible] (200[illegible]) [illegible]

[illegible]

Kimber, R. M. (1980). *The traffic capacity of roundabouts*. TRRL Laboratory Report 942. Transport and Road Research Laboratory, Crowthorne, United Kingdom.

Kimber, R. M. and Hollis, E. M. (19[illegible]) [illegible] relationships [illegible]

[illegible] R. M. and [illegible] (19[illegible]) [illegible]

Kimber, R. M. and [illegible] (19[illegible]) [illegible] queues and delays [illegible]

Ko[illegible] [illegible]

Wu, N. (1999) An approximation [illegible] the distribution of queue lengths [illegible] In *Proceedings* [illegible] Vol. 2, pp. [illegible]

Chapter 6

Roundabouts analysis

Orlando Giannattasio[1]

I love roundabouts. … Will I go? Will I not go? The other car might go in my lane. There's a bit of a dance going. It's like a samba.

Craig Taylor

Outline. *This chapter introduces basic definitions and notations common to the roundabouts. Some basic elements concerning the roundabout design are provided, and main details are shown with respect to the capacity and delay methods.*

As described before, at-grade junctions can be classified into:

- linear;
- roundabout.

The two types are differentiated as follows:

- In the first case (Linear junction) is applied the rule of precedence on the right, with the introduction of appropriate road signs (stop or precedence);
- In the second case (Roundabout) the vehicles that are circulating within the ring have priority compared to those who have to access.

The main reasons why roundabouts are preferred with respect to Linear junctions (with or without traffic light regulation) are tied to safety and capacity; regarding safety, the roundabouts reduce the number of conflict points thanks to the one-way ring. In particular, on each arm there is an entry conflict point and a diversion point: these manoeuvres create the greatest inconvenience but are partly controlled thanks to the reduced speed guaranteed by the deflection of the trajectories. Crossing manoeuvres, which represent the cause of the greatest concerns, are eliminated due to the configuration of the roundabouts.

It is important to note that the accident rate due to queue conflicts cannot be eliminated completely but is still contained as it has been verified that roundabout junctions generally increase the capacity of the junction, also taking advantage of the dead times for insertions and decreasing the queue length.

[1]Department of Civil, Computer Science and Aeronautical Technologies Engineering, Roma Tre University, Italy

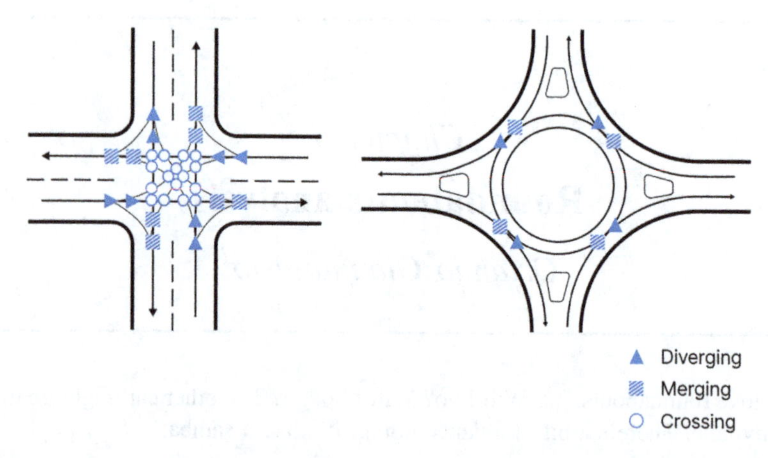

Figure 6.1 Points of conflict in at-grade junctions (Courtesy: Texas Department of Transportation)

Figure 6.1 shows the huge difference between a four-arm junction and a single-lane roundabout: the linear junction has 32 points of conflict compared to the eight of the roundabouts. If the number of lanes on the arms or on the ring is increased, the conflict points will also increase, but they will always be lower than those of Linear junctions.

6.1 Design of the geometry

The design of a roundabout is a process that interests different aspects. Due to their design, the roundabouts can increase safety and simultaneously capacity requirements of the junction. Traffic safety translates into a reduction in entry and circulation speeds in the roundabout by deflecting vehicle trajectories, but unfortunately, this reduction in speed negatively affects the capacity of the junction.

The design process of the roundabout is characterised by a first step that evaluates the feasibility of this junction. The next step is to define the geometry, knowing its positioning, the deflection of the vehicle trajectories and all the geometric elements necessary for the correct design. In the end, must be verified two important elements: the circulation of heavy vehicles and the safety of pedestrians and cyclists.

To summarise, the geometrical elements to consider for the roundabout design are:

- The centre of the roundabout,
- External diameter of the roundabout,
- Entry and Exit width;
- Rotatory Crown;
- Central Island;
- Entry and Exit curve;
- Splitter Island width;
- Possible Crossing (pedestrian, cycling,..).

In the next paragraph, each design element will be briefly analysed and discussed.

6.1.1 Centre of the roundabout

The positioning of the centre of the roundabout represents a key element for the design of this type of junction. The optimal condition is obtained when the road axes converge towards the centre of the roundabout (Figure 6.2); this design rule guarantees reduced speed in the entry, crossing and exit phases within the roundabout and ideal condition for insertion and deflection.

However, if the condition described is not possible to achieve, a slight eccentricity towards the right can be allowed, to at least safeguard the entrance which is a more critical point than the exit. It must be noted that alignment to the right should be avoided because it favours entry at very high speeds within the roundabout, increasing the possibility of a car crash. If the roundabout is part of T-junction, the arms of the roundabout must be directed towards the central island, avoiding the possibility of having roads that enter the roundabout and lapping it tangentially.

The successive branches of a roundabout form an angle of attack which should be close to the right angle. If it is not possible to obtain this value, it would be advisable for the angle to be no less than 30° to avoid visibility and safety problems. If the angle is less than 30°, one of the branches can be deviated (Figure 6.3).

After the evaluation of the centre of the roundabout, it is possible to design the remaining elements that are part of the roundabout.

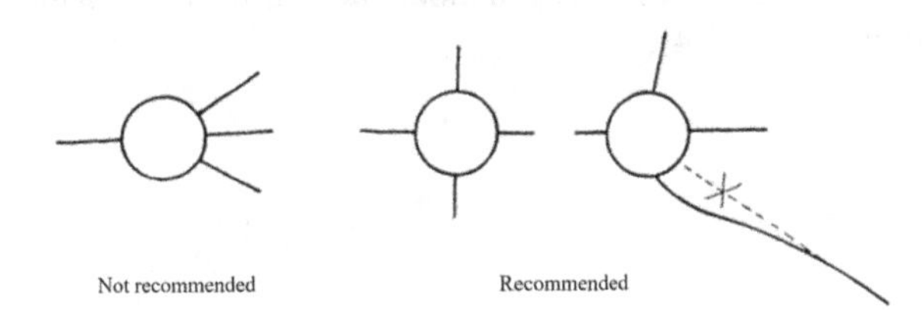

Figure 6.2 Suggested positioning for Roundabout's branches

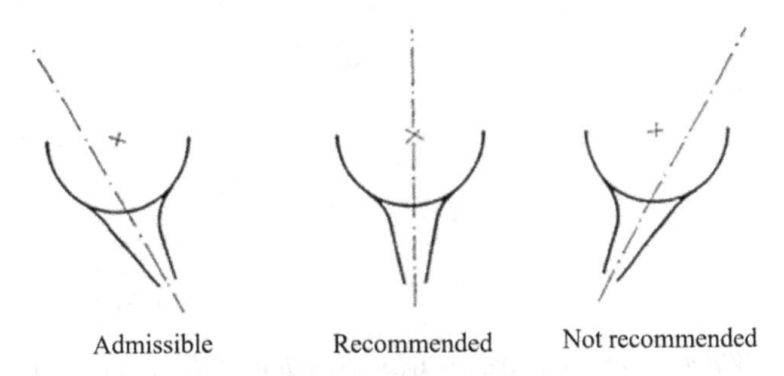

Figure 6.3 Line-up evaluation for roundabout's branches

6.1.2 External diameter

The external diameter of the roundabout is the maximum distance between two external points of the roundabout crown and is therefore equal to the sum of the width of the crown counted twice and the diameter of the central island. The ratio between the diameter of the inscribed circle and the diameter of the central island is of fundamental importance for the transit of heavy vehicles, especially in smaller roundabouts where, for example, a diameter of 18 m of the central island requires an external diameter minimum of 36 m to avoid the possibility of overtopping.

6.1.3 Enter and exit width

The entry width is one of the most important parameters to evaluate the roundabout's capacity; it measures the actual width of the entrance, near the stop line, starting on the left from the junction of the stop line with the rotary crown and on the right perpendicularly on the other side of the arm. For safety reasons, the entry width must be the shortest possible and must be between 3 m and 4 m. The number of entry lanes can be increased to satisfy the entry flow. Must be noticed that the entry point with more than two lanes must be avoided because it can reduce drastically the safety.

The width of the rotary crown must be equal to or greater than the maximum value of the entry's width and so therefore it must be sized according to it. Furthermore, if the verification of the curvature radius for heavy vehicles is not satisfied, the entrance width can be increased accordingly.

As an example, to increase the width of the entrance it is possible to:

- add a new lane upstream of the junction, starting from a distance of approximately 60 m (Figure 6.4);

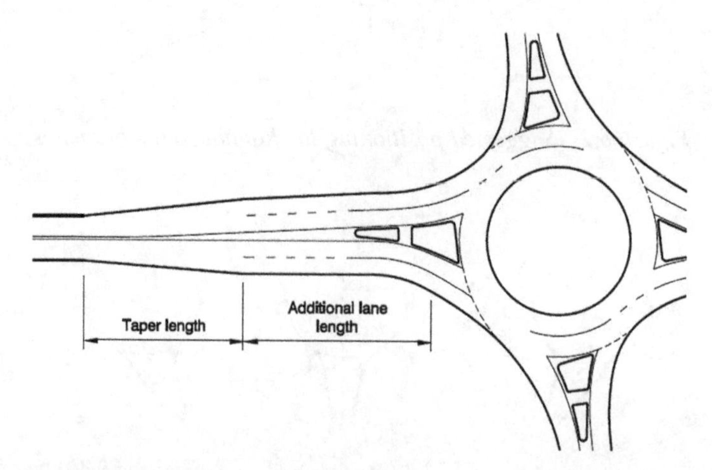

Figure 6.4 Widening of the entrance with insertion of the second entry lane upstream

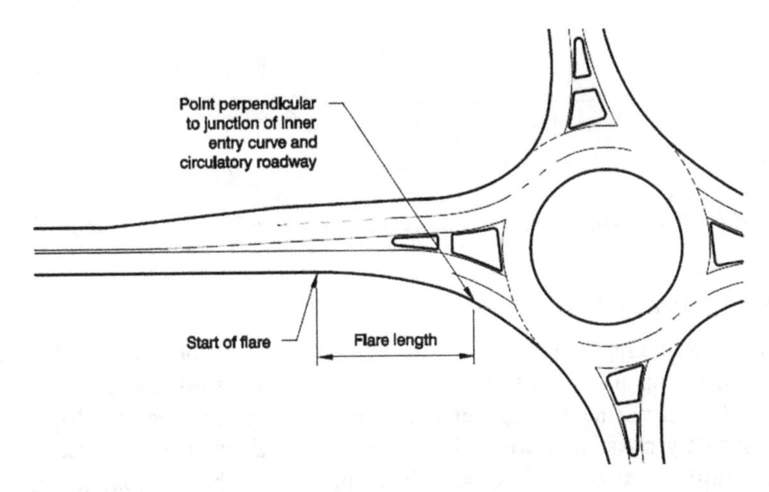

Figure 6.5 Widening of the entrance with adjustment of the entry curve

- create a flare of the entrance, starting from a distance of 25 m in urban areas and 40 m in extra-urban areas, in order to obtain a new lane before reaching the junction (Figure 6.5).

As regards the width of the exit lane, it is between 4 m and 5 m, while in the case of double lanes, the overall width is 7 m. If there is a significant flow of right turns, it is possible to insert a special lane reserved for right turns outside the roundabout.

6.1.4 Rotatory crown

The rotatory crown (or circulatory ring) represents the roadway with one or more lanes that surround the central island, it is travelled by vehicles in an anti-clockwise direction. The size of the rotary crown varies depending on the width of the entrances and the radii of curvature for the possible movements. The width of the ring must be equal to or slightly greater than the widest inlet width, without undergoing variations. Generally, the width to be adopted for the ring, excluding the platforms, is 7 m; if there are two entry lanes it can reach a width of 10 m.

If the width of the crown necessary for the transit of heavy vehicles is so large as to compromise the deflection of the trajectory of light vehicles, the construction of a passable platform around the central island is envisaged, with a consequent reduction in safety. Generally, the width is 1.50 m for mini roundabouts. For other roundabouts, it is sufficient to provide a 0.5 m paved internal shoulder. The slope of the curb varies between 3% and 6% to discourage light vehicles from travelling, but it must not be too high otherwise there is a risk of heavy vehicles overturning.

The transversal slope of the ring is 1.5%–2% and must be directed towards the outside of the roundabout, to control the speed of vehicles, improve visibility, and allow the collection and disposal of rainwater. Also, in this case, the main problem is the possibility of heavy vehicles overturning (Figure 6.6).

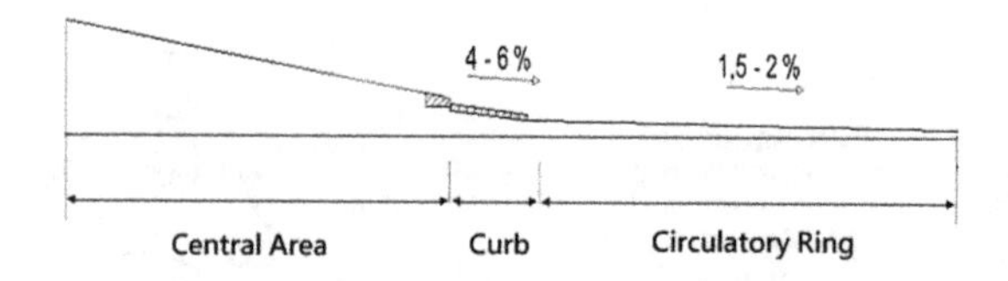

Figure 6.6 Cross-section of the roundabout crown and the surmountable platform

6.1.5 Central island

The central island represents the non-passable part of the roundabout. Its shape must be circular to guarantee constant traffic speeds and have sufficient deviation of the vehicles that cross the roundabout diameter. Elliptical or irregular shapes could augment safety risks, as it would be easier to have accelerations or decelerations that are difficult to control. As already mentioned in the previous paragraph, a surmountable dock can be provided to facilitate the manoeuvring of heavy vehicles.

The size of the central island is determined by the difference between the external diameter and the rotary crown. It is also defined as a function of the deflection angle; in fact, if this angle does not have a suitable value, the diameter of the central island and therefore the external diameter of the roundabout must be increased. Alternatively, the entrance can be moved to the left, with an increase in inclination and a reduction in width, ensuring that the turning movement checks are satisfied.

The central island is often used to carry out decorative or advertising interventions. Excavations under the ground level must be avoided because the roundabout must be clearly visible and recognisable. The platform must not be accessible to pedestrians and must not have walls, high edges or structures in general that could limit visibility and consequently safety.

6.1.6 Entry and exit curves

The entrance curves are the set of curves (one or more) along the right curb or the right shoulder that lead from the entrance to the roundabout (as shown in Figures 6.7 and 6.8).

The curvature radius of the entrance slows down the speed of the incoming vehicles and together with the dimensions of the parameters discussed so far (width of the entrances, roundabout crown, and central island) affects capacity and safety.

The entrance curve must have a curvilinear path tangent to the external point of the roundabout crown for the right side and tangential to the central island for the left side. The radius must be less than or equal to that of the outer circumference of the roundabout. The entrances can be single or double-lane.

The exit curves are designed to free the circulation ring in the shortest possible time, favouring the exit flow. The radius of the exit curve must be larger than that of the roundabout (in the external part) so as not to create difficulties for the vehicles in the exit phase. It must be curvilinear tangential to the rotary crown on the right side and to the central island on the left side.

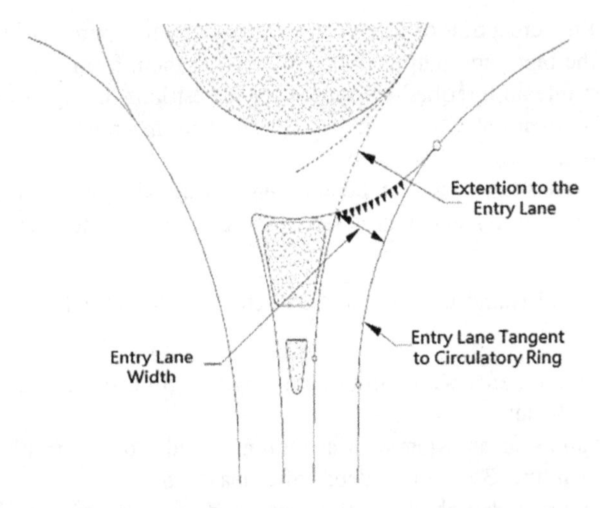

Figure 6.7 Design of entry curves for single-lane entry

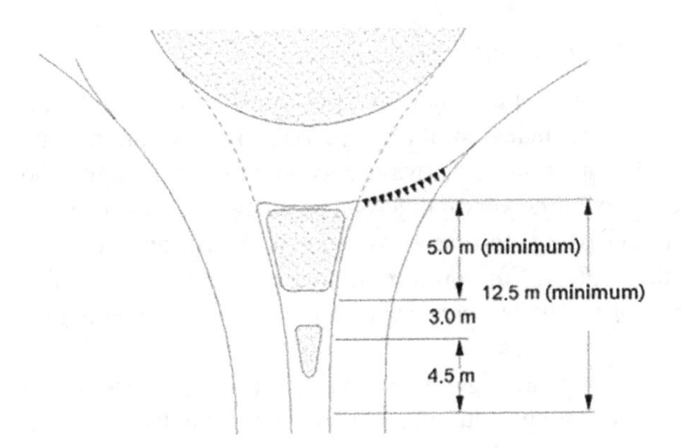

Figure 6.8 Design of exit curves for single-lane exit

Even if you want to encourage the flow of exiting vehicles, speed values must still be kept low to protect pedestrians on the exiting arch.

6.1.7 Divisional islands

The divisional islands are created at the entrance to the roundabouts and must always be present unless they represent an obstacle to circulation and visibility of the central island (for example in mini roundabouts).

The divisional islands perform various functions:

- improve the perception of the junction by users who approach it;
- separate the input and output flows, preventing them from colliding;
- they constitute an intermediate refuge for pedestrians during crossing;
- facilitate correct entry and exit vehicle circulation, preventing incorrect left-turn manoeuvres.
- create a space suitable for the positioning of vertical signs, provided this does not reduce visibility and facilitates slowing down and directing towards the desired itinerary.

The divisional island is confined between the external entry and exit curves and is made up of:

- an accumulation area with a minimum length of 5 m, sized so as to contain at least one vehicle;
- a pedestrian/cycle crossing with a minimum width of 2.5 m plus 0.25 m on each side (in total 3 m) and a depth of at least 1.5 m;
- an approach area, which closes the island and protects pedestrians.

It is important to remember that the divisional islands should be clearly visible both day and night, using light-coloured materials that contrast with the flooring or that reflect light.

6.1.8 Possible crossing

During the design phase, particular attention must be paid to the needs of vulnerable users. In fact, the inclusion of a roundabout involves an increase in pedestrian routes and the distances are not always accepted by pedestrians who tend to make faster but at the same time very dangerous crossings. For this reason, it is necessary to avoid placing crossings too far away, so that pedestrians can prefer these routes designed for their safety. Furthermore, it may be useful to insert obstacles (such as vases or plants) or pedestrian barriers with the aim of channelling pedestrian flows more on the prefixed routes.

Pedestrian crossings are highlighted on the roadway by means of zebra stripes with white stripes (or alternating contrasting colours) parallel to the direction of travel of the vehicles, with a length of no less than 2.50 m on local roads and urban neighbourhood roads, and 4 m on other roads; the width of the stripes and intervals is 0.50 m.

Cycle crossings must be provided only to ensure the continuity of cycle paths in the junction areas. They are highlighted on the roadway by two discontinuous white stripes, 0.50 m wide, with segments and intervals 0.50 m long.

6.2 Analysis

After design, functional tests are an important step for the designer due to the necessity of verifying the roundabout. In this paragraph will be presented:

- Methods for roundabout capacity;
- Methods for the delay calculation in a roundabout.

6.2.1 *Capacity of the roundabout*

The capacity C of the arm of a roundabout is the smallest value of the flow on the arm that determines the permanent presence of vehicles waiting to get in the roundabout.

The calculation of the capacity of a roundabout is carried out by considering the junction as a succession of T-junctions, in which the vehicles circulating on the ring have priority over those entering and considering the right turn as the only permitted manoeuvre.

Considering the entry to a, the capacity C can be expressed as a function of:

- geometric characteristics of the junction or its configuration;
- disruptive traffic Q_d which makes entering the roundabout difficult and which depends on the incoming flows Q_e, outgoing Q_u, and circulating Q_c;
- psychotechnical times, relating to user behaviour (usually the critical interval T_c and the sequence time T_f);
- numerical constants.

Other than the entry capacity, it is possible to define some capacity indices:

- Generic entry reserve capacity $(RC)_i$ is calculated as $(RC)_i = C_i - Q_{ei}$
- Percentage reserve capacity RC is calculated as $RC = (RC)_i \,/\, C_i \times 100$
- Percentage capacity rate TC is calculated as $TC = Q_{ei} \,/\, C_i \times 100$
- Simple Capacity CS identifies, with respect to a given traffic flow distribution scenario, the maximum flow value that can be had incoming from each branch when the congestion begins for one of those.
- Total Capacity CT represents, with respect to a given traffic distribution scenario, the sum of the values of the entering flows from each branch which simultaneously determine the congestion of the branches themselves.

The determination of capacity can be done with:

- empirical methods: are based on the observation of existing roundabouts and concern the definition of relationships between geometric characteristics, traffic flows, and capacity through regression techniques;
- theoretical methods: are based on the gap-acceptance theory, in which a time interval accepted by the driver while waiting to enter the roundabout is identified among the vehicles traveling along the ring.

In fact, when the vehicles arrive at a junction, having to carry out a manoeuvre to merge from a secondary flow into a main flow (without having priority), they carry out the manoeuvre according to the position and speed of the vehicles in the main flow. The critical interval T_c represents the smallest time interval between vehicles of the mainstream accepted by a user of the secondary stream to execute the desired crossing or entry manoeuvre. The critical interval represents a random variable. The gap represents the temporal spacing between two consecutive vehicles of the mainstream, considered sufficient for entry and can be greater than or equal to the critical interval.

The empirical-statistical methods differ slightly from each other because the capacity calculation is carried out as a function of one or more characteristics of the roundabout (geometry, configuration, and user behaviour).

Capacity assessment reports have been developed by various authors, all listed below. It should also be noted that within the square brackets has been reported the country that uses this formulation at a regulatory level.

- Kimber Formulation (or TRL) [Great Britain];
- HCM 2000 Formulation [USA];
- Brilon & Bonzio Formulation [Germany];
- Brilon & Wu Formulation [Germany];
- Bovy & Coll Formulation [Switzerland];
- Girabase and CERTU Formulation [France].

In the following, the reference unit of measurement for entry traffic and entry capacity is the unit of passenger vehicles per hour (or [uvp/h]). To homogenise different vehicle types incoming flows, buses and heavy vehicles are set at 2 uvp, while two-wheeled vehicles (motorcycles,...) are equal to 0.5 uvp.

According to shared regulatory guidelines, the number of lanes on the ring represents the number of rows of circulating vehicles that can be contained in the ring, without being marked with horizontal signs.

In all of the following examples, stationarity conditions are assumed, i.e., the entrances are under-saturated and the traffic demand at each arm remains unchanged for a time period T of such magnitude that the operating conditions at the junction can stabilise at average values in T of constant state parameters, with specific values of the latter little dispersed around the same average values.

Kimber formulation (or TRL)

The Transport Research Laboratory (TRL) method considers the geometry of the roundabout as a fundamental element for defining the capacity. In this formulation, obtained from a series of data collected on roundabouts in operation in England, the calculation of the inlet capacity is carried out considering the geometric sizes of the arm, the ring, and the circulating flow Q_c at the entrance itself.

The mathematical relation can be expressed as:

$$C = K \cdot (F - f_c \cdot Q_c) \tag{6.1}$$

with

- $F = 303 \cdot x_2$
- $f_c = 0.210 \cdot t_D \cdot (1 + 0.2 \cdot x_2)$
- $k = 1 - 0.00347 \cdot (\Phi - 30) - 0.978 \cdot (1/(r - 0.05))$
- $t_D = 1 + 1 / (2 \cdot [1 + \exp((D - 60)/10)]$
- $x_2 = v + (e - v)/ (1 + 2 \cdot S)$
- $S = 1.6 \cdot (e - v)/ l' = (e' - v')/l'$

The geometric parameters and the corresponding ranges of variability are indicated in Table 6.1. They refer to the left-hand driving circulatory regime in force in the United Kingdom, which involves driving around the ring of roundabouts in an anti-clockwise direction. In the right-hand driving circulatory regime, the homologous specular quantities are considered.

The geometrical parameters of the roundabout (Figure 6.9) that must be considered for capacity calculation are:

Table 6.1 Geometrical parameters for capacity C calculation

Parameter	Definition	Value range
e	Entry width	3.6–16.5 [m]
V	Lane width	1.9–12.5 [m]
e'	Previous entry width	3.6–15 [m]
V'	Previous lane width	2.9–12.5 [m]
U	Ring width	4.9–22.7 [m]
l, l'	Flaring medium length	1–∞ [m]
S	Flaring accuracy	0–2-9
R	Entry curvature radius	3.4–∞ [m]
Φ	Entry comer	0–77 [°]
$D = D_{ext}$	Inscribed circle diameter	13.5–171.6 [m]
w	Exchange lane width	7–26 [m]
L	Exchange lane length	9–86 [m]

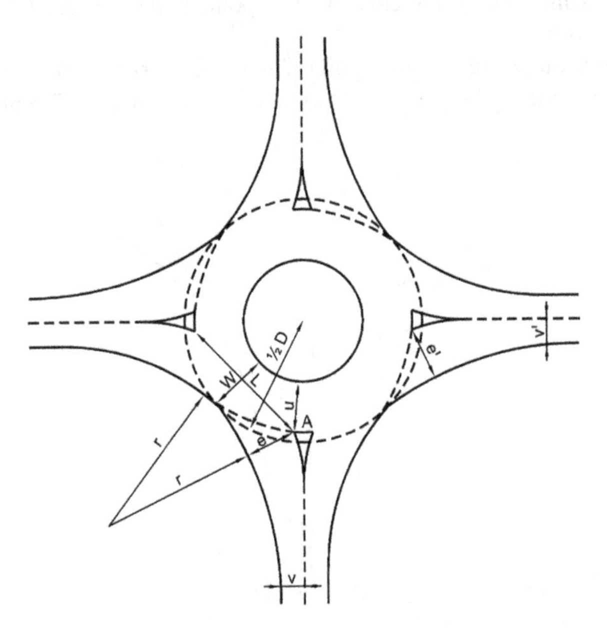

Figure 6.9 Geometrical parameters (Mauro, 2010a)

- the width of the entrance e, measured along the perpendicular taken from point A to the external edge;
- the width of the entrance lane v, measured upstream of the widening of the arm at the entrance along the perpendicular conducted from the axis of the roadway to the external edge;
- the width of the ring u represents the distance between the separating island at the arms (point A) and the central island;
- the entry radius r is the smallest radius of curvature of the external edge near the entrance;
- the width of the exchange trunk w is the smallest distance between the central island and the external edge in the section between an entrance and the next exit;
- the length of the exchange trunk L is defined as the smallest distance between the separating islands at the arms of two consecutive entrances.

In the figure ahead the medium length of the flare is indicated, which can be identified using two parameters (l or l'). The geometric construction for identifying the two parameters differs slightly.

In both cases, a parallel to the HA curve is drawn at a distance "v" from it (by "v" we mean the width of the lane previously defined) and the GD curve is identified, which intersects the segment AB (which represents the width of the entrance) at point D. The segment BD is then considered, and the midpoint C is identified.

At this point, to identify the length l, from C draw the perpendicular to the segment BD, which meets the curve GD at point F. The length CF represents l (Figure 6.10).

To identify the length l', from point C draw the curve parallel to the external edge BG, which meets the curve GD at point F'. The length CF' represents the l' (Figure 6.11).

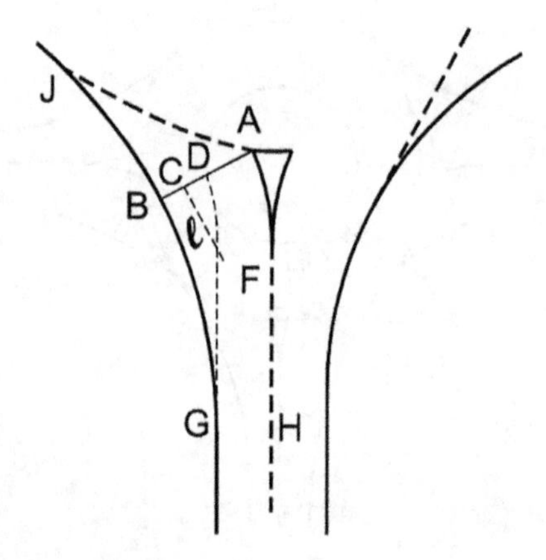

Figure 6.10 Geometrical construction for l evaluation (Mauro, 2010a)

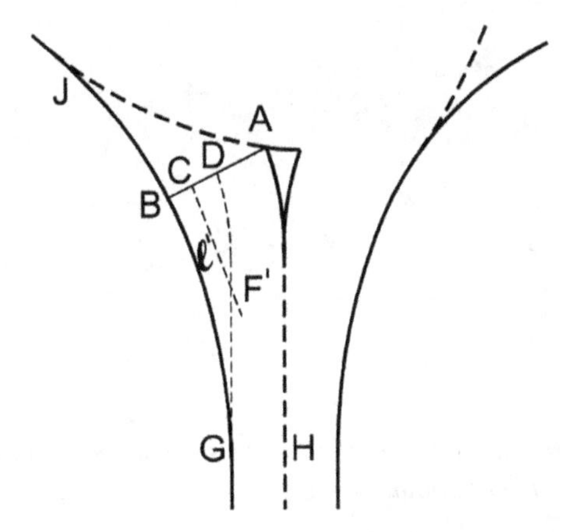

Figure 6.11 Geometrical construction for l' evaluation (Mauro, 2010a)

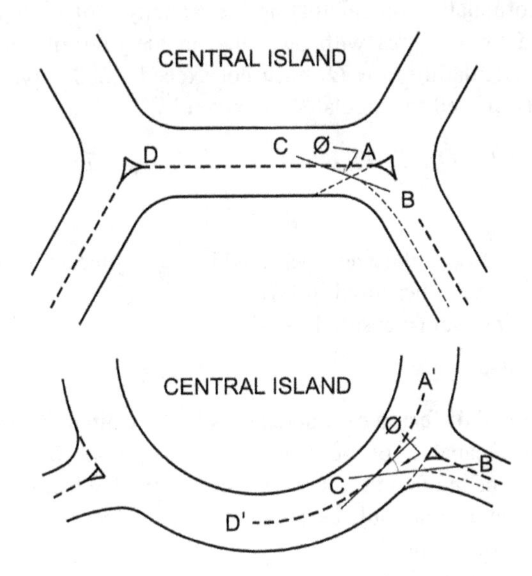

Figure 6.12 Geometrical construction for the evaluation of entry angle (Mauro, 2010a)

Between l and l' the following approximative formula is applied $l' = 1.6 \cdot l$.

The entry angle Φ represents the conflict angle between the flows entering and those circulating in the ring and is identified by tracing the straight line tangential to the incoming flow and the straight line tangential to the circulating flow (Figure 6.12).

In the case of a roundabout of a smaller dimension, the conflict angle can be obtained by crossing the lines tangential to the incoming and outgoing flow and dividing this angle by 2 (Figure 6.13).

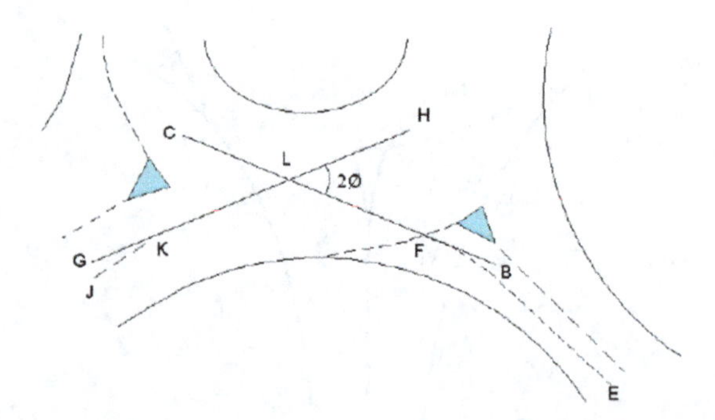

Figure 6.13 Geometrical construction for the evaluation of the entry angle within smaller roundabouts

HCM formulation

The HCM 2000 formulation for calculating the capacity C of a roundabout entrance can only be used for schemes with one lane on the ring and one on the arms. Furthermore, the circulating flow Q_c must not exceed 1,200 uvp/h.

The relationship used to calculate C is given by:

$$C = (Q_c \cdot e^\wedge((-Q_c \cdot T_c)/3,600)/(1 - e^\wedge((-Q_c \cdot T_f)/3,600) \tag{6.2}$$

where

Q_c is the circulating flow in the ring (measured in [uvp/h]) near the entry considered;
T_c is the critical interval (measured in [s]);
T_f is the sequence interval (measured in [s]).

It is worth noting that:

- due to the limited diffusion of roundabouts in the United States, there is a lack of information capable of guaranteeing the optimal functioning of these types of junctions in operation. For this reason, the Highway Capacity Manual (HCM) procedure contemplates a range of capacity values obtained with suitable values of T_c and T_f.
- the HCM formulation does not include any information for the delay estimation in the roundabout.
- To calculate the upper and lower boundaries for the capacity, the following values are used:
- Upper boundary $T_c = 4.1$ s and $T_f = 2.6$ s;
- Upper boundary $T_c = 4.6$ s and $T_f = 3.1$ s.

Brilon-Bondzio formulation

For the Brilon-Bondzio formulation was developed based on the configuration of the scheme, through the number of lanes at the entrances and the ring. The capacity relationship of an entry is:

$$C = A - B \cdot Q_c$$

where the value of the parameters A and B ranges depending on the number of lanes at the entrance and on the ring (Table 6.2). The relationship is valid for circular schemes with external diameters ranging between 28 m and 100 m.

In this formulation, the disturbance traffic Q_d coincides with that circulating in the ring Q_c at the entrance for which C is evaluated.

The formula is valid for the geometric configurations indicated in Figure 6.14.

Brilon-Wu formulation

In Germany, the Brilon and Wu's capacity calculation formula was developed based on the configuration of the scheme and user behaviour. The parameters considered are:

Table 6.2 Parameters for Brilion-Bondzio formulation

Number of lane entering the ring	Number of entry lane	A	B	Sample numerousness
3	2	1,409	0.42	295
2	2	1,380	0.5	4,574
2–3	1	1,250	0.53	879
1	1	1,218	0.74	1,504

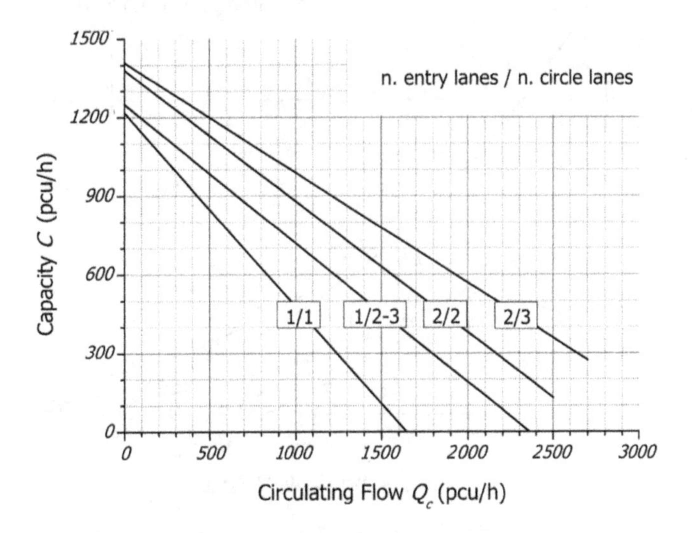

Figure 6.14 Trend of C as a function of Q_c according to the Brilon-Bondzio capacity formulation (Mauro, 2010a)

- Number of entry lane n_e
- Number of ring lane n_c
- Flow circulating in the ring at the entry Q_c
- Critical range T_c
- Sequence time T_f
- Minimum distance between vehicles circulating on the ring Δ

The relationship for calculating the capacity of a roundabout entrance (adopted in the German standard HBS 2001) is used in the case of schemes with one lane at the ring and one lane at the entrances is given by:

$$C = 3{,}600 \cdot (1 - (\Delta \cdot Q_c/3{,}600)/n_c)^\wedge(n_c) \cdot n_e / T_f \cdot \exp(-Q_c / 3{,}600 \cdot (T_c - T_f/2 - \Delta) \qquad (6.3)$$

For the calibration of the model to the German case study (using the experimental data), the following parameters have the corresponding value:

- $T_c = 4.1$ s
- $T_f = 4.1$ s
- $\Delta = 4.1$ s

In Figure 6.15 it is possible to see the relation capacity trends and traffic circulating on the ring by varying the geometric configuration of the scheme.

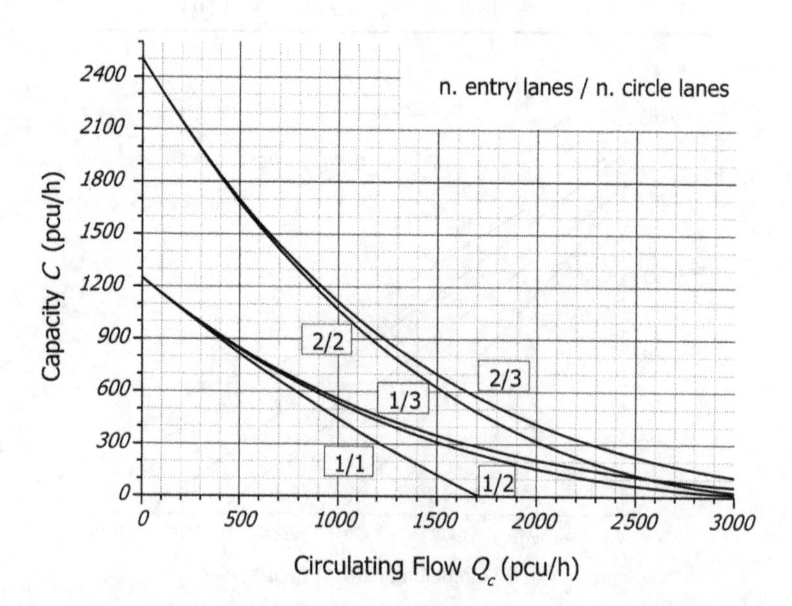

Figure 6.15 Trend of the capacity of the entrances of a roundabout (Mauro, 2010a)

Brion-Wu formulation (HBS 2001)

In Figure 6.16 the capacity trends as a function of the flow circulating in the ring of the American HCM method and the German one are compared.

The trend of the curves highlights that with the same geometric, traffic, and psychotechnical conditions, for Q_c values higher than 500–550 uvp/h, the American formula overestimates the capacity compared to the values obtained with the German formulation.

In the case of roundabouts that have rings that can be driven on in two rows, with an external diameter between 40 m and 60 m and a ring width of 8–10 m, the capacity calculation is carried out using the following relationship:

$$C = 3{,}600 \cdot n_e/T_f \cdot \exp(-Q_c/3{,}600 \cdot (T_c - T_f/2 - \Delta)) \tag{6.4}$$

with

- $T_c = 4.3$ s
- $T_f = 2.5$ s
- n_e with value
  - o = 1 for single entry lane
  - o = 1.4 for two entry lane

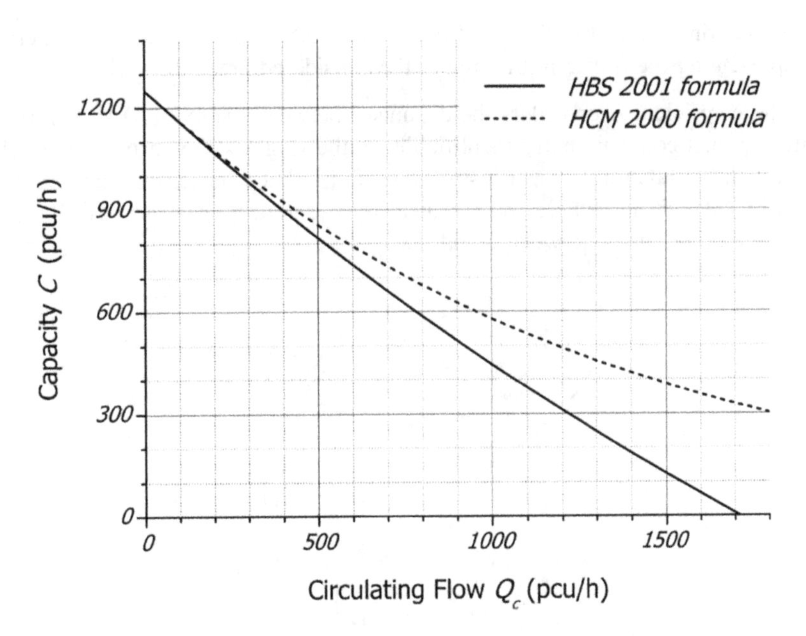

Figure 6.16 Confrontation between HBS 2001 and HCM 2000 capacity relation (Mauro, 2010a)

Bovy and Coll formulation

In this case, the capacity calculation report depends on the number of lanes at the entrances and on the ring. Roundabout junction schemes must have an impassable central island with a maximum internal diameter of 18–20 m and an external diameter of the inscribed circle varying between 24 m and 34 m. The entrances must be flared, that is, with multiple lanes corresponding to the stop line to facilitate reaching the direction considered.

$$C = 1/\gamma \cdot (1{,}500 - 8/9 \cdot Q_d) \tag{6.5}$$

γ is the parameter that takes into account the number of incoming lanes and can take on different values:

- = 1 for one lane;
- = 0.6–0.7 for two lanes (depending on the smaller or larger width of the entrance), it is generally assumed equal to 0.667;
- = 0.5 for three lanes.

Distributed traffic Q_d is calculated with the following equation:

$$Q_d = \alpha \cdot Q_u + \beta \cdot Q_c \tag{6.6}$$

with

Q_u representing the exit traffic from the roundabout to the considered bracket;
Q_c representing the traffic in the ring at the considered bracket.

The coefficient α considers the distance l between the exit (A) and entry (B) conflict points conventionally identifiable on the ring (as shown in Figure 6.17). Through the simulation, it was possible to establish that α decreases with l until, for $l > 28$ m, the exiting vehicles do not disturb the entering ones ($\alpha = 0$). The coefficient therefore varies between 0 and 0.8.

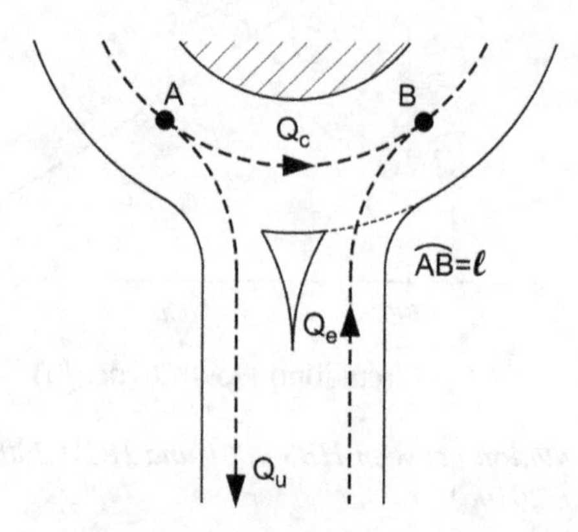

Figure 6.17 Distance l between the conflict point in exit (A) and in entry (B) (Mauro, 2010a)

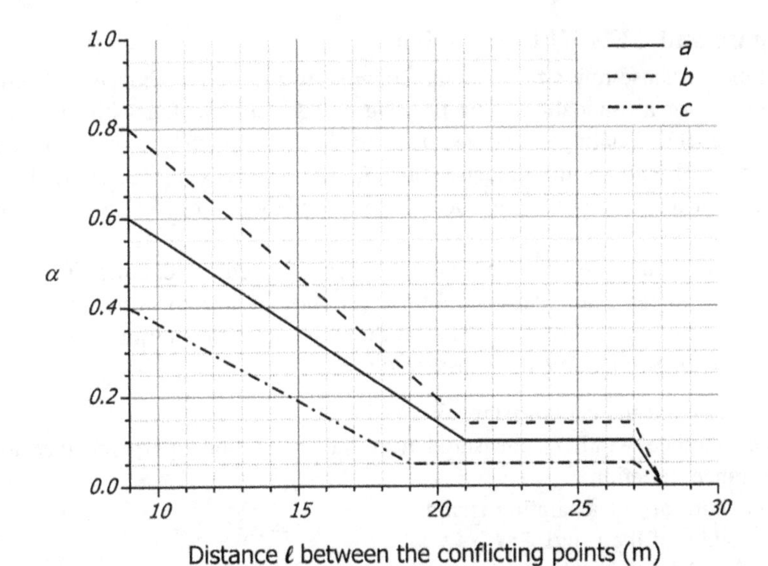

Figure 6.18 Values of the parameter α as a function of the distance l between the outgoing and incoming conflict points (Mauro, 2010a)

The trend of α as a function of distance l is shown in the graph (Figure 6.18) and refers to different speed values (V = 20–25 km/h for curve a, $V >$ 20–25 km/h for curve b and $V <$ 20–25 km/h for curve c).

The coefficient β is expressed as a function of the number of lanes on the ring. Its value decreases as the number of lanes increases, with a reduction in the minor contribution of the circulating flow to disturbing traffic. The values of β are:

- β = 0.9–1.0 for one lane;
- β = 0.6–0.8 for two lanes;
- β = 0.5–0.6 for three lanes.

For this formulation, passenger car unit (or [pcu]) are calculated using the following conversion coefficients:

- a cycle or motorcycle on the ring = 0.8 pcu;
- an incoming cycle or motorcycle = 0.2 pcu;
- a heavy vehicle or a bus = 2 pcu.

Also, this formulation provides the following two capacities index:

- Capacity used rate at Entry, TCU_e, is calculated as: $TCU_e = ((\gamma \cdot Q_e)/C_e) \cdot 100$
- Rate of Capacity used at Point of Conflict, TCU_c, is calculated as: $TCU_c = ((\gamma \cdot Q_e + 8/9 \cdot Q_d))/1500 \cdot 100$

Girabase and CERTU formulation

The name Girabase refers to the commercial software used in France to determine the capacity of a roundabout. The procedure was implemented with reference to some parameters relating to the geometry of the roundabout and by introducing a predetermined value of the sequence time T_f and its processing was carried out by collecting a large amount of data from roundabouts operating in saturated conditions.

The procedure can be used for all types of circular schemes (from mini-roundabouts to large roundabouts) with a variable number of arms from three to eight and with one, two or three lanes at the ring and at the entrances.

The geometric parameters considered are:

- the radius of the central island R_i;
- the entrance width L_e measured near the roundabout, perpendicular to the entrance direction;
- the width of the separating island L_i;
- the width of the ring LA;
- the width of the outlet L_u.

Table 6.3 shows the value range of the variable.

In addition to the geometric parameters, traffic flows are also considered (Figure 6.19) (all the following values are measured in [uvp/h]):

Table 6.3. Variability fields of geometric parameters in the Girabase software

Parameter	Definition	Value range [m]
L_e	Entry width	3–11
L_i	Dividing island width	0–70
L_u	Exit width	3.5–10.5
LA	Ring width	4.5–17.5
R_i	Central island radius	3.5–87.5

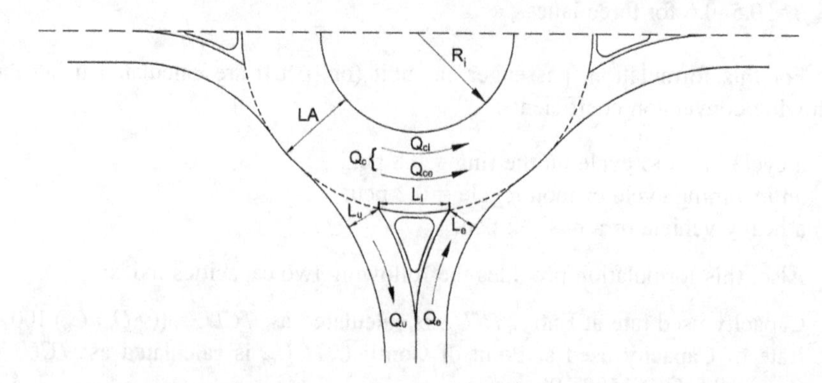

Figure 6.19 Traffic flows and geometric quantities considered in the Girabase procedure (Mauro, 2010a)

- Q_u is the traffic flow exiting the considered entry;
- Q_{ci} is the amount of traffic Q_c on the internal semi-roadway of the ring;
- Q_{ce} represents the amount of traffic Q_c on the external semi-roadway of the ring (close to the entrance);
- $Q_c = Q_{ci} + Q_{ce}$ constitutes the traffic flow circulating on the ring at the entrance considered.

The capacity of an input (measured in [uvp/h]) is based on the exponential regression technique and is evaluated using the following relationship:

$$C = A \cdot e^\wedge(-C_B \cdot Q_d) \tag{6.7}$$

with the following specifications:

- C_B is a coefficient that is 3.525 for urban areas and 3.625 for extra-urban areas;
- $A = (3{,}600/T_f) \cdot (L_e/3.5)^\wedge 0.8$ with T_f representing the sequence time equal to 2.05 s;
- $Q_d = Q_u \cdot k_a \cdot (1 - Q_u /(Q_c + Q_u)) + Q_{ci} \cdot k_{ti} + Q_{ce} \cdot k_{te}$ with:
  - Q_d is the distributed traffic near the entrance considered;
  - $k_a = (R_i/(R_i + LA)) - (L_i/L_{i\ max}) =$ with $L_i < L_{i\ max}$ otherwise 0;
  - $L_{i\ max} = 4.55 \cdot radq(R_i + LA/2)$;
  - k_{ti} is equal to the minimum between $160/(LA \cdot (R_i + LA)$ and 1;
  - k_{te} is equal to the minimum between $(1- ((LA - 8)/LA)) \cdot (R_i/(R_i + LA))^2$ and 1.

For urban roundabouts, there is also another simplified capacity formulation (CERTU), valid for schemes with a diameter of the central island varying between 20 m and 60 m, with one lane at the entrances, symmetrical placement of the arms and demand for incoming traffic balanced.

The formula for calculating the capacity of an input is given by:

$$C = 1500 - 5/6 \cdot Q_d \tag{6.8}$$

It is expressed as a function of the disturbance flux Q_d:

$$Q_d = a \cdot Q_c + b \cdot Q_c \tag{6.9}$$

with

- a variable depending on the radius of the central island, between 0.9 for $R_i <$ 15 m and 0.7 for $R_i >$ 30 m;
- b variable depending on the width of the separating island at the arms, between 0 for $L_i >$ 15 m and 0.3 for $L_i =$ 0 m.

The Girabase formulation has long since replaced the other procedures developed by the various road research institutes in France, and therefore also the SETRA formulation which is still widely used in Italy.

6.2.2 *Methodology for the calculation of delay at a roundabout*

The delay represents the time that is lost during the journey due to junctions along the route. In the case of the roundabout, the delay is made up of three rates which are represented in Figure 6.20; the diagram of the delay can be approximated in Figure 6.21 for simplicity.

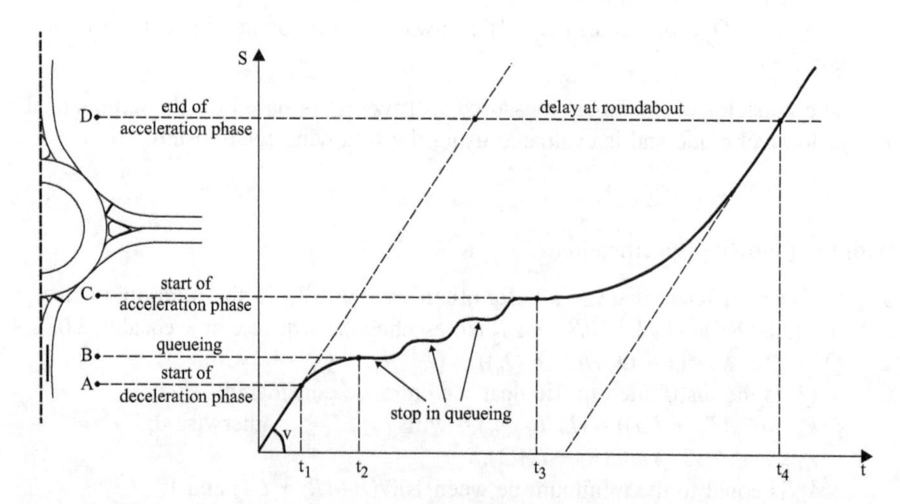

Figure 6.20 Delay caused by the presence of a roundabout on the travel (Mauro, 2010b)

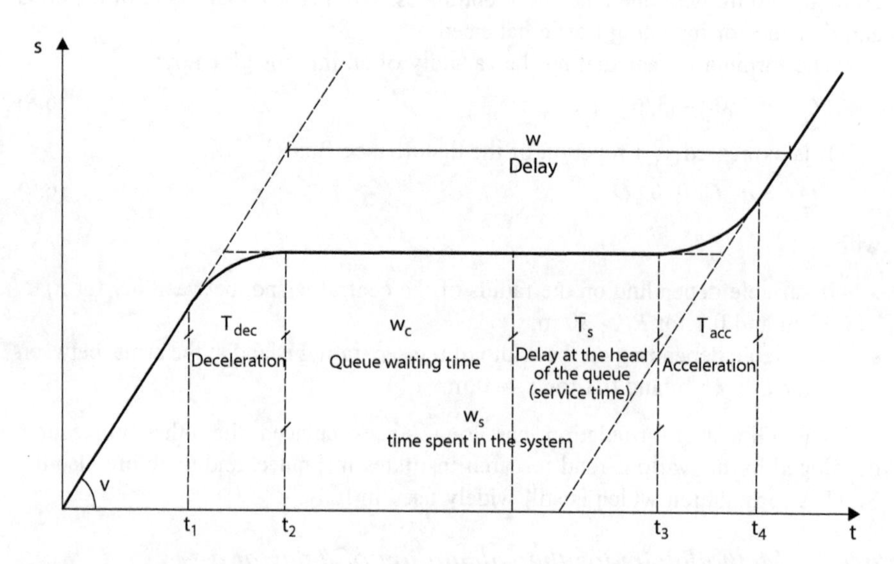

Figure 6.21 Simplified representation of delays due to the presence of a roundabout along the route (Mauro, 2010b)

Analysing Figure 6.21, it is possible to identify the overall delay of a vehicle as the sum of three rates:

- the time spent in the system or waiting time w_s given by the sum of the waiting time in the queue w_c, i.e., the time that a single vehicle loses from queuing until it positions itself at the entrance to the junction (i.e., at the line of give precedence) to be able to use it and of the service time T_s, that is the time that passes between when the vehicle reaches the head of the queue (i.e., the aforementioned giving way line) and the instant from which it starts its entry into the ring to carry out the desired manoeuvre;
- the time T_{dec} due to the deceleration of the vehicle approaching the roundabout;
- the time T_{acc}, referring to the acceleration phase to exit the roundabout.

So, it is possible to obtain the following formula for the delay:

$$w = w_s + T_{dec} + T_{acc} \tag{6.10}$$

The calculation of the overall delay of w is useful for example when carrying out comparative analyses of various types of junctions, choosing the most suitable one; instead, when you want to evaluate the circulation functionality of a junction you can directly use the residence time in the system w_s equal to the sum of two rates:

$$w_s = w_c + T_s \tag{6.11}$$

In addition to calculating the delay per vehicle, the number of vehicles waiting at an entrance to the roundabout, including the service vehicle, can be defined by indicating it with L_s; the number of vehicles in line waiting behind the vehicle in service is defined as queue length and indicated by L_c (Figure 6.22). For example, in the case of a roundabout with one lane at the entrances, we have:

$$L_c = L_s - 1 \tag{6.12}$$

The variables relating to queue length and delay are random. As regards the delays, the average values $E[w_s]$ of w_s and $E[w_c]$ of w_c are of interest, while for the queue both the average values $E[L_c]$ for L_c and $E[L_s]$ for L_s and the percentile values must be calculated $L_{c,p}$ and $L_{p,c}$. The determinations of $E[w_s]$ and L_s require the characterisation of the state of the system in terms of the presence or absence of stationarity.

In stationary conditions, $E[w_s]$ represents the average of the time spent in the roundabout w_s relating to the waiting vehicles. In case of the absence of stationarity, the average waiting time must be calculated considering that the average queue length varies over time due to the variation of incoming flows. For example, if the system is saturated or oversaturated for a period T, the queues will be shorter at the beginning of the period and will increase at the end. In this case, the average waiting time in queue $E[w_c]$ relating to this period is given by:

$$E[w_c] = (w_c(t_0) + w_c(t))/2 \tag{6.13}$$

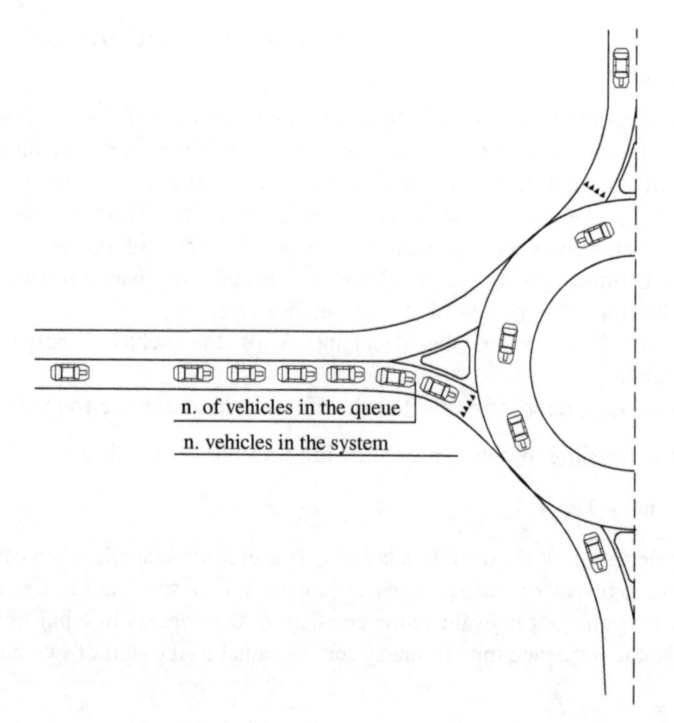

Figure 6.22 Representation of the number of vehicles in the system and in the queue (Mauro, 2010b)

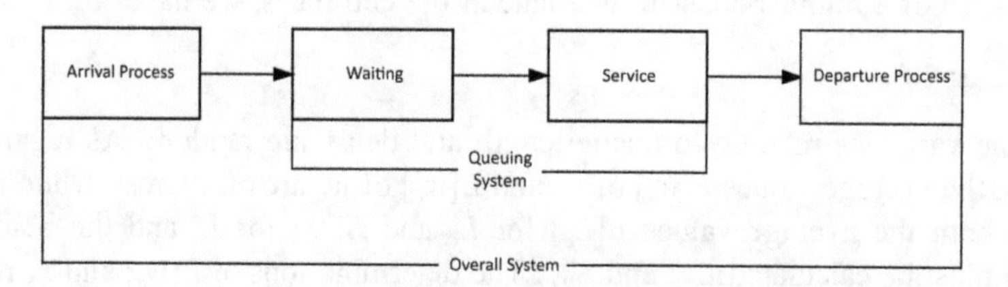

Figure 6.23 Representation of the waiting and its component

where $w_c(t_0)$ and $w_c(t)$ are the waiting times in the queue of the two vehicles arriving in the queue respectively at the beginning t_0 and $t = t_0+T$ of the observation period T.

The calculation of average delays, queue lengths, and the number of vehicles entering is carried out considering the entrance lanes to the arms of a roundabout as channels of an ordinary system object of the mathematical theory of queues (Figure 6.23).

Table 6.4 Service level limit values according to the HCM

Service level	Medium delay per vehicle [sec]
A	≤ 10
B	10–15
C	15–25
D	25–35
E	35–50
F	$d > 50$

6.2.2.1 Delay and queue lengths evaluation

The delay is made up of various rates. The most used delay calculation formula is that of Akçelic and Troutbeck:

$$d = 3{,}600/C + 900 \cdot T \cdot [Q_i/C - 1 + radq((Q_i/C - 1)^\wedge 2 + (3{,}600/C)(Q_i/C)/(450 \cdot T))] \tag{6.14}$$

with

d is the average delay per vehicle (measured in [s]);
C is the input capacity to the branch (measured in [veh/h]);
T is the period considered ($T = 1$ for a period of 1 h and $T = 0.25$ for an interval of 15 min);
Q_i is the flow entering the arm (measured in [veh/h]).

Based on the delay thus calculated, the HCM establishes six service level classes which are defined in Table 6.4.

In addition to the formula for calculating the delay, a relationship can also be defined for estimating the queue at the 95th percentile due to W_u:

$$L_{95} = C/3,600 \cdot 900 \cdot T \cdot [(Q_i/C) - 1 + radq((Q_i/C - 1)^\wedge 2 + ((3,600/C) \cdot (Q_i/C)/(150 \cdot T)))] \tag{6.15}$$

with
L_{95} is the tail of the 95th percentile (measured in [veh]);
C is the input capacity to the branch (measured in [veh/h]);
T is the length of the considered period (measured in [h]);
Q_i is the entry flow in the bracket (measured in [veh/h]).

6.3 Summary

This chapter presents the basic definitions and notations associated with roundabouts. It outlines fundamental aspects of roundabout design and provides key details regarding methods for assessing capacity and delay.

This chapter deals with the roundabouts with a main focus on the methods for the analysis, considering that they can be included as sub-models in larger models for Transportation Systems Analysis and Design. Furthermore, there are several applications in which the roundabouts are also integrated with traffic lights; most common cases are referred to the public transport (e.g., bus) signals prioritisation.

References

Some suggestions for further readings are reported below among the huge literature available on these topics.

Di Gangi, M., and Mussone, L. (2010). *Linee guida per la progettazione e verifica funzionale delle intersezioni non semaforizzate* (pp. 1–201). Maggioli.

Esposito, T., and Mauro, R. (2003). *Fondamenti di infrastrutture viarie: La geometria stradale.* Hevelius.

Esposito, T., and Mauro, R. (2003). *Fondamenti di infrastrutture viarie: La progettazione funzionale delle strade.* Hevelius.

Kimber, R. M. (1980). The traffic capacity of roundabouts, TRRL Laboratory Report 942. Transport and Road Research Laboratory, Crowthorne, United Kingdom, 63.

Kimber, R. M., and Hollis, E. M. (1977). Flow/delay relationships for major/minor priority junctions. *Traffic Engineering & Control*, 18(Analytic).

Kimber, R. M., and Hollis, E. M. (1978). Peak-period traffic delays at road junctions and other bottlenecks. *Traffic Engineering & Control*, 19(N10).

Kimber, R. M., and Hollis, E. M. (1979). Traffic queues and delays at road junctions (No. LR909 Monograph).

Krogscheepers, C., Robinson, B., and Rodegerdts, L. (2001). Roundabout operations: a summary of FHWA's – '*Roundabouts: An Informational Guide*'. *20th South African Transport Conference*, 2001.

Mauro, R. (2010a). Capacity evaluation. In: *Calculation of Roundabouts: Capacity, Waiting Phenomena and Reliability* (pp. 15–57). Berlin: Springer-Verlag.

Mauro, R. (2010b). Calculation of roundabouts: Problem definition. In: *Calculation of Roundabouts: Capacity, Waiting Phenomena and Reliability* (pp. 1–13). Berlin: Springer-Verlag.

Wu, N. (1994). An approximation for the distribution of queue lengths at unsignalized intersections. In *Proceedings of the Second International Symposium on Highway Capacity* (Vol. 2, pp. 717–736). Victoria, Australia: Australian Road Research Board Ltd.

Chapter 7
Analysis and control

Giulio Erberto Cantarella[1] *and Orlando Giannattasio*[2]

If you don't enjoy screaming, on a running motorcycle you do not make great conversations. Instead, you spend time to perceive things and meditate on it.

Zen and the art of motorcycle maintenance – R.M. Pirsig

Outline. *This chapter reviews the basic definitions and notations for single signalised junction analysis and control. It introduces green timing and green timing and scheduling and reviews some common optimisation methods for signalised single junction static control.*

Basic elements of Signalised Junction Analysis (SJA) are reviewed below and in Chapter 8. As said in the introduction a complete discussion of this topic is out of the scope of this book; a recent comprehensive presentation of most of these topics is in Roess *et al.* (2010), see also HCM (2020), Chapter 2 in Cascetta (2009) and Di Pace (2019).

Aiming at developing methods for signalised junction control description and design, it is better distinguishing:

- **Single junctions**: control of adjacent junctions is not taken into account, as discussed below;
- **Systems of junctions**: Control of adjacent junctions is to be considered, as discussed in Chapter 8:
  - (**Junction**) **Arterials**, are simple sequences of junctions connected without loops, usually the word "junction" is omitted;
  - (**Junction**) **Networks**, are junctions connected by a network with loops.

Aiming at developing (macroscopic) models for delay computation, as a performance index, it is better distinguishing:

- **Isolated junctions**: no approach interacts with any of the upstream junctions, effects of adjacent junctions are negligible as discussed below;

[1]Department of Civil Engineering, University of Salerno, Italy
[2]Department of Civil, Computer Science and Aeronautical Technologies Engineering, Roma Tre University, Italy

- **Interacting junctions**: (at least) some approaches interact with any of the approaches of the upstream junction, the effects of adjacent junctions are not negligible, as discussed in Chapter 8.

As it will be clearer in the following text, single and isolated junctions have different definitions, as well as systems of junctions (arterials or networks) and interacting junctions.

Basic definitions and notations for single junctions are described in Section 7.1 with reference to the Webster (1958) modelling framework; then unitary delay for an isolated approach is discussed in Section 7.2, including the two-term Webster (1958) formula. Afterward, Sections 7.3 and 7.4 describe some methods for the signal setting design of a single junction including the simple Webster method in closed form (Webster, 1958) (see also Webster and Cobbe, 1966) as well as some optimisation methods.

7.1 Basic definitions and notations for single junctions

As already said in Chapter 4, some streets merging in a common area define a junction. Vehicles (or pedestrians) can cross the junction following different paths, called manoeuvres, defined by the entering-exiting points, such as straight-on, right or left turns (In the examples reported throughout this paper, vehicles move according to the right-hand rule.). According to the lane allocation (junction layout), two (or more) manoeuvres may share the same lanes. Such a set of manoeuvres should be given the right of way at the same time and forms a *stream*, as well as a single manoeuvre with one or more reserved lanes. Let

f_i be the arrival flow, or in-flow, of stream i, the average number of users arriving at the junction in the unitary time,
s_i be the saturation flow of stream i, the number of users that can cross the junction in the unitary time when a queue is present, not to be confused with the capacity introduced below;
$y_i = f_i / s_i$ be the flow ratio of stream i.

Arrival and saturation flows will be assumed as known and measured in passenger car units (pcu) per unit of time for vehicular streams. The arrival flow can be directly measured or can be the result of a network analysis (for instance through a demand assignment method), whilst the saturation flow mainly depends on junction layout and geometrical characteristics (more details are reported in the Appendix).

At signalised junctions, different streams are given the right to cross the junction according to the signals displayed by the traffic lights: *green*, *red*, and *amber* in a cyclic order (In some countries, a red-amber signal precedes the green signal.). Let

c be the cycle time, the least time for a complete succession of signals,
G_i be the displayed green time for stream i,
R_i be the displayed red time for stream i,

A_i be the (displayed) amber time for stream i, preceding the red signal for safety reasons,
h_i be the (displayed) green-amber time for stream i, such that:

$$h_i = G_i + A_i \ \forall i \tag{7.1}$$

Remark. *Some of the above variables (and others below) are commonly denoted by capital Roman letters even if they do not denote matrices or sets.*

The green and red time are the result of the signal setting, whilst the amber time (usually in the range of 3 to 4 seconds) mainly depends on junction geometrical characteristics and vehicle performances, such as approaching speed, braking deceleration, and so on, and will be assumed as known (more details are reported in Appendix). The cycle time is assumed common to all the streams, thus for each stream i *the following equation holds:*

$$c = R_i + G_i + A_i = h_i + R_i \quad \forall i \tag{7.2}$$

7.1.1 Effective green time

The analysis of a signalised junction is usually carried out with reference to effective green (and red times), after Webster (1958). The cycle time is assumed divided into an *effective red* period during which no departure occurs and an *effective green* period during which users cross the junction at a constant rate $1/s$ (until there are users waiting in a queue). Furthermore, the *lost time* (usually in the range of 2 to 3 seconds) depends on junction layout and geometry and models the part of the green-amber time that cannot be used for departure due to starting and stopping transients (more details are reported in Appendix). Let

g_i be the effective green time for stream i,
r_i be the effective red time for stream i,
$l_i = l_{1i} + l_{2i}$ be the lost time for stream i, due to starting l_{1i} and stopping transients l_{2i}.

The displayed and effective green times and the saturation flow for each stream i satisfy the following equations, see Figure 7.1:

$$\begin{aligned} g_i &= (1/s)\int_0^{Gi+Ai} \phi(t)\, dt \quad \forall i \\ h_i &= g_i + l_i \qquad\qquad\qquad\quad \forall i \end{aligned} \tag{7.3}$$

where $\phi(t)$ is the departure flow against the time t, provided that a queue is waiting to cross the junctions.

The cycle time, c, and the green-amber times, h_i, allow computing all other relevant variables for each stream i, assuming the amber and lost times, A_i and l_i, known:

$$g_i = h_i - l_i \quad \forall i \tag{7.4}$$

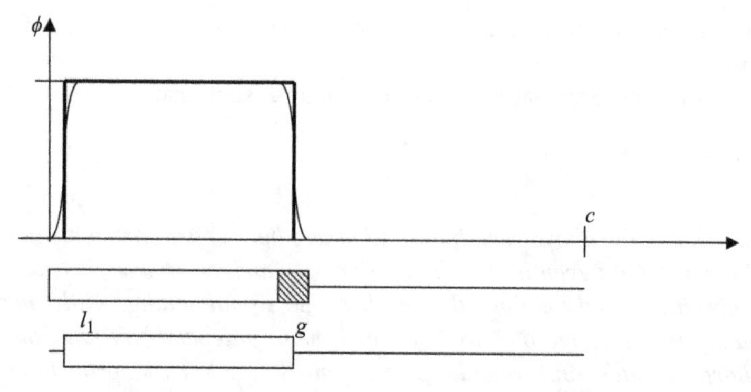

Figure 7.1 *Effective green, red, and lost time at a signalised junction approach*

$$r_i = c - g_i = c - (h_i - l_i) \quad \forall i \tag{7.5}$$

$$G_i = g_i + l_i - A_i = h_i - A_i \quad \forall i \tag{7.6}$$

$$R_i = c - (g_i + l_i) = c - h_i \quad \forall i \tag{7.7}$$

Minimum green time constraints may be introduced for practical implementations:

$$g_i \geq g_{MINi} > 0 \quad \forall i \tag{7.8}$$

Under undersaturation conditions, there is no queue at the beginning of the effective red period. Moreover, during the effective red period, no vehicle may leave the queue, thus at the end of this period the queue length n_i is:

$$n_i = f_i r_i = f_i(c - g_i) > 0 \tag{7.9}$$

Minimum green time constraints are also useful if the queue that can be stored along a street entering the junction is a major concern. Let

n_{MAXi} be the storage or longitudinal, street capacity for stream i.

Then the longitudinal queue length constraints are given by:

$$\begin{aligned} &f_i(c - g_i) \leq n_{MAXi} \\ &\text{or } g_i \geq c - n_{MAXi}/f_i > 0 \quad \forall i \end{aligned} \tag{7.10}$$

- Degree of saturation and capacity factor

A stream i is *under-saturated* if the number of users arriving during a cycle, f_i c, is less than the maximum number of users that can depart, s_i g_i, *over-saturated* otherwise. If the flow ratio of a stream i is larger than 1, (i.e., $y_i > 1$), this stream is over-saturated regardless of the green timing (see remark at the end of the chapter). With this assumption, oversaturation could be avoided by duly designed green timing. Let

$f_{MAXi} = s_i\, g_i / c$ be the capacity of stream i, $f_i < f_{MAXi}$ meaning that stream i is under-saturated, $f_i > f_{MAXi}$ meaning that stream i is over-saturated.

The arrival flow and capacity of the stream i can be compared through two indicators:

$x_i = f_i / f_{MAXi} = (f_i\, c) / (s_i\, g_i) = y_i / (g_i/c)$ the *degree of saturation* of stream i, $x_i <$ 1 stream i is under-saturated, $x_i > 1$ stream i is over-saturated;

$z_i = f_{MAXi} / f_i = (s_i\, g_i) / (f_i\, c) = (g_i/c) / y_i$ the *capacity factor* of stream i, $f_i > 1$ stream i is under-saturated, $f_i < 1$ stream i is over-saturated; it is the maximum rate by which the arrival flow can be increased still avoid oversaturation ($f_i = z_i\, f_{MAXi}$) and measures how far is over-saturation for stream i.

The above two indices are simply related since one is the reciprocal of the other, $z_i\, x_i = 1$.

The *junction degree of saturation* x is the maximum degree of saturation among all streams:

$$x = \max_i x_i = \max_i (f_i c)/(s_i g_i) \tag{7.11}$$

Similarly, the minimum capacity factor among all streams gives the *junction capacity factor*, z (sometimes also denoted by μ). It is the maximum rate by which the arrival flows of all the streams can be increased, while still avoiding over-saturation for all the streams:

$$z = \min_i z_i = \min_i (s_i g_i)/(f_i c) \tag{7.12}$$

The two indicators are clearly equivalent, since $z \cdot x = 1$. The junction capacity factor seems more intuitive, as it increases when capacity increases, and vice versa.

From the above equations, some streams, called *critical streams*, have a degree of saturation and a capacity factor equal to the junction degree of saturation or the capacity factor, respectively. The other streams have a smaller degree of saturation and a larger capacity factor.

The junction capacity factor is not affected by arrival flow patterns; thus, it is not affected by interaction with adjacent junctions and only depends on arrival and saturation flows, effective green times, lost times, and cycle time. For highly congested junctions, it can be worth to analyse signal setting that maximises the capacity factor. As noted in the introduction, several methods have been proposed to compute green times that maximise capacity factor, based on optimisation models (see Chapter 8).

7.1.2 *Signal plan of a single junction*

Green-amber scheduling is feasible if the green-amber periods of two streams that cannot safely cross the junction at the same time do not overlap; such a couple of streams are called *incompatible* (see section 2.1, for instance, streams *1* and *3* in Figure 7.2). Usually, two crossing or merging streams are considered incompatible (However sometimes a left turn is considered compatible with the opposing straight-on manoeuvre, in this case, its saturation flow is reduced accordingly). Given the cycle time, the signal plan of a single junction, aiming at capacity factor

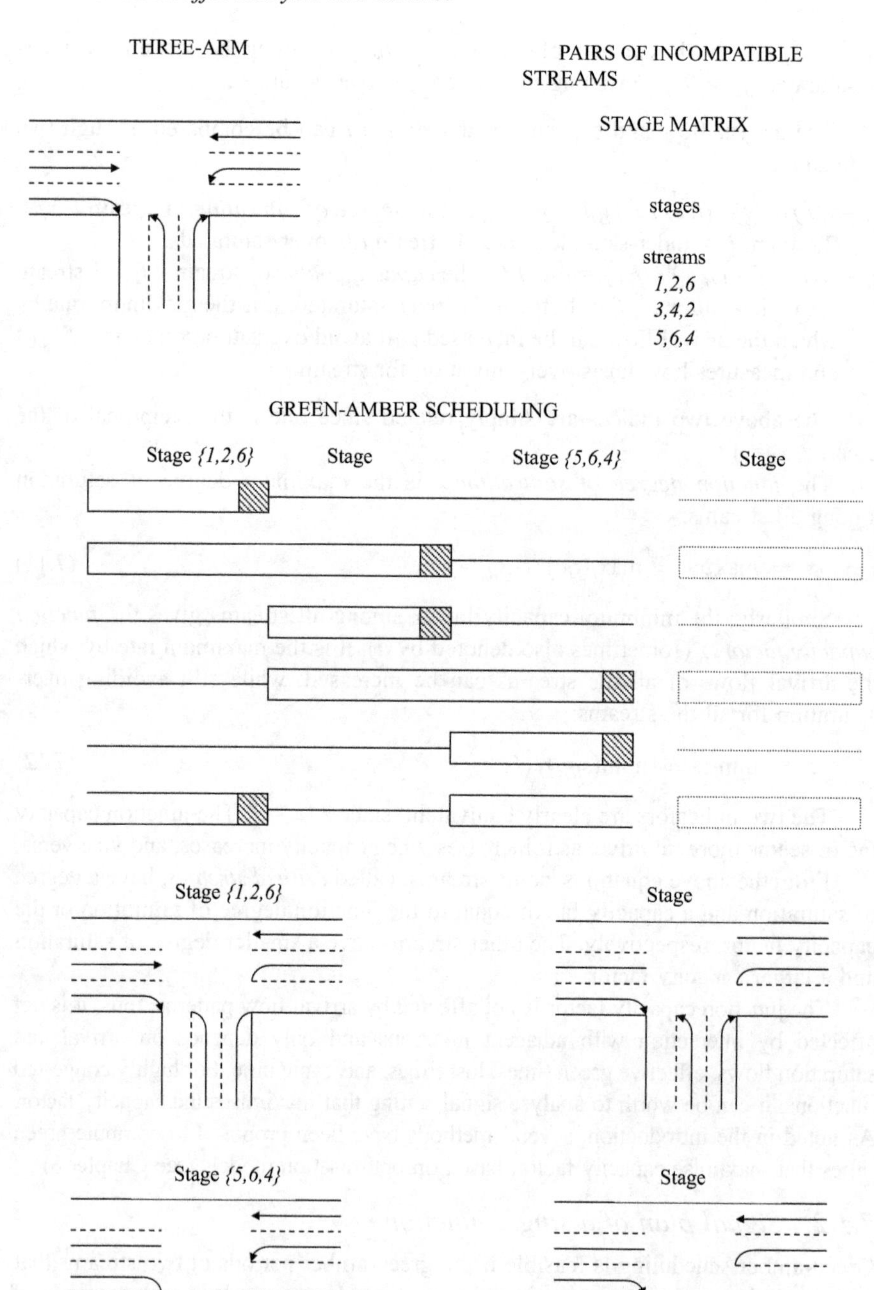

Figure 7.2 Example of green-amber scheduling and stage sequence (from Cantarella, 1998)

maximisation or total delay minimisation, is completely specified by one of the two approaches described in the following sections:

- **Green timing** is defined by the length of each stage knowing the stage matrix, as described in Section 7.3;
- **Green timing and scheduling**, describe how the green-amber periods are located within the cycle, defined by the beginning and the end of each green-amber period, as described in Section 7.4.

Remark. *The so-called all-red period (or stage) at the end of each stage to allow for the safe clearance of the junction is assumed to be* t_{SG} *(see Appendix for more details); in the following, for simplifying notations, the all-red period is assumed included, as an additional term, within each stage length* t_j *and each green-amber time* h_i*. This period for each approach having green also in the successive stage is green, it is often called inter-green.*

7.2 Delay for isolated junctions

Modelling delay for each stream or better for the corresponding approach of an isolated junction is based on the general framework described in Section 2 of Chapter 2 for queuing links under *steady-state conditions*, that is assuming that the arrival flow, f_i, and the saturation flow, s_i, are constant during the time analysis interval given by [0, c].

Two main conditions are distinguished below (see also Akcelik, 1980).

7.2.1 Undersaturation

The in-flow is less than capacity, $f_i < f_{MAXi}$, or $(f_{MAXi} - f_i) > 0$.

In this case, for each approach i, the evolution over time of the queue length is equal for each cycle time, made up of an effective red period followed by an effective green period (Figure 7.3).

7.2.1.1 Deterministic delay

In this sub-section, we describe a delay function developed under the assumption that the arrival flow and the saturation flow are represented by deterministic variables, constant over time. Remembering Equation (7.9):

$$n_i = f_i r_i = f_i(c - g_i) \tag{7.9}$$

the total delay d_{TRi} accumulated during the effective red period $[0, r_i]$ is given by:

$$d_{TRi} = n_i \cdot r_i / 2 = f_i \cdot (c - g_i)^2 / 2 \tag{7.13}$$

See case B.1 in Section 2 of Chapter 2, comparing Equations (7.12) and (7.13) with (7.17a) and (19) assuming $f_{MAXi} = 0$ and $T = c$.

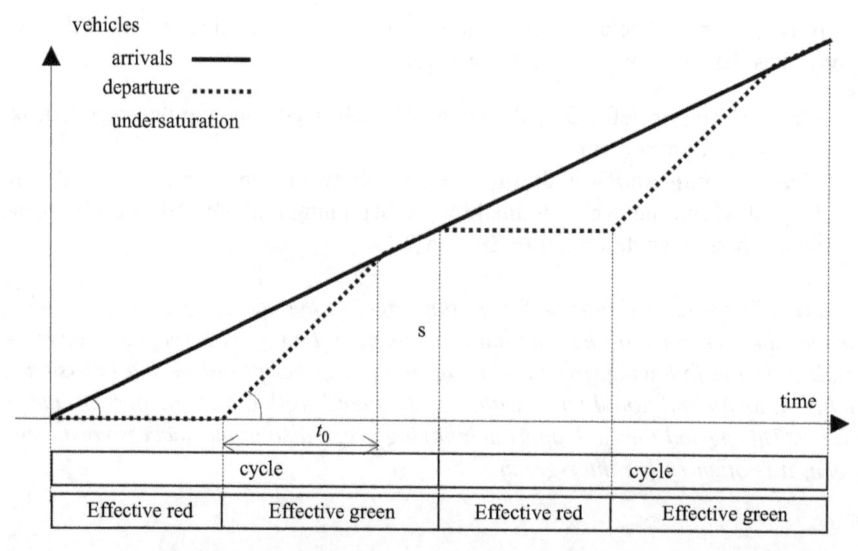

Figure 7.3 Arrivals to and departures from an undersaturated approach (adapted from Cantarella and Vitetta, 2010, see also Akcelik, 1980)

Due to the undersaturation hypothesis, the queue n_i at the end of the effective red period vanishes at time t_0 before the end of the effective green period:

$$t_0 = n_i / (s_i - f_i) = f_i(c - g_i) / (s_i - f_i) \tag{7.14}$$

The total delay d_{Ti} accumulated during the effective green period $[r_i, c]$ is:

$$d_{TGi} = n_i \cdot t_0 / 2 = n_i^2 / (2 (s_i - f_i)) = f_i^2 (c - g_i)^2 / (2(s_i - f_i)) \tag{7.15}$$

See case A.2.1a in Section 2 of Chapter 2, comparing Equations (7.14) and (7.15) with (7.10) and (12) assuming $f_{MAX} = s_i$.

The total delay d_{Ti} accumulated during the whole cycle $[0, c]$ is given by the sum of (13) and (15):

$$d_{Ti} = (c - g_i)^2 f_i s_i / (2 (s_i - f_i)) \tag{7.16}$$

The average unitary deterministic undersaturation delay, d_{DUi}, accumulated during the whole cycle $[0, c]$ is given by the ratio between the total delay (16) and the number of users, $f_i\, c$, arriving at the approach i during the whole cycle $[0, c]$:

$$d_{DUi}(f_i) = (c - g_i)^2 s_i / (2c(s_i - f_i)) = c(1 - g_i/c)^2 / (2(1 - f_i/s_i)) \tag{7.17}$$

This function is defined over the interval $[0, f_{MAXi}]$ and takes values in the interval $[c\,(1 - g_i/c)^2 / 2, c\,(1 - g_i/c) / 2]$ (Figure 7.4).

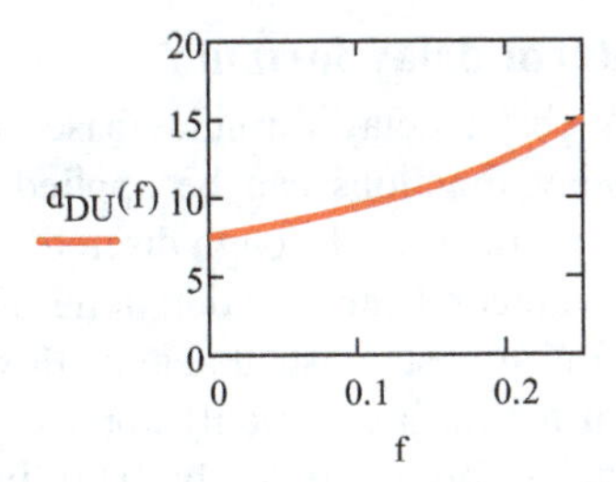

Figure 7.4 Undersaturation deterministic delay function against flow, c = 60, g = 30, g/c = 0.5, s = 0.5

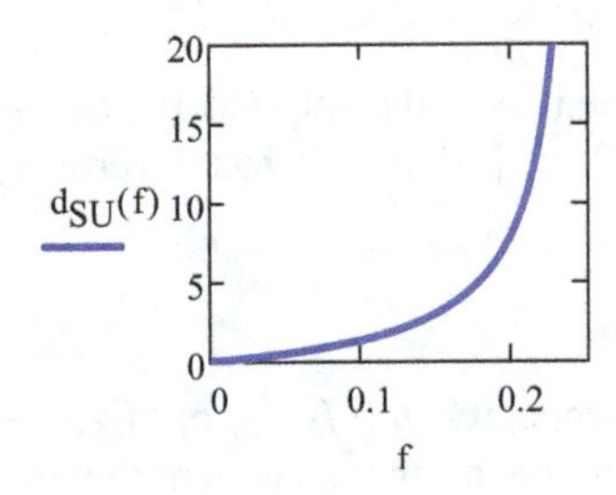

Figure 7.5 Undersaturation stochastic delay function against flow, c = 60, g = 30, g/c = 0.5, s = 0.5

7.2.1.2 Stochastic delay

In this sub-section, we describe an additional delay function developed under the assumption that the arrival flow and the saturation flow are represented by random variables, with distribution constant over time.

Indeed, the arrival flow and the saturation flow may show fluctuations over time whose effect can be described by the (Stochastic) Queueing Theory briefly recapped in Section 2 of Chapter 2. Assuming time lag between two arrivals and two departures distributed as an exponential random variable, case: M/M/1; ((+∞), FIFO) in Section 2 of Chapter 2, and assuming, as above, that:

$$f_{MAXi} = s_i g_i \,/\, c \text{ is the capacity of approach } i$$

$$x_i = f_i / f_{MAXi} = (f_i c) \,/ (s_i g_i) \text{ is the degree of saturation of approach } i$$

the average unitary stochastic undersaturation delay, d_{SUi}, accumulated during the whole cycle $[0, c]$ is given by Equation (7.24) in Section 2 of Chapter 2 (Figure 7.5):

$$d_{SUi}(f_i) = x_i / \,(2 f_{MAXi}(1 - x_i)) = 1 \,/\, (2(s_i g_i /\, c)((s_i g_i /\, f_i c) - 1)) \qquad (7.18)$$

This function is defined over the interval $[0, f_{MAXi}]$ and takes values in the interval $[0, \infty]$, that it has a vertical asymptote for $f_i = f_{MAXi}$ (Figure 7.4).

7.2.1.3 Webster two-term delay formula

In practical applications Webster delay formulas based on the sum of the deterministic and stochastic delay functions can be applied. In the three-term delay function a third negative term, obtained through discrete simulation, is added to the other two given by the above theoretical considerations. This third term has a rather complicated expression but is almost constant against flow and roughly 10% of the other two; therefore, one of the most commonly applied delay functions is the so-called Webster two-term delay formula given by 0.90 times the sum of the deterministic and stochastic delay functions:

$$d_{WUi}(f_i, g_i) = 0.90/2(c(1 - g_i/c)^2/ (1 - f_i/s_i) + 1 /((s_i g_i/ c)((s_i g_i/(f_i c)) - 1))) \quad (7.19)$$

This function is defined over the interval $[0, f_{MAXi}]$ and takes values in the interval $[c\ (1 - g_i\ /c)^2\ /\ 2, \infty]$, that is it has a vertical asymptote for $f_i = f_{MAXi}$ (Figure 7.6).

7.2.2 *Oversaturation*

The in-flow is larger than capacity, $f_i > f_{MAXi}$, or $(f_{MAXi} - f_i) < 0$.

In this case for each approach i the queue length increases with time within the time analysis interval $[0, T]$, assuming that T is longer than the cycle time, $T > c$. For simplicity's sake, T is assumed an integer multiple of c; so that the interval $[0, T]$ begins at the beginning of an effective red period and ends at the end of an effective green period.

7.2.2.1 Deterministic delay

In this sub-section, we describe a delay formula developed under the assumption that the arrival flow and the saturation flow are represented by deterministic variables, constant over time.

The deterministic oversaturation delay, d_{DOi}, is given by two additional terms (Figure 7.7) to be added to the undersaturation delay from Equation (7.17) with $f_i = f_{MAXi}$:

$$d^{(1)}_{DUi}(f_i) = c\ (1 - g_i/c)\ /\ 2 \quad (7.20)$$

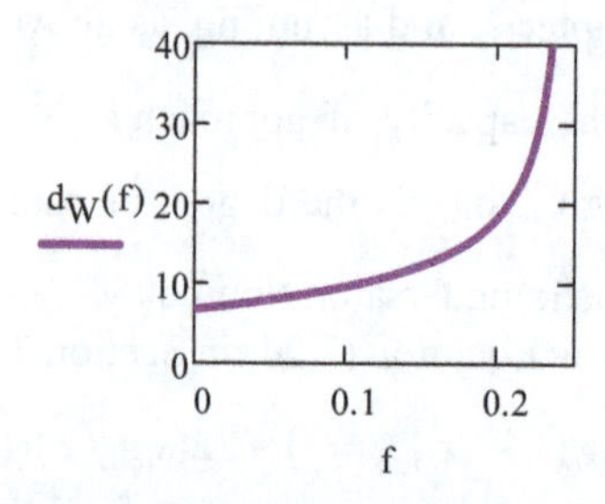

Figure 7.6 Webster two-term delay function against flow, c = 60, g = 30, g/c = 0.5, s = 0.5

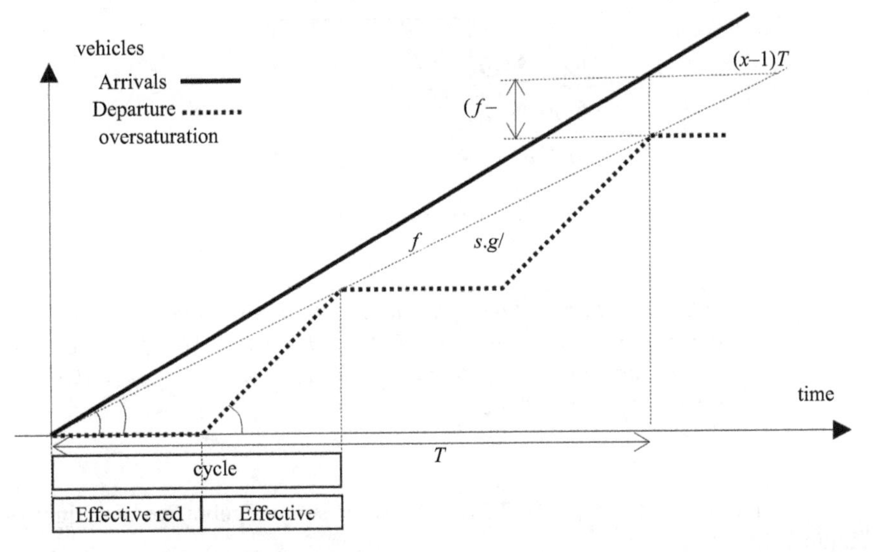

Figure 7.7 Arrivals to and departures from an oversaturated approach (adapted from Cantarella and Vitetta, 2010)

A term depending on the arrival flow f_i and the length of interval T describing the delay due to the oversaturation experienced by the users leaving the approach before time T. It is given by the total delay expressed by Equation (7.19) in Chapter 1.2 divided by the number of users who arrived up to time T, $f_i \cdot T$:

$$d^{(2)}_{DOi}(f_i) = ((f_i - f_{MAXi}) \cdot T^2/2)/(f_i \cdot T) = (1 - s_i g_i/(f_i c)) \cdot T/2$$
$$\text{or} d^{(2)}_{DOi}(f_i) = (1 - 1/x_i)T/2 = ((x_i - 1)/x_i) \times T/2 \tag{7.21a}$$

A term depending on the arrival flow f_i and the length of interval T describing the delay due to the oversaturation experienced by the users leaving the approach after time T.

In this case, the total delay is given by the queue length at the end of interval T, $(f_i - f_{MAXi})\ T$, multiplied by the average delay time $(f_i / f_{MAXi} - 1)\ T/2$:

$$(f_i - f_{MAXi})T \cdot (f_i/f_{MAXi} - 1)T/2$$

Thus, the term is given by the total delay divided by the number of users who arrived up to time T, say $f_i \cdot T$:

$$d^{(3)}_{DUi}(f_i) = (1 - s_i g_i/(f_i c)) \cdot (f_i c/(s_i g_i) - 1) \cdot T/2$$
$$\text{or } d^{(3)}_{DUi}(f_i) = ((x_i - 1)^2/x_i) \cdot T/2 \tag{7.21b}$$

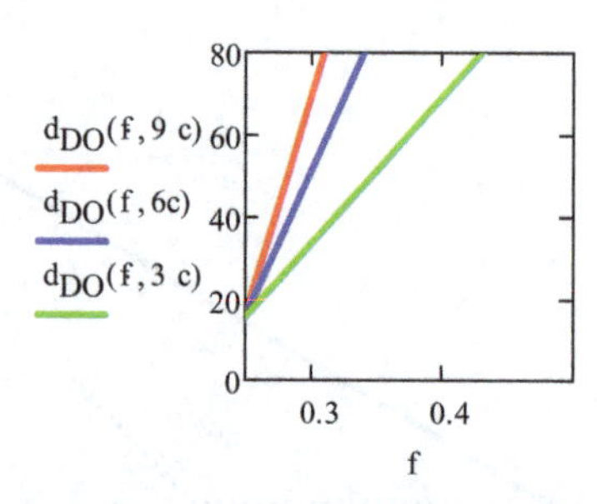

Figure 7.8 Undersaturation and oversaturation deterministic delay function against flow in the range [0, fMAX] or [fMAX, 2 fMAX] respectively, T = 3c (red continuous), 6c (blue dash and point), 9c (green dashed), with c = 60, g = 30, g/c = 0.5, s = 0.5

Combining Equations (7.20, 7.21, 7.22) after some algebra we get (right of Figure 7.8):

$$d_{DOi}(f_i) \ = c(1 - g_i/c) \ / \ 2 \ + (f_i c/ \ (s_i g_i) - 1) \cdot T/ \ 2 \qquad (7.21c)$$

defined over the interval range $[f_{MAX}, \infty]$.

7.2.2.2 Stochastic delay

In this sub-section, we describe an additional delay function developed under the assumption that the arrival flow and the saturation flow are represented by random variables, with distribution constant over time.

Indeed, the arrival flow and the saturation flow may show fluctuations over time as already noted above. The application of the (Stochastic) Queueing Theory leads to so complicated formulas that they are never been used in practice.

A heuristic method is based on a transformation of the Undersaturation Stochastic delay formula so that it has an oblique asymptote given by the Oversaturation Deterministic delay instead of the vertical one. The result is a function that includes both the undersaturation and oversaturation stochastic delay functions as well as the oversaturation deterministic delay function, the undersaturation delay function being given by Equation (7.16). Details are not reported for brevity's sake (see Akcelik, 1980, for details).

A simpler method is based on the tangent approximation of the Webster undersaturation delay formula for values of in-flow close to capacity. The point where the tangent is applied is chosen trying to approximate the oversaturation deterministic delay formula. Figure 7.9 shows some examples of this method.

7.3 Methods for Green Timing

According to the Gren Timing approach the green-amber scheduling is described by dividing the cycle time into *stages* (Figure 7.2), periods during which the signal

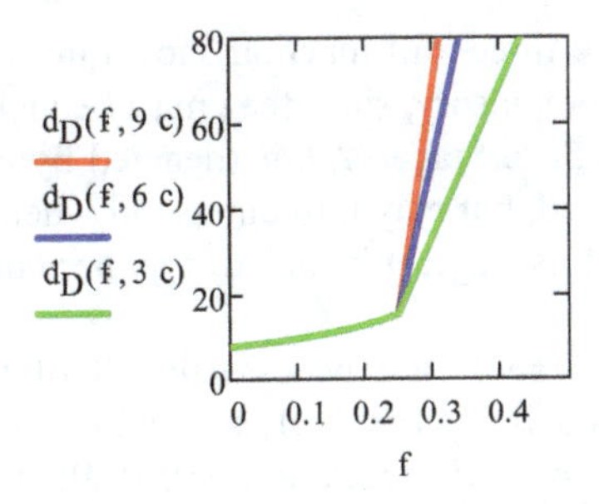

Figure 7.9 Tangent approximation (dotted) at $f = \alpha\ s\ g\ /\ c$, α =(0.90,0.85,0.80), T = 3c (red continuous), 6c (blue dash and point), 9c (green dashed), with c = 60, g = 30, s = 0.5

received by each stream does not change from red to green or from amber to red (the change from green to amber being not relevant to stage definition). Each stage is associated with a set of streams having green (or amber) signal, the others receiving red signal of course (set *{1,2,6}* in Figure 7.2). It is generally assumed that each access *i* has only one green period in each cycle. The design of the signal plan of a single junction through the **green timing** approach requires that the stage matrix is known, it may be defined as described below.

7.3.1 Stages and stage matrix

A *stage* is a set of *approaches* that have green at the same time. For safe operations all the approaches in a stage must be mutually compatible, namely, they may have green without any conflict – *compatibility requirement*. Usually, it is also required that each stage be maximal – *completeness requirement*, meaning that no further approach may be added to any of them without violating the compatibility requirements (adapted from Cantarella *et al.*, 2015). With reference to the three-arm junction shown in Figure 7.2, there are four stages *{1,2,6}*, *{3,4,2}*, *{5,6,4}*, and *{2,4,6}* meeting both those requirements.

[All candidate stages satisfying both the compatibility and completeness requirements can easily be generated by the Bron-Kerbosch algorithm (see Wikipedia) for finding all maximal cliques of the compatibility graph, with a node for each approach and an edge for each pair of compatible approaches. In other words, the adjacency matrix of the graph is the (square symmetric) compatibility matrix with as many rows and columns as the number of approaches and '0/1' entry for each 'incompatible/compatible' pairs of approaches.]

Let $n \geq 2$ be the number of candidate stages ($n = 1$ meaning there is no pair of incompatible approaches, and there is no need for traffic control). A candidate stage that contains an approach not included in any other stage is called *compulsory*, otherwise it is called *optional*. Let n_c and $n_o = n - n_c$ be the number of compulsory and optional candidate stages, respectively.

With reference to the three-arm junction shown in Figure 7.2, stages *{1,2,6}*, *{3,4,2}*, and *{5,6,4}* are compulsory, since they must be included in any sequence to give green to streams *1, 3, 5*. Stage *{2,4,6}* (denoted by dotted lines) is optional, since it may not be included, but may turn out useful when curb-side streams show heavy traffic conditions; this stage, if included, may be scheduled between any pair of the other stages.

A set of candidate stages (not necessarily all of them) in an order is a *sequence*. A sequence is called *feasible* if each approach belongs to at least one stage in the sequence – *feasibility requirement.* If all the compatible and complete stages are considered, such a condition surely holds (in the limited case each stage contains only one approach). Moreover, a compulsory candidate stage, as defined above, must be included in each sequence since it contains an approach not included in any other stage. The number of feasible sequences may be very large.

If there is no optional stage, $n_o = 0$ and $n = n_c$, the number of feasible sequences is given by $n!$, say the number of permutations of the n stages. However, it is worth noting that any periodic rotation of a sequence, for instance, such that the sequence (1,2,3) becomes (2,3,1), then (3,1,2), then (1,2,3,) again, does not affect optimal green and performance indicators, while optimal offsets change in an easily predictable way (see Chapter 8).

Thus, for each junction, all the sequences may be considered grouped into $(n_i! / n_i) = (n_i - 1)!$ equivalence classes and only one sequence for each equivalence class is further analysed. Thus,

- if only two stages are available, two possible sequences can be built up {(1,2); (2,1)} which are equivalent, thus, only one equivalence class exists;
- if three stages are available two equivalence classes exist: {(1,2,3); (2,3,1); (3,1,2)} and {(3,2,1); (2,1,3); (1,3,2)}; either may be obtained from the sequence (1,2) by positioning stage 3 after stage 1 or after stage 2; thus a binary variable is enough to define the optimal sequence;
- if four stages are available six equivalence classes exist: {(1,2,3,4); (2,3,4,1); (3,4,1,2); (4,1,2,3)}; {(1,3,2,4); (3,2,4,1); (2,4,1,3); (4,1,3,2)}; {(1,4,2,3); (4,2,3,1); (2,3,1,4); (3,1,4,2)}); {(1,4,3,2); (4,3,2,1); (3,2,1,4); (2,1,4,3)}; {(1,3,4,2); (3,4,2,1); (4,2,1,3); (2,1,3,4)}; {(1,2,4,3); (2,4,3,1); (4,3,1,2); (3,1,2,4)}); each may be obtained by changing the position of stage 4 with respect to sequences (1,2,3) or (3,2,1); thus a binary variable and a ternary variable are enough to define the optimal sequence;
- and so on.

If there is at least one optional stage, $n_o > 0$, $n > n_c$, the stages can be grouped into 2no sub-sets, each including all the n_c compulsory stages and some (or none at all) of the n_o optional stages; stages belonging to any of such sets may be arranged in a number of feasible sequences equal to the factorial of its size minus 1, as described above. Thus,

- if only one compulsory, 1, and one optional, 2, stages were available it would imply that stage 2 is a sub-set of stage 1, therefore, stage 2 violates the completeness requirement, even though two equivalent classes exist: {(1)}; {(1,2); (2,1)}, thus this case it is not further considered;
- if two compulsory, 1 and 2, and one optional, 3, stages are available, three equivalent classes exist: {(1,2); (2,1)}; {(1,2,3), (2,3,1); (3,1,2)} and{(3,2,1); (2,1,3); (1,3,2)}; each may be obtained from sequence (1,2) by not including stage 3 or by positioning it after stage 1 or after stage 2; thus a ternary variable is enough to define the optimal sequence;
- if three compulsory, 1 and 2 and 3, and one optional,4, stages are available, eight equivalent classes exist: {(1,2,3), (2,3,1); (3,1,2)} and {(3,2,1); (2,1,3); (1,3,2)}; {(1,2,3,4); (2,3,4,1); (3,4,1,2); (4,1,2,3)}; {(1,3,2,4); (3,2,4,1); (2,4,1,3); (4,1,3,2)}; {(1,4,2,3); (4,2,3,1); (2,3,1,4); (3,1,4,2)}); {(1,4,3,2); (4,3,2,1); (3,2,1,4); (2,1,4,3)}; {(1,3,4,2); (3,4,2,1); (4,2,1,3); (2,1,3,4)}; {(1,2,4,3); (2,4,3,1); (4,3,1,2); (3,1,2,4)}; each may be obtained by not including stage 4 or by changing its position within the sequences (1,2,3); thus three binary variables are enough to define the optimal sequence;
- and so on.

Quite often it is also required that each approach has green in consecutive stages within the sequence if more than one – *consecutiveness requirement*. This requirement is effective only for sequences containing four or more stages.

The stage sequence can be described by the following matrix (see Figure 7.2):

$\Delta = [\delta_{ij}]$, the *stage matrix*, a binary matrix with a row for each stream i and a column for each stage j, and entries, $\delta_{ij} = 0$ 1 if stream i has green in stage j, = 0 otherwise.

The stages during which a stream has green are assumed consecutive, considering the last stage as preceding the first one (the green-amber period of stream 6 in Figure 7.2). Let

t_j be the length of stage j.

The sum of the stage lengths is equal to the cycle time:

$$\sum_j t_j = c \tag{7.22}$$

The consistency among stage lengths t_j and green-amber time h_i of the stream i is given by:

$$h_i = g_i + l_i = \sum_j \delta_{ij} t_j \quad \forall i \tag{7.23}$$

7.3.2 *Webster method*

The widely used Webster method (Webster 1958; Webster and Cobbe 1966) allows green timing through a simply closed formulation, once the stage matrix has been defined [This method is somehow equivalent to the analysis in Chapter 21 in Roess *et al.* (2010)].

The application of this method requires that each stage j be associated with a *representative stream i(j)*, which may have green in stage j only; other streams may have green in more than one stage; this assumption may well affect the choice of the stages, indeed the number of stages in stage matrix may not be larger than the number of streams. With reference to the three-arm junction shown in Figure 7.2, the three stages *{1,2,6}*, *{3,4,2}*, *{5,6,4}* may be considered with representative streams 1, 3, 5, respectively; the stage *{2,4,6}* may not be used. Let m be the number of considered stages, equal to the number of representative streams.

Thus, the stage lengths are equal to the green-amber times of the representative streams:

$$t_j = h_{i(j)} = g_{i(j)} + l_{i(j)} \; \forall j = 1, \cdots, m \tag{7.24}$$

Combining Equations (7.24) and (7.22), yields:

$$c = \sum\nolimits_{j=1,m} (g_{i(j)} + l_{i(j)}) = \sum\nolimits_{j=1,m} g_{i(j)} + \sum\nolimits_{j=1,m} l_{i(j)} = \sum\nolimits_{j=1,m} g_{i(j)} + L \tag{7.25}$$

where $L = \Sigma_{j=1,m} \, l_{i(j)}$ is the total lost time over the representative streams. Thus

$$\sum\nolimits_{i=1,m} g_{i(j)} = c - L \tag{7.26}$$

7.3.2.1 Equisaturation principle

The Webster method is based on the *equisaturation principle*: all representative streams have a *common capacity factor* z^*, and saturation degree $x^* = 1/z^*$:

$$z_{i(j)} = g_{i(j)}/(y_{i(j)}c) = z* \; \forall j = 1, \ldots, m \tag{7.27}$$

where $y_i = f_i / s_i$ is the flow ratio of the representative stream I, introduced above.

Combining Equations (7.27) with (7.26) yields:

$$\sum\nolimits_{i=1,m} (y_{i(j)} \cdot c \cdot z*) = c - L$$

Therefore:

$$z* = (c - L) / (cY) = (1 - L/c)\,(1/Y) \tag{7.28}$$

where $c > L$, and $Y = \Sigma_{j=1,m} \, y_{i(j)}$ is the total flow ratio over the representative streams.

7.3.2.2 Cycle time

The common capacity factor increases with the cycle length up to the upper bound $(1/Y)$ for infinite cycle length, thus for $(1/Y) \leq 1$, or $Y \geq 1$, whichever the cycle length, undersaturation may never occur.

If $Y < 1$, undersaturation, $z^* = 1$ (for practical application a value less than 1 is used, say 0.95), is assured for any cycle length larger than:

$$c_{MIN} = L/\ (1 - Y)\ > L$$

On the other hand, delay tends to increase with large values of cycle time, but for very small values. The cycle time, c^*, that minimises the total delay can be approximated by the following equation, obtained through some simulations (Webster 1958; Webster and Cobbe 1966):

$$c_D = (1.5\,L + 5)\ /\ (1 - Y)\ > c_{MIN}[\text{seconds}]$$

In practice, a cycle length of less than 45" is used only in very particular applications.

7.3.2.3 Green timing

According to Equations (7.27) and (7.28), the green times for representative streams are:

$$g_{i(j)} = (y_{i(j)}/Y)\ (c - L)\ \forall j = 1, \ldots, m \tag{7.29}$$

In other words, the difference between the cycle time and the total lost time is allocated as effective green times (of the representative streams) proportionally to the flow ratios.

Hence, remembering Equation (7.24), the stage lengths are

$$t_j = y_{i(j)}(/Y)\ (c - L) + l_{j(j)}\ \forall j = 1, \ldots, m \tag{7.30}$$

The green-amber times of all (other) streams may be computed through the consistency (Equation (7.23)). Examples of the application of the Webster method are reported in Tables 7.1 and 7.2 below.

Table 7.1 Comparison between Webster and SIGCAP methods – example 1, in tdis case, tdere is no need for stage {2,4,6} (from Cantarella, 1998)

Data c = 120.0 seconds					**Webster's method z = 1.765**			**SIGCAP method z = 1.917**		
i	***f$_i$***	***s$_i$***	***y$_i$***	***l$_i$***	***h$_i$***	***g$_i$***	***z$_i$***	***h$_i$***	***g$_i$***	***z$_i$***
1	300	1,800	0.167	3.0	47.4	44.4	2.220	41.3	38.3	**1,917**
2	500	1,800	0.278	2.0	87.4	85.4	2.562	80.3	78.3	2,348
3	250	1,800	0.139	3.0	40.0	37.0	2.220	39.0	36.0	2,156
4	600	1,800	0.333	2.0	72.6	70.6	**1.765**	78.7	76.7	**1,917**
5	200	1,800	0.111	3.0	32.6	29.6	2.220	39.7	36.7	2,755
6	550	1,800	0.306	2.0	80.0	78.0	2.127	81.1	79.1	2,156
Stages					t_j			t_j		
{126}					47.4			41.3		
{342}					40.0			39.0		
{564}					32.6			39.7		
{246}					00.0			00.0		

Table 7.2 Comparison between Webster and SIGCAP methods – example 2, in this case adding stage {2,4,6} lead to better results (from Cantarella, 1998)

Data c = 120.0 seconds					**Webster's method z = 1.875**			**SIGCAP method z = 2.156**		
i	f_i	s_i	y_i	l_i	h_i	g_i	z_i	h_i	g_i	z_i
1	150	1,800	0.083	3.0	30.8	27.8	2.775	26.0	23.0	2.300
2	500	1,800	0.278	2.0	80.0	78.0	2.340	84.1	82.1	2.464
3	250	1,800	0.139	3.0	49.2	46.2	2.775	38.9	35.9	**2.156**
4	600	1,800	0.333	2.0	89.2	87.2	2.181	94.0	92.0	2.300
5	200	1,800	0.111	3.0	40.0	37.0	2.775	35.9	32.9	2.464
6	550	1,800	0.306	2.0	70.8	68.8	**1.875**	81.1	79.1	**2.156**
Stages					t_j			t_j		
{126}					30.8			26.0		
{342}					49.2			38.9		
{564}					40.0			35.9		
{246}					00.0			19.2		

Note: Minimum values in bold.

7.3.2.4 Equisaturation and junction capacity factor maximisation

Any green timing different from that described by Equations (7.30) and (7.23) can increase the capacity factor of some representative streams but will also decrease the capacity factor of other representative streams. Hence, the equisaturation principle assures the maximisation of the common capacity factor of the representative streams, only; other streams may have larger or smaller capacity factors.

In other words, when one (or more) streams may have green in more than one stage, the Webster's method may fail to find the green timing that maximises the junction capacity factor. This is the case of a three-arm junction, as the one shown in Figure 7.2, where some streams may have green in two (or three) stages. In this case, optimisation method described in Chapter 8 may be applied for green timing.

A special case occurs if each stream has green in one stage and the representative stream $i(j)$ of each stage j is the *critical stream*, say the one with the highest flow ratio among all streams in stage j. In this case, all non-critical streams have a larger capacity factor, and the equisaturation principle assures the maximisation of the minimum capacity factor for the whole junction.

7.3.2.5 Minimum green time constraints

Against common knowledge, the Webster method may cope with minimum green time constraints. Indeed, if a stream i receives a green time g_i less than the minimum value g_{MINi}, it suffices to increase the lost time of that stream to the value $l_i + g_{MINi}$ and decrease flow ratio to 0, then re-apply the Webster method. This way the junction capacity factor may decrease.

7.3.3 Optimisation methods for Green Timing

The main optimisation methods for Green Timing are based on models with linear constraints and convex or linear objective functions for total delay minimisation or capacity factor maximisation, respectively. The methods described below are very close to those introduced in the early 70s by the late R.E. Allsop (Allsop 1971, 1972b), and named SIGSET and SIGCAP (Allsop, 1972a). All the methods below can easily be implemented in a spreadsheet.

The main **decisional variables** are the stage lengths (for total delay minimisation the cycle time c may also be a decisional variable, see below):

$$t_j \geq 0 \quad \forall j$$

The main **constraint** concerns stage lengths cycle time consistency, Equation (7.22) below:

$$\sum\nolimits_j t_j = c \tag{7.22}$$

For Equation (7.22) there is no need to introduce a constraint stating that the stage lengths are upper bounded by the cycle time, $t_j = c \ \forall j$.

The green-amber time and the effective green time for each stream i are given by Equation (7.23). Minimum stage length as well as minimum effective green time constraints may straightforwardly be added (see Section 7.1):

$$t_j \geq t_{MIN} \quad \forall j \tag{7.31}$$

$$\sum\nolimits_j \delta_{ij} t_j - l_i^3 g_{MINi} \quad \forall i \tag{7.32}$$

- SIGCAP capacity factor maximisation – the cycle time c is an input data

  At this aim, the junction capacity factor is given by Equation (7.12):

$$z = \min_i z_i = \min_i (s_i \, g_i)/(f_i c) \tag{7.12}$$

The values of stage lengths maximising the capacity factor can be obtained by adding a further **decisional variable**:

$$z \geq 0$$

and further constraints, as many as the streams are:

$$z \leq (s_i g_i) \ / \ (f_i c) \ \ \forall i$$
$$\text{or } z \leq s_i(\textstyle\sum_j \delta_{ij} - l_i))/(f_i c) \ \ \forall i$$

thus

$$\sum\nolimits_j \delta_{ij} s_i t_j - (f_i c) z \geq l_i s_i \ \ \forall i \tag{7.33}$$

The whole optimisation model is:

$$\text{maximise } z \tag{7.34}$$

subject to constraints: (22), (31), (32), and (33)

with respect to decisional variables: $t_j \geq 0 \ \forall j, z \geq 0$

The model (34), SIGCAP, is a linear optimisation model with non-negative decision variables, which may be solved through the simplex algorithm. Due to the greater than/equal to constraints (33) the model (34) may not have the solution. Indeed, from the pure mathematical point-of-view, this can be the case for very low values of cycle time c, depending on the values of the lost times l_i, but this case never occurs for practical values, say $c > 30$ seconds, and $l_i < 5$ seconds.

The optimal value of the capacity factor increases as the cycle time goes to infinity (not considering constraints 31 and 32), thus the cycle time c is to be considered an input data.

If the optimal value of the capacity factor is less than 1, no signal plan exists assuring undersaturation conditions.

SIGCAP vs. Webster method. The solutions of the linear optimisation model (34) SIGCAP, are consistent with those of the Webster method when the latter is applicable, otherwise, if some streams have green in more than one stage better values of the junction capacity factor may be obtained through SIGCAP.

With reference to the three-arm junction in Figure 7.2, and considering three or four stages, the results of the application of both methods to two examples are briefly reported in Tables 7.1 and 7.2. Example 1 in Table 7.1 shows that in this case there is no need to include the optional stage {2,4,6}; whilst example 2 in Table 7.2, obtained with different arrival flows, shows that in this case, adding the optional stage {2,4,6} leads to a better solution.

7.3.4 Total delay minimisation

At this aim, any formula for the unitary delay $d_U(f, g)$ may be used, provided that it is convex with respect to the effective green time g, such as the Webster two-term delay formula, or any extension of it. There the total delay is given by:

$$d_T = \sum\nolimits_i f_i d_U(f_i, g_i) \tag{7.35}$$

where $g_i = \Sigma_j\, \delta_{ij}\, t_j - l_i$ according to Equation (7.23).

Two slightly different models can be defined.

~ SIGDEL total delay minimisation – the cycle time *c* is an input data, as for SIGCAP.

The whole optimisation model is:

$$\text{minimise} \sum\nolimits_i f_i d_U(f_i, S_j d_{ij} t_j - l_i) \tag{7.36}$$

subject to constraints: (22), (31), and (32)

with respect to decisional variables: $t_j \geq 0\ \forall j$,

The model (36), SIGDEL introduced in this book, is a convex optimisation model, say convex objective function with linear constraints, with non-negative decision variables, which may be solved through the convex simplex algorithm. The above considerations about solution existence still hold.

The solutions of the convex optimisation model (36) are generally different from those provided by the model (34) SIGCAP. It might even occur that the junction capacity factor is less than 1. Therefore, it is better to first solve model (34) and check whether the optimal junction capacity factor is greater than 1, $z > 1$, before applying model (36).

When the optimal junction capacity factor is not much greater than 1, it may be worth ensuring that the solution of model (34) does not lead to oversaturation conditions by adding a further constraint given by (33) with $z = 1.10$ for safety.

~ **SIGSET total delay minimisation – the cycle time *c* is a decisional variable:**

$$c \geq 0$$

It is usually lower and upper bounded, such as:

$$30 \text{ seconds} \leq c \leq 120 \text{ seconds}$$

To keep the convexity of the objective function, the stage lengths as decisional variables should be scaled to the cycle time, duly amending Equations (7.5, 7.6):

$$t_j = t_j / c \geq 0 \; \forall j$$

Thus, the stage lengths cycle time consistency, Equation (7.2) above, becomes:

$$\sum\nolimits_j \tau_j = 1 \tag{7.37}$$

Furthermore, instead of the cycle time *c* its reciprocal should be considered as the decisional variable:

$$q = 1/c \geq 0$$

with a slight modification of the definition of the unitary delay function as $d_U(f, g/c)$.

Therefore Equations (7.31) and (7.32) become:

$$\tau_j \geq t_{MIN} q \; \forall j \tag{7.38}$$

$$\sum\nolimits_j \delta_{ij} \tau_j - l_i q \geq g_{MINi} q \; \forall i \tag{7.39}$$

The whole optimisation model is:

$$\text{minimise} \sum\nolimits_i f_i d_U(f_i, S_j d_{ij} t_j - l_i q) \tag{7.40}$$

subject to constraints: (37), (38), and (39)

with respect to decisional variables: $\tau_j \geq 0 \; \forall j$, $q \geq 0$

Considerations made above for model (36) also apply to model (40) SIGSET. The optimal value of the cycle time is generally within the range [30, 120].

7.4 Methods for Green Timing and Scheduling

Main optimisation methods for Green Timing and Scheduling are based on models with linear constraints and some binary variables or with non-linear constraints and real variables only. All the methods described below, named SICCO (after Improta and Cantarella, 1984; Cantarella and Improta, 1983) have not been implemented in commercial software. Similar methods are implemented in OSCADY (TRL, 2021).

The description of SICCO methods below is consistent with those existing in literature, but somehow original for this book. It is assumed that each stream i has only one green amber (g-a) period within each cycle; however, depending on the time reference point, the g-a period of a stream may begin in a cycle and end in the subsequent cycle, see for instance the g-a period of stream 6 of the three-arm junction in Figure 7.2.

As already said, the main variables for each stream i are:

$h_i \geq 0$ is the (displayed) green-amber time,
$u_i \geq 0$ the beginning of the (displayed) green-amber time,
$v_i \geq 0$ the end of the (displayed) green-amber time.

The values of variables u_i and v_i $\forall i$ allow us to define the stage matrix, which in this case is not an input data but an output of the post-processing of the solution, see the three-arm junction in Figure 7.2. A consistency equation holds among the above variables:

$$v_i = u_i + h_i \ \forall i \tag{7.41}$$

Thus, the main **decisional variables** for each stream i are only two:

$$h_i \geq 0 \ \forall i$$

$$u_i \geq 0 \ \forall i$$

both being upper bounded by the cycle time

$$h_i \leq c \tag{7.42}$$

$$u_i \leq c \tag{7.43}$$

On the other hand, as noted in Section 7.2, may well occur $0 \leq v_i = u_i + h_i \leq 2\,c$.

With no loss of generality, the beginning of the g-a time of a stream, for instance 1, can be used as a reference to avoid redundant solutions by translation or mirroring:

$$u_1 \geq 0$$

This way the number of actual decisional variables is reduced by 1. Indications to choose this stream are given below.

Minimum stage length, as well as minimum effective green time constraints, may straightforwardly be added:

$$h_i \geq g_{MIN} + l_i \quad \forall i \tag{7.44}$$

The following constraints ensure that the g-a times of a pair of incompatible streams (see Chapter 4) j and k do not overlap.

– **Either** the g-a period of stream j begins first, thus it must end before the beginning of the g-a period of stream k, and in the next cycle, the g-a period of stream j must begin after the end of the g-a period of stream k.

$$u_j + h_j \leq u_k \quad \forall j, k \text{ incompatible} \tag{7.45}$$

$$u_k + h_k \leq u_j + c \quad \forall j, k \text{ incompatible} \tag{7.46}$$

– **or** the ga period of stream k begins first:

$$u_k + h_k \leq u_j \quad \forall j, k \text{ incompatible} \tag{7.47}$$

$$u_j + h_j \leq u_k + c \quad \forall j, k \text{ incompatible} \tag{7.48}$$

The two constraints (45) and (46) exclude the two constraints (47) and (48) and vice versa. Two ways can be used to express this XOR (exclusive OR) condition.

- SICCO 1

A further binary **decisional variable** is introduced for each pair of incompatible streams j and k:

$$w_{jk} = 0/1 \quad \forall j, k \text{ incompatible}$$

with $w_{jk} = 1$ the g-a period of stream j begins first,
$w_{jk} = 0$ the g-a period of stream k begins first.

Thus constraints (45) and (46) EXOR (47) and (48) can be expressed as:

$$u_j + h_j \leq u_k + (1 - w_{jk}) \cdot c \quad \forall j, k \text{ incompatible} \tag{7.49}$$

$$u_k + h_k \leq u_j + w_{jk} \cdot c \quad \forall, k \text{ incompatible} \tag{7.50}$$

Indeed for $w_{jk} = 1$ constraints (49) and (50) become (45) and (46), whilst for $w_{jk} = 0$ they become (47) and (48). As far as the g-a period of stream 1 is used as a reference its g-a period surely begins before that of any stream incompatible with it, therefore:

$$w_{1k} = 1 \quad \forall k$$

In this case, the following two equations hold instead of (7.49) and (7.50)

$$u_1 + h_1 \leq u_k \quad \forall 1, k \text{ incompatible} \tag{7.51}$$

$$u_k + h_1 \leq u_k + c \quad \forall 1, k \text{ incompatible} \tag{7.52}$$

This way the number of binary decisional variables is reduced. So far it is useful to choose as reference stream one of those with the maximum number of incompatibilities.

- SICCO 2

The following two non-linear constraints can be used to describe (45) and (46) EXOR (47) and (48).

$$(u_k - u_j - h_j) \cdot (u_j - u_k - h_k) \leq 0 \ \forall j, k \text{ incompatible} \tag{7.53}$$

$$(u_k + c - u_k - h_j) \cdot (u_j + c - u_{kj} - h_k) \leq 0 \ \forall j, k \text{ incompatible} \tag{7.54}$$

As far as the g-a period of stream 1 is used as a reference its g-a period surely begins before that of any stream incompatible with it, therefore constraints (51) and (52) can be used instead of (53) and (54) for $j = 1$. This way the number of non-linear constraints is reduced. So far it is useful to choose as reference stream one of those with the maximum number of incompatibilities.

- SICCO **capacity factor maximisation** – the cycle time c is known, say it is an input data

At this aim, the junction capacity factor is given by Equation (7.12):

$$z = \min_i z_i = \min_i (s_i g_i)/(f_i c) \tag{7.12}$$

The values of stage lengths maximising the capacity factor can be obtained by adding a further **decisional variable**:

$z \geq 0$and further constraints as many as the streams are:

$$z \leq s_i(h_i - l_i) \ / \ (f_i c) \ \forall i \tag{7.55}$$

remembering that $g_i = h_i - l_i \ \forall I$, from Equation (7.4).

The whole optimisation model is:

$$\text{maximise } z \tag{7.56}$$

subject to constraints: (42), (43), (44), and (55)

SICCO 1: (49) and (50) as well as (51) and (52)

SICCO 2: (53) and (54) as well as (51) and (52)with respect to decisional variables: h_i, $u_i \geq 0 \ \forall j$, $z \geq 0$, and

SICCO 1 only: $w_{jk} = 0/1 \quad \forall j, k$ incompatible

The model (56) SICCO 1 is a linear mixed-binary optimisation model with some non-negative real decision variables and some binary decision variables, which may be solved for instance by Branch&Bound or Branch&Cut methods or by one of the several exact algorithms available for a small number of variables. The model (56) SICCO 2 is a non-linear optimisation model with non-negative real decision variables, which may be solved for instance through the Nelder and Meade algorithm or other more advanced methods. Post-processing of the solution of SICCO 2 gives the values of binary variables w_{jk}.

Remark. *SICCO –* ***max*** *z may provide the non-critical streams (those which may have a capacity factor greater than optimal value) a g-a time less than it was possible. Therefore, after the solution of the model, it is worthwhile the application of a reduced version of SICCO1 –* ***min*** *z with values of variables* w_{jk} *given by the optimal solution of SICCO –* ***max*** *z, and objective function maximise* $\sum_i h_i$*; the solution of this linear optimisation model does not affect the optimal junction capacity factor.*

SICCO – max z vs. SIGCAP. The solutions of the optimisation model (56) SICCO – **max z** are consistent with those from SIGCAP. SICCO avoids generating each feasible stage and tests each feasible stage matrix (see Section 7.1) generated by any combination of optional stages as is the case for SIGCAP. This feature is of the utmost relevance for complex junctions with many optional stages. Anyhow the optimal stage matrix can be obtained *a posteriori* by computing the end of the g-a times v_i through Equation (7.41).

With reference to the three-arm junction in Figure 7.2, the results of the application of SICCO – **max z** method to the two examples in Tables 7.1 and 7.2 are briefly reported in Tables 7.3 and 7.4 below; results are consistent with those in Tables 7.1 and 7.2.

Table 7.3 Application of SICCO – ***max*** *z, example 1, the optimal stage matrix computed a posteriori contains three stages*

Data c **= 120.0 seconds**					**SICCO – max z method** z **= 1.917**		
i	f_i	s_i	y_i	l_i	h_i	u_i	z_i
1	300	1,800	0.167	3.0	41,3	00.0	**1,917**
2	500	1,800	0.278	2.0	80,3	00.0	2,348
3	250	1,800	0.139	3.0	39.0	41.3	2,156
4	600	1,800	0.333	2.0	78,7	41.3	**1,917**
5	200	1,800	0.111	3.0	39,7	80.3	2,755
6	550	1,800	0.306	2.0	81,1	80.3	2,156

Note: Minimum values in bold.

Table 7.4 Application of SICCO – ***max*** *z, example 2, the optimal stage matrix computed a posteriori contains four stages*

data c **= 120.0 seconds**					**SICCO – max z method** z **= 2.156**		
i	f_i	s_i	y_i	l_i	h_i	u_i	z_i
1	150	1,800	0.083	3.0	26.0	00.0	2.300
2	500	1,800	0.278	2.0	84.1	100.8	2.464
3	250	1,800	0.139	3.0	38.9	26.0	**2.156**
4	600	1,800	0.333	2.0	94.0	26.0	2.300
5	200	1,800	0.111	3.0	35.9	64.9	2.464
6	550	1,800	0.306	2.0	81.1	64.9	**2.156**

Note: Minimum values in bold.

• SICCO total delay minimisation

At this aim, any formula for unitary delay $d_U(f, g)$ may be used, provided that it is convex with respect to the g-a time h, such as the Webster two-term delay formula, or any extension of it. There the total delay is given by Equation (7.57):

$$d_{\mathrm{T}} = \sum_i f_i d_U(f_i, g_i) \tag{7.57}$$

where $g_i = h_i - l_i$ according to Equation (7.3).

7.4.1 SICCO total delay minimisation

At this aim, any formula for unitary delay $d_U(f, g)$ may be used, provided that it is convex with respect to the g-a time h, such as the Webster two-term delay formula, or any extension of it. There the total delay is given by Equation (7.57):

$$d_T = \sum_i f_i d_U(f_i, g_i) \tag{7.57}$$

where $g_i = h_i - l_i$ according to Equation (7.3).

Two slightly different models can be defined, as for SIGDEL and SIGSET above.

∼ **the cycle time c is known, say it is an input data, as for model (56).**

The whole optimisation model is:

$$\text{minimise} \sum_i f_i d_U\left(f_i, \sum_j d_{ij} t_j - l_i\right) \tag{7.58}$$

subject to constraints: (42), (43), and (44)

SICCO 1: (49) and (50) as well as (51) and (52)
SICCO 2: (53) and (54) as well as (51) and (52)

with respect to decisional variables: h_i, $u_i \geq 0 \; \forall j$, $z \geq 0$, and
SICCO 1 only: $w_{jk} = 0/1 \forall j, k$ incompatible

∼ **the cycle time c is unknown, say it is a decisional variable.**

The whole optimisation model is quite similar to model (58) see also model (40).

As noted above, It is better to first solve model SICCO – **max z** and check whether the optimal junction capacity factor is greater than 1, $z > 1$, before applying SICCO – **min d**, and/or adding a further constraint $z \geq 1.10$ for safety to SICCO – **min d**.

When the optimal junction capacity factor is not much greater than 1, it may be worth ensuring that the solution of model (34) does not lead to oversaturation conditions by adding a further constraint given by (33) with $z = 1.10$ for safety.

SICCO – min d vs. SIGDEL/SIGSET. The solutions of the SICCO – **min d** are consistent with those of the SIGDEL/SIGSET methods, but it avoids generating each feasible stage and test each feasible stage matrix (see Section 7.1) generated by any combination of optional stages as is the case for SIGCAP. This feature is of the utmost relevance for complex junctions with many optional stages. Anyhow the optimal stage matrix can be obtained *a posteriori* by computing the end of the g-a times v_i through Equation (7.41).

7.5 Summary

This chapter introduces basic variables and methods for the single signalised junction analysis and control, including methods for Green Timing, such as Webster, SIGCAP, SIGDEL, SIGSET, as well as Green Timing and Scheduling, such as SICCO. Methods for Green Timing can easily be implemented in a spreadsheet; SICCO is not available in commercial software. A method based on graph theory useful to solve SICCO – **max z** is described by Cantarella and Improta (1988).

Similar methods are the basis of the commercial software OSCADY (Burrow, 1987; TRL, 2021); others are in SIDRA (Akcelik and Besley, 1992; Akcelik, 1993); see remarks below.

All the considerations in this chapter can almost straightforwardly be applied to pedestrian streams, as well as to bus streams if distinguished from car ones.

Optimal control of oversaturated signals is usually based on queue allocation strategies, such as the store and forward method introduced by Michalopoulos P. G. and Stephanopoulos G. (1977a) for single junctions, (1977b) for systems of junctions (see Chapter 8), see also Chang and Lin (2000).

Junctions is a software package by Transport Research Laboratory. It incorporates the previously separate programs PICADY, for priority junction analysis (see Chapter 5), ARCADY for roundabouts analysis (see Chapter 6), and OSCADY for singles signalised junction analysis and control (this chapter). The latest version, Junctions 10, was launched on 3 February 2021 (Wikipedia 2024, 10, 31). https://en.wikipedia.org/wiki/Junctions_(software).

Sidra Intersection (styled **SIDRA**, previously called **Sidra** and **aaSidra**) is a software package used for intersection (junction), interchange and network capacity, level of service and performance analysis, and signalised intersection, interchange and network timing calculations by traffic design, operations and planning professionals. Version 9.0 was released in May 2020 (Wikipedia 2024, 10, 31). https://www.sidrasolutions.com.

A.1 Appendix

This appendix describes how the values of some parameters relevant to signalised junction control can be obtained from observations or from some simple models. Before starting the analysis of a junction, possibly in a system, it is relevant to collect some infrastructural characteristics of the junction:

- the number of lanes for each direction of travel and the relative width and slope;

- the presence of parking near the junction;
- the presence of bus stops near the junction;
- the presence of pedestrian crossings near the junction;
- the location of the junction.

A.2 Arrival flow data

Arrival flow data can be collected through several available technologies. All flows are measured in vehicles per unit of time, for example, vehicles/second, or better in passenger car units (pcu) per time unit using appropriate homogenisation coefficients for types of vehicles other than cars, such as buses, medium and heavy vehicles, etc.

A.3 Lost time and saturation flow data

The lost time can be measured by observing the actual time interval between the lag between two successive vehicles crossing the stop line during the green time. It is generally observed that this value is maximum at the beginning of the green period, then decreases until a constant value is generally close to 2 s (in the presence of a queue). A similar phenomenon occurs during the amber period. The lost time is given by the sum of all time intervals exceeding the constant value. The saturation flow can also be estimated, remembering that it is equal to the reciprocal of the constant value.

A.4 Lost time and saturation flow method

HCM (2022) and previous editions (see also Chapter 2 in Cascetta, 2009) describe a widely used method for computing the saturation flow s of an approach or stream, expressed in pcu per time unit (see also Kimber *et al.*, 1986):

$$s = s_0 \cdot N \cdot F_w \cdot F_{HV} \cdot F_g \cdot F_p \cdot F_{bb} \cdot F_a \cdot F_{RT} \cdot F_{LT}$$

where

s is the saturation flow in pcu per time unit;
s_0 is the ideal saturation flow per lane, usually 1,900 pcphgpl (passenger cars per hour of green time per lane);
N is the number of lanes in the lane group;
F_w is the adjustment factor for lane width (12 ft or 3.66 m lanes are standard);
F_{HV} is the adjustment factor for heavy vehicles in the traffic flow;
F_g is the adjustment factor for approach grade;
F_p is the adjustment factor for the existence of a parking lane adjacent to the lane group and the parking activity in that lane;
F_{bb} is the adjustment factor for the blocking effect of local buses that stop within the junction area;
F_a is the adjustment factor for the area type;
F_{RT} is the adjustment factor for right turns in the lane group;
F_{LT} is the adjustment factor for left turns in the lane group.

A.5 Amber and all-red

The lengths of the amber and all-red periods are determined allowing vehicles to stop before reaching the stop line, by the end of the amber period, or to cross the junction before the end of the all-red period. It is appropriate to preliminarily specify that the amber period is specific to each access approach, while the all-red period is common to all approaches that do not have green in the subsequent stage.

All the parameters used below (for example speed, deceleration, reaction time, . . .) should be considered as the results of statistical analyses that account for the variability in driver characteristics (e.g., reaction times), vehicle features (e.g., presence of driver assistance systems), infrastructure conditions (e.g., pavement quality), and environment factors (e.g., weather conditions). Either the average value or a safety-favored percentile may be used.

Consider a vehicle moving with speed $v > 0$, at the beginning of the amber period, the driver of the vehicle, at the beginning of the amber period, decides to carry out one of the following two actions:

- brake and stop the vehicle before the stop line uniformly decelerating before the end of the amber period;
- proceed beyond the stop line and cross the junction (before the end of the all-red period) if there is not enough space to carry out the first action moving at a constant speed.

In both cases, a reaction and decision time, t_R, is considered, during which the vehicle, moving at a constant speed, covers a distance $v \cdot t_R$. After the time t_R, stopping the vehicle requires, a further time:

$$t_F = v/a$$

with a positive value (of the average over time) of the braking deceleration. The distance d_S travelled during this action is:

$$d_S = v^2/(2a)$$

This distance coincides with the minimum distance necessary to brake before the stop line. Vehicles that stop before the stop line cannot collide with other vehicles entering the junction. Therefore, the length of the amber period is not influenced by the stop time.

A vehicle that during the amber period is at a distance d from the stop line less than d_S, however, it does not have enough space to stop without colliding with other vehicles, therefore it must necessarily cross the junction. The crossing requires a time made up of the sum of two rates:

- time needed to reach the stop line:

$$t_{NS} = d_S/v = \left(v^2/(2a)\right)/v = v/(2a)$$

- time required to cross the junction after passing the junction or *clearance time*:

$$t_{SG} = d_{SG}/v$$

with d_{SG} being the sum of the length of the vehicle and the distance to be crossed between the stop line and the limit of the junction where the manoeuvre ends (maximum distance between the possible trajectories).

The minimum length of the amber period, A_{min}, is therefore given by:

$$A_{min} = t_R + t_{NS} = t_R + v/(2a)$$

It must be noted that in the absence of other indications is assumed $A = A_{min}$. Under these conditions, the length of the all-red period is equal to the clearance time, t_{SG}:

$$t_{SG} = d_{SG}/v$$

The length of the amber period can be extended by consistently reducing that of the all-red period. Therefore, if regulation without an all-red period is adopted, the amber period is given by the sum of A_{min} and t_{SG}. The following values are generally accepted:

- Reaction time: 1 second;
- Deceleration for motor vehicles: 3.0–3.5 m/s^2;
- Deceleration for buses: 2.0–2.5 m/s^2;
- Deceleration for trams: 1.5 m/s^2.

References

A huge literature is available on these topics, some relevant papers are in the list of references below, besides those quoted in the main text.

Akçelik, R. (1980). Time-Dependent Expressions for Delay, Stop Rate and Queue Length at Traffic Signals.

Akcelik, R., and Besley, M. (1992). SIDRA user guide, Part 2, Input. ARRB Transport Research Ltd., Report No. WDTE91/012B, Victoria, Australia, 143p.

Akcelik, R. (1993). Traffic signals: capacity and timing analysis. Australian Road Research Board Ltd., Research Report No. 123, Fifth Reprint, Victoria, Australia.

Allsop, R.E. (1971). Delay-minimising settings for fixed-time traffic signals at a single road junction. *Journal of the Institute of Mathematics and Its Applications*. 8 (2), 164–185.

Allsop, R.E. (1972a). SIGSET: A computer program for calculating traffic capacity of signal-controlled road junctions. *Traffic Engineering & Control* 12, 58–60.

Allsop, R.E. (1972b). Estimating the traffic capacity of a signalized road junction. *Transportation Research* 6 (3), 245–255.

Allsop, R.E. (1972c). Delay at fixed time traffic signal I: theoretical analysis. *Transportation Science* 6 (3), 260–285.

Allsop, R.E. (1976). SIGCAP: A computer program for assessing the traffic capacity of signal-controlled road junctions. *Traffic Engineering & Control* 17, 338–341.

Allsop, R. E. (1977). Treatment of opposed turning movements in traffic signal calculations. *Transportation Research* 11 (6), 405–411.

Allsop, R.E. (1992). Evolving application of mathematical optimisation in design and operation of individual signal controlled road junctions. In: Gri.ths, J.D. (Ed.), *Mathematics in Transport and Planning and Control*. Clarendon Press, Oxford, pp. 1–24.

Burrow, I.J. (1987). OSCADY: a computer program to model capacities, queues and delays at isolated traffic signal junctions, TRRL Report, Vol. 105, *Transport and Road Research Laboratory*, Crowthorne.

Cantarella G.E. (1998). Capacity Maximization at Three-Arm Junctions: An Extension of Webster's Method. In *Proceedings of the Third International Symposium on Highway Capacity*, R. Rysgaard ed., 279-296, Road Directorate (Copenhagen, Denmark, June 1998).

Cantarella G.E. and Improta G. (1983). A Non-Linear Model for Control System Design of an Individual Signalized Junction. In Atti delle Giornate di Lavoro AIRO 83, 709-722. Guida Editore. (Napoli, settembre 1983.)

Cantarella G.E. and Improta G. (1988). Capacity Factor Optimization for Signalized Junctions: a Graph Theory Approach. *Transportation Research Part B* 22, 1–23.

Cantarella G.E., and Vitetta A. (2010). La regolazione semaforica di intersezioni stradali: metodi ed applicazioni. Collana Trasporti - Franco Angeli Editore.

Cascetta E. (2009). *Transportation Systems Analysis: Models and Applications*. Springer.

Di Pace R. (2019) Introduction to the Traffic Flow Theory, appendix 2 in Cantarella G.E., D.P. Watling, S. de Luca, and R. Di Pace. 2019. *Dynamics and Stochasticity in Transportation Systems: Tools for Transportation Network Modelling*. Elsevier.

Cronje, W.B. (1983). Optimization model for isolated signalized traffic intersections. *Transportation Research Record 905*, 80–83.

Chang, T.-H., and Lin, J.-T. (2000). Optimal signal timing for an oversaturated intersection. *Transportation Research Part B* 34, 471–491.

Gallivan, S., and Heydecker, B.G. (1988). Optimising the control performance of traffic signals at a single junction. *Transportation Research* 22B (5), 357–370.

HCM. (2022). Highway Capacity Manual 7th Edition: A Guide for Multimodal Mobility Analysis. National Academies of Sciences Engineering and Medicine, Transportation Research Board.

Heydecker, B.G. (1992). Sequencing of traffic signals. In: Gri.ths, J.D. (Ed.), *Mathematics in Transport and Planning and Control*. Clarendon Press, Oxford, pp. 57–67.

Heydecker, B.G., and Dudgeon, I.W. (1987). Calculation of signal settings to minimise delay at a junction. In: *Proceedings of 10th International Symposium on Transportation and Traffic Theory, MIT*. Elsevier, New York, pp. 159–178.

Kimber, R.M., McDonald, M., and Hounsell, N. (1986). The prediction of saturation flows for road junctions controlled by traffic signals, TRRL Report, Vol. RR 67, Transport and Road Research *Laboratory*, Crowthorne.

Improta G. and Cantarella G.E. (1984). Control System Design for an Individual Signalized Junction. *Transportation Research Part B* 18, 147–167.

Michalopoulos P. G. and Stephanopoulos G. (1977a). Oversaturated signal systems with queue length constraints-I. Single Intersection. *Transportation Research* 11(6), 413–421.

Michalopoulos P. G. and Stephanopoulos G. (1977). Oversaturated signal systems with queue length constraints-II. Systems of intersections. *Transportation Research* 11(6), 423–428.

Roess R. P., Prassas E. S., and McShane W.R. (2010). *Traffic Engineering.* Pearson College *Div.*

TRL. (2021). *Junctions is a software package by Transport Research Laboratory*, see Section 7.5.3.

Webster F.V. (1958). *Traffic Signal Settings. Road Research Technical Paper No. 39, Road Research Laboratory*, London, UK.

Webster F.V., and Cobbe B.M. (1966). Traffic Signals. *Road Research Technical* Paper No. 56, London UK.

Yagar S. (1974). Capacity of a signalized road junction: critique and extensions. *Transportation Research* 8(1974), pp. 137–147.

Yagar, S. (1975). Minimizing delay at a signalized intersection for time-invariant demand rates. *Transportation Research* 9(2), 129–141.

Yagar S. (1977). Minimizing delays for transient demands with application to signalized road junctions. *Transportation Research* 11(1977), 53–62.

Chapter 8

Systems of signalised junctions analysis and control

Giulio Erberto Cantarella[1] *and Orlando Giannattasio*[2]

You look at where you're going and where you are and it never makes sense, but then you look back at where you've been and a pattern seems to emerge.

Zen and the art of motorcycle maintenance – R.M. Pirsig

Outline. *This chapter reviews the basic definitions and notations for systems of signalised junction analysis and control. It introduces coordination and sychronisation and reviews some common optimisation methods for systems of signalised junction static control.*

Basic elements of Signalised Junction Analysis (SJA) are reviewed below and in this chapter. As said in the introduction a complete discussion of this topic is out of the scope of this book; a recent comprehensive presentation of most of these topics is in Roess *et al.* (2010), see also HCM (2020), Chapter 2 in Cascetta (2009) and Di Pace (2019).

For the reader's convenience, from Chapter 7, which aims at developing methods for describing and designing signalised junction control, it is better distinguishing:

- **Single junctions**: control of adjacent junctions is not taken into account, as discussed in Chapter 7;
- **Systems of junctions**: Control of adjacent junctions is to be taken into account, as discussed below:
  - **(Junction) Arterials**, simple sequences of junctions connected without loops, usually the word 'junction' is omitted;
  - **(Junction) Networks**, junctions connected by a network with loops.

Aiming at developing (macroscopic) models for delay computation, as a performance index, it is better distinguishing:

[1]Department of Civil Engineering, University of Salerno, Italy
[2]Department of Civil, Computer Science and Aeronautical Technologies Engineering, Roma Tre University, Italy

- **Isolated junctions**: no approach interacts with any of the upstream junctions, effects of adjacent junctions are negligible, discussed in Chapter 7;
- **Interacting junctions**: (at least) some approaches interact with any of the approaches of the upstream junction, the effects of adjacent junctions are not negligible, as discussed below.

Basic definitions and notations for systems of junctions are described in Section 8.1; then unitary delay for an interacting approach is discussed in Section 8.2. Afterward, Sections 8.3 and 8.4 outline some methods for the signal setting design of an arterial including the simple green wave method in closed form as well as some optimisation methods. For a detailed description of the later methods see the references at the end of the chapter.

8.1 Basic definitions and notations for systems of junctions

This section introduces the basic notations for the analysis and signal-setting design of

- **Arterials**, simple sequences of junctions connected without loops,
- **Networks**, junctions connected by a network with loops.

The junction arterial or network may be modelled through an undirected connected graph with a node for each junction and an edge for each pair of adjacent junctions, say two junctions connected by a street [If the graph were not connected the analysis below would be carried out for each connected sub-graph in it.]. Let

n_J be the number of junctions;
n_S be the number of streets connecting them.

An arterial contains $n_S = n_J - 1$ streets and can be modelled by a graph with $(n_J - 1)$ edges. On the other hand, a (connected) network contains $n_S > n_J - 1$ streets and can be modelled by a graph with more than $n_J - 1$ edges; it contains $n_S - (n_J - 1)$ (independent) loops.

It is assumed that:

- the cycle time $c > 0$ is common to all junctions (or sub-multiple of it, say $c/2$);
- the signal plan of each single junction k is identically repeated each cycle;
- the signal plan of each single junction k is described by
  - the stage matrix Δ_k and the stage length t_{ki}, of each stage j, or
  - the beginning u_{ki} and the end v_{ki} of the green-amber period of each stream i;
- the signal plan of each single junction k is given a reference starting time, say:
  - the beginning of the stage numbered 1, or
  - the beginning u_{k1} of the green-amber period of the stream numbered 1.

Thus the signal plan of the arterial or the network is completely defined by the schedule of each single junction signal plan with respect to the common reference time, usually the reference starting time of junction 1. Let

$\phi_k \geq$ be the *node offset* for junction k, say the time lag between the reference starting time of its signal plan and the common reference time.

To avoid redundant solutions by translation or mirroring let $\phi_1 = 0$ be a reference time point, all the other $(n_J - 1)$ node offsets be independent and can be decisional variables for control.

On the other hand, delay models for interacting junctions, see next Section 8.2, are more easily defined by

$\phi_{hk} = \phi_k - \phi_h$ the *arc offset* from junction h to junction k, say the time lag between the beginning of the signal plan of junction h and that of junction k; arc offsets may be positive or negative, depending on which signal plan starts earlier (Figure 8.1).

It is worth noting that the *arc offset* between junctions (k, h) is the opposite of that of between junctions (h, k), $\phi_{kh} = -\phi_{hk}$; thus one arc offset is enough for each pair of adjacent junctions. The direction of the chosen arc offset is independent of the actual direction of the street connecting them.

Since the signal plan of each single junction is identically repeated, each cycle, node, and arc offsets may be re-defined within range [0, c[, through the **modulo operation** applied to the offsets with respect to c, with no loss of generality.

With reference to an arterial, all the $n_S = (n_J - 1)$ arc offsets are independent and can be chosen as decisional variables for junction control, see Section 3.1 for a simple method, and Section 3.2 for a general method, using node offsets.

On the other hand, a network contains $n_S > (n_J - 1)$ arc offsets, but only $(n_J - 1)$ of them are independent, thus it is not worth choosing them as decisional variables for junction control, see Section 8.4 for the outline of a general optimisation method using node offsets.

Remark. *Indeed, for each independent loop within the graph modelling the junction network, there exists a set of consistency equations among all the arc offsets of the edges in the loop, analogous to Kirchoff's second law or loop rule for an*

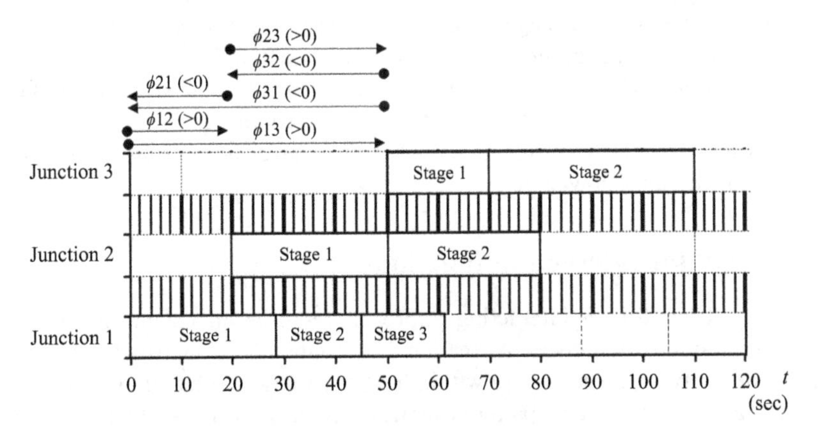

Figure 8.1 Example of arc offsets (adapted from Cantarella and Vitetta, 2010)

electric circuit. Therefore the use of arc offsets as decisional variables requires that:

- a set of $n_S - (n_J - 1)$ independent loops is found;
- $n_S - (n_J - 1)$ constraints, one for each loop, are included in the optimisation method (with discrete variables too to take into account that the consistency equation is to be restricted to the range $[0, c]$).

8.1.1 Signal plan of a system of junctions (arterial or network)

Given the cycle time common to all junctions, the signal plan of a system of junctions is completely specified by the signal plan of each junction *k*, (Chapter 7) say:

1. **Green timing** – defined by the stage matrix Δ_k and the length of each stage t_{kj};
   or
   Green timing and scheduling – describing how the green-amber periods are located within the cycle, defined by the beginning, u_{ki}, and the end, v_{ki}, of each green-amber period, such that $h_{ki} = v_{ki} - u_{ki}$;
   and
2. the values of $(n_J - 1)$ node offsets ϕ_k, with $\phi_1 = 0$ as a reference time point. With regard to arterials only, the signal plan may also be specified by the $(n_J - 1)$ arc offsets instead.

Several methodologies are available:

Coordination – the values of the offsets are determined after the signal plan of each single junction has been specified, as said above;
Synchronisation – the values of the offsets and the length of each stage t_{ki} of each junction *k* are determined at the same time, after the stage matrix Δ_k of each junction *k* has been defined, as discussed in Section 3 of Chapter 7;
Scheduled synchronisation – the values of the offsets, the length of each stage t_{ki}, and the stage matrix Δ_k of each junction *k* are determined at the same time after a set of feasible possible stages is defined for each junction *k* [a method addressing this problem has recently been presented in Cantarella *et al.*, 2015];
Complete synchronisation – the values of the offsets and the signal plan of each junction *k* are determined at the same time given the compatibilities among streams for each junction [this methodology has never been discussed in the literature to authors' knowledge].

8.2 Delay for interacting junctions

Modelling delay for an interacting approach requires an enhancement of the modelling framework outlined in Section 2 of Chapter 7 for an isolated approach. Indeed, the arrival flow may vary within the time analysis interval given by [0, c], with average f_i, but the same pattern is considered each cycle, while the saturation flow, s_i, is constant during the time analysis interval given by [0, c]. Under this assumption the undersaturation deterministic delay analysis in Section 2 of Chapter 7 no longer applies; on the other hand, the stochastic and/or oversaturation delay may still be computed as in section in Section 2 of Chapter 7.

The undersaturation deterministic delay is the result of several phenomena occurring when vehicles move from an upstream junction to a downstream one:

- inter-**division**: the flow pattern of vehicles arriving at the upstream junction is divided among all available approaches, each corresponding to a different stream, possibly including several manoeuvres leading to different exits;
- intra-**division**: after inter-division, the flow pattern of vehicles leaving an approach is divided among all manoeuvres (if more than one) in the corresponding stream;
- **distortion**: after intra-division, the arrival flow pattern of vehicles leaving an approach following a manoeuvre is changed by the sequence of red/green periods before entering the downstream street;
- **merging**: after distortion, the flow patterns of vehicles leaving different approaches leading to the same street merge together;
- **dispersion**: after merging, the flow patterns of vehicles moving along the street tend to become closer and closer to a uniform pattern due to speed distribution among vehicles. Indeed the longer the travel time is the closer to a uniform profile the dispersed flow pattern is. This consideration is the main element in distinguishing isolated (see Chapter 7) vs. interacting approaches.

The above framework is implemented in an operative model based on the notion of a (discrete) cyclic flow profile within the TRANSYT method described in Section 8.4.

8.3 Methods for arterial coordination

Designing the signal plan of an arterial is simpler than that of a network with loops. Indeed, in this case, both the node offsets or the arc offsets may be used as decisional variables, as discussed in Section 8.1. Moreover, even if rarely noted in the literature, methods for arterials may be used for any network without loops, even though it is not a sequence of streets.

The main methods aim at maximising the bandwidth, that is the period of time (within the green period) during which a vehicle can travel along the arterial without encountering a red, assuming a speed known to users or given by duly signals. Two long-established methods are described below, the simple Green-Wave method for a one-way arterial and the MaxBand/MultiBand method for two-way ones. In both cases, the cycle time c common to all junctions and the stage matrix of each junction are assumed input data.

8.3.1 Green-Wave method for arterial coordination

The Green-Wave method allows the coordination of a one-way arterial through a very simple closed formulation, once the stage matrix and green timing of each junction have been defined (recent reviews in Warberg, 2008; Wu *et al.*, 2014). Assuming that the arterial contains $n_S = n_J - 1$ streets, let

L_{hk} be the length of a street between each two adjacent junctions h and k;
v be the travel speed suggested to users moving along the arterial (or known by them);
tt_{hk}/v be the travel time between each two adjacent junctions h and k.

The signal plan is specified by the $(n_J - 1)$ arc offsets given by:

$$\phi_{hk} = tt_{hk} \tag{8.1}$$

In this way, a vehicle leaving junction h at the beginning of the green period and travelling at speed v along the street will encounter the beginning of the green period upon arriving at the downstream junction k, and so on, as long as it keeps moving along the arterial. Methods for two-way arterial coordination are described in Section 8.3.2 and in Section 8.4.

8.3.2 MaxBand and MultiBand methods for arterial control

The MaxBand method allows the bandwidth maximisation of a two-way arterial through an optimisation method (Little, 1966, Little *et al.* 1981).

The main notations are described below and shown in Figure 8.2. To be consistent with Figure 8.2, the notations used in this section differ from those used elsewhere in this book.

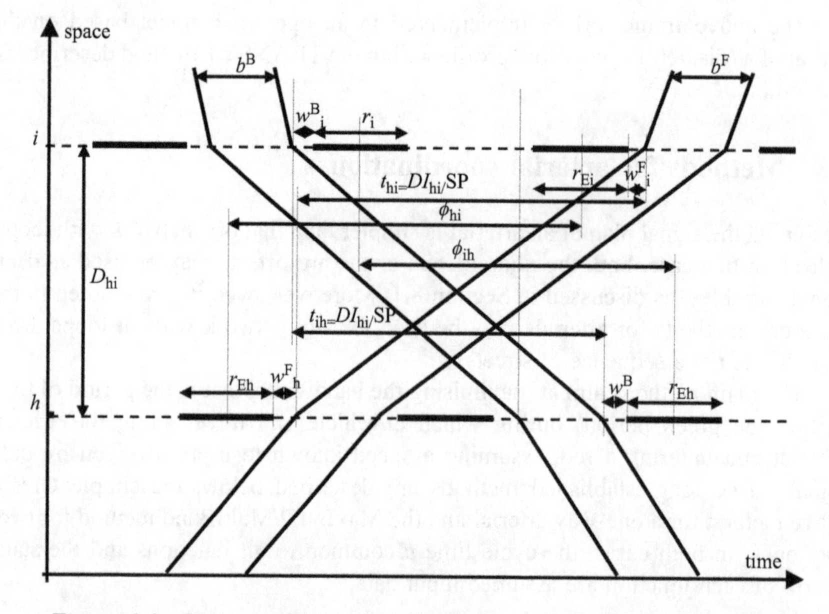

Figure 8.2 Main variables of MAXBAND method (Little et al. 1981)

$DI_{hi} = DI_{ih}$ is the distance between the junction h and I, considered input data;
SP_{hi} is the travel speed for users moving from junction h towards junction i (SP_{ih} on the opposite side);
t_{hi} is the travel time from junction h to junction i (t_{ih} for the opposite way);
r_{Ei} is the effective red time for the manoeuvre of the junction i which follows the arterial from i towards the consecutive junction h (r_{Eh} for the opposite way);
$g_{Ei} = c - r_{Ei}$ is the corresponding effective green time ($g_{Eh} = c - r_{Eh}$ for the opposite side);
$w^F{}_i$ is the time interval between the end of the red phase and the beginning of the band for the flow moving from junction h towards junction i;
$w^B{}_i$ is the time interval between the end of the band and the beginning of the red phase of the flow moving from junction i towards junction h;
ϕ_{hi} is the arc offset between the junction h and i, say the time interval between the midpoints of the red periods preceding the band at the junction h and i (ϕ_{ih} for the opposite way).

The sum of forward ϕ_{hi} and the backward ϕ_{ih} offsets between the junction h and i is equal to the cycle length c or any integer multiple of it, see Section 8.1:

$$\phi_{hi} + \phi_{ih} = m_{hi}c \quad \forall h, i \ adjacent \tag{8.2}$$

with m_{hi} is an integer non-negative number.

The whole optimisation model is outlined below.

The main **input data** are:

the cycle time c, common to all junctions,
the distance between each pair of adjacent junctions
the red proportions r_{Ei}/c,
the stage matrix for each junction.

Moreover, the progression speed SP between each pair of adjacent junctions, and/or the length of each stage of each junction may be an output of the method or input data.

The **objective function** is the maximisation of the weighted sum of the bandwidths in the two travel directions:

$$\text{maximise}\left(b^F/c\right) + \gamma\left(b^B/c\right) \tag{8.3}$$

where

b^F is the bandwidth for the flow moving from intersection h towards intersection i (or direction *Forward*), being a decisional variable;
b^B is the bandwidth for the flow moving from intersection i to intersection h (or direction *Backward*), being a decisional variable;
γ is the weight parameter between the bands (it is equal to 1 in conditions of equal weight between the two flow streams), being an input data.

The main **constraints** are, besides (2):

a minimum-maximum interval for the progression speeds,
a minimum-maximum interval for the difference of the speeds of two adjacent streets,

a minimum-maximum interval for the difference of the speeds of the directions of the same street.

The main **output data**, that is the decisional variables beside integer values m_{hi} in Equation (8.2), are:

the bandwidths for the two ways, b^F/c and b^B/c
the position of the two bands, w^B_i/C and w^F_i/C for each junction i
the arc offset ϕ_{hi} for each pair of adjacent junctions h and i.
the (progression) speeds SP_{hi} and SP_{ih} between each pair of junctions h and i, if they are not an input data.

The application of this method also allows for defining the order of left-turn phases (in all junctions) based on the weighted maximisation of the width of the green bands in the two directions of travel, along the signalised arterial.

The optimisation problem obtained is a mixed-integer linear programming problem, since some decisional variables are non-negative real, others, m_{hi}, non-negative integer, and all constraints are linear. It may be solved for instance by Branch & Bound or Branch & Cut methods or by one of the several exact algorithms available for a small number of variables.

A specific case arises when the values of the progression speeds are input data. In this case, the arc offsets are given by:

$$\phi_{hi} = w_h^F + DI_{hi}/SP_{hi} - w_i^F \quad \forall h, i \text{ adjacent} \tag{8.4}$$

The MAXBAND method can be extended to account for the variation in flow along the artery due to vehicle entry and diversion to/from the main flow, as in the MULTIBAND method (Gartner *et al.*, 1990).

8.4 Methods for network coordination or synchronisation

One of the most used methods for designing the signal plan of a network, or an arterial, is TRANSYT (introduced by Robertson, 1969; now version 16, TRL, 2024). It is based on an implementation of the general traffic model described in Section 8.2. The TRANSYT method, just **TRANSYT** in the following, is implemented in a commercial software fully integrated with methods in **Junctions** software for single junctions, see Chapter 7. TRANSYT 7F, the americanised version, is briefly commented on in the final remarks.

The basic elements of TRANSYT are outlined below, for details see TRL (2024); the general structure is in Figure 8.3. It is assumed that:

all the junctions have the same cycle time c (or a sub-multiple of it, for instance, $c/2$), considered an input data;
the signal plan of each junction and that of the whole network is repeated each cycle.

TRANSYT may be used in three different ways.

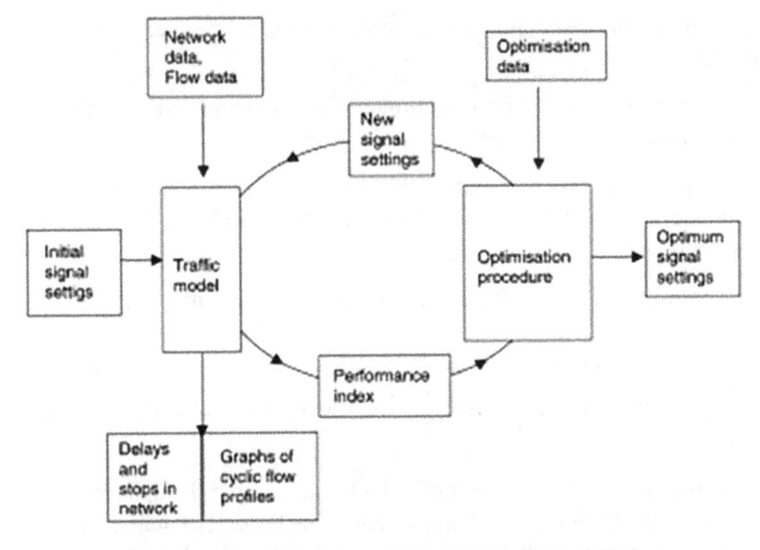

Figure 8.3 Structure of TRANSYT (TRL, 2024)

8.4.1 TRANSYT traffic model – analysis

Knowing the signal plan of each junction and the node offsets

8.4.1.1 Computation of the total delay

As already noted in Section 8.2, the undersaturation deterministic delay analysis in Section 2 of Chapter 7 for an isolated approach no longer applies to an interacting approach. The stochastic and/or oversaturation delay may still be computed as in Section 2 of Chapter 7. It is assumed that:

- the arrival arc flows entering the network are constant over time and known;
- the departing arc flows exiting the network are constant over time and known;
- either the route flow from each entry point to each exit point is constant over time and known,
- or the turning proportions at each approach are constant over time and known.

The main traffic model available in TRANSYT is the Platoon Dispersion model described below, others are commented on at the end of this section.

Platoon Dispersion Model (PDM)

This model is suitable for any kind of junction network and well describes the dispersion phenomenon along a street between two junctions, as commented in Section 8.2. The travel time along a street does not depend on the flow, nor on the length of the queue. This model does not describe the backward effects in the upstream junctions of the downstream queuing phenomena constraining the entering capacity of the street.

According to Platoon Dispersion Model (PDM), the undersaturation deterministic delay is the result of several phenomena occurring when vehicles move along a street from an upstream junction to a downstream one, see Section 8.2. It is assumed that:

- the cycle time c is divided into small intervals with length $\Delta\tau$, usually 1 s or less,
- $c/\Delta\tau$ is a positive integer for simplicity's sake.

Omitting approach/stream index, the flow pattern over time at a point during one cycle cab modelled by a discrete **cyclic flow profile** (CFP), described by a vector **q**, with entries:

$q(i)$ is the number of vehicles moving during interval i, with $i = 1, \ldots, c/\Delta\tau$, such that:
$f = \sum_{i=1,\ldots,c/\Delta\tau} q(i)/c$ is the corresponding total flow.

The cyclic flow profiles entering the network are assumed known, that is an input data.

- inter-**division**: the flow pattern of vehicles arriving at the upstream junction is divided among all available approaches, each corresponding to a different stream j, possibly including several manoeuvres leading to different exits according to known proportions.

Flow conservation is assured by inter-division. No delay occurs during inter-division.

- intra-**division**: after inter-division, the flow pattern of vehicles leaving an approach with more than one manoeuvre is divided among all exit manoeuvres, according to known proportions; otherwise, the approach has one manoeuvre and there is no intra-division.

Flow conservation is assured by intra-division. No delay occurs during inter-division.

- **distortion**: after intra-division, the arrival flow pattern of vehicles leaving an approach following a manoeuvre is changed due to the sequence of red/green periods before entering the downstream street. Omitting approach/stream index, let

$q_{D1}(i)$ $\forall i$ be the cycle profile after division, before distortion, waiting for green time;
$q_{D2}(i)$ $\forall i$ be the cycle profile after distortion, exiting the junction during the green period.

In undersaturation, the queue due to the red period vanishes at time t_o before the end of the cycle, (see case A.2.1, in Section 2 of Chapter 2, see also Section 2 in Chapter 7):

a) red period $[0, r]$, the values $q(i)$ of the exit value of the cyclic flow profile are zero:

$$q_{D2}(i) = 0 \;\; \forall i : i \cdot \Delta\tau \in [0, r]$$

b) green period up to the time of the end of the queue t_o, the values $q(i)$ of the exit value of the cyclic flow profile are equal to the saturation flow s time the sub-interval length:

$$q_{D2}(i) = s \cdot \Delta\tau \ \ \forall i : i \cdot \Delta\tau[r, r + t_o]$$

c) green period from the time of the end of the queue t_o to the cycle time c the values $q(i)$ of the exit value of the cyclic flow profile are equal to the entering values:

$$q_{D2}(i) = q_{D1}(i) \ \ \forall i : i \cdot \Delta\tau[r, r + t_o]$$

In the case of oversaturation the queue does not vanish before the end of the cycle, therefore case c) (mentioned above) does not occur, and there is a residual queue at the end of the cycle.

The total flow f is unchanged by distortion. A delay occurs during intra-division.

delay: after intra-division, the arrival flow pattern of vehicles leaving an approach

The number of arrivals over time can be obtained from the CFP waiting for green time, $q_{D1}(i)\ \forall i$, while the number of departures over time can be obtained from the CFP exiting the junction during the green period, $q_{D1}(i)\ \forall i$; this way the evolution of the queue over time can be defined, hence the unitary undersaturation deterministic delay for an interacting approach $d_{DU^*}(f)$ can be computed, as a function of the total flow f (see Section 2 in Chapter 2). Therefore the unitary delay $d_{U^*}(f)$ is:

$$d_{U*}(f) = d_{DU*}(f) + d_{SUO}(f, g) \tag{8.5}$$

where $d_{SUO}(f,\ g)$ is the unitary stochastic and/or oversaturation delay, as in Section 2 of Chapter 7. With a uniform CFP after division, Equation (8.5) gives results consistent with results in Section 2 of Chapter 7 for isolated approaches.

- **merging**: after distortion, the flow patterns of vehicles leaving different approaches leading to the same street merge together; the corresponding CFPs are summed up, to get the CFP entering the street to move toward the downstream junction.

Flow conservation is assured by merging. No delay occurs during inter-division.

- **dispersion**: after merging, the whole flow pattern of vehicles moving along the street from junction j to junction k tends to become closer and closer to a uniform pattern due to the travel speed distribution among users. Let

$q_{Mj}(i)\ \forall i$ be the cycle profile at the upstream junction j just after merging;
$q_{D3k}(i)\ \forall i$ be the cycle profile at the downstream junction k after dispersion;
t_{jk} be the travel time between upstream junction j and downstream junction k;
$\Delta i = \text{int}(0.5 + 0.8 \cdot t_{jk}/\Delta\tau)$ be the closest integer to the travel time of the vehicles moving with the highest speed, assumed $0.8 \cdot t_{jk}$, a coefficient different from 0.80 may be used;
$\eta \in [0,1]$ (often denoted by F) be a dispersion convex combination coefficient such that:

$\eta = 0$: the cyclic flow profile is equal to that at the upstream junction shifted by Δi sub-intervals, that is by the travel time of the vehicles moving with the highest speed, this case occurs for very short streets;

$\eta = 1$: the cyclic flow profile is uniform, and all values are equal to

$$q_{D3k}(i) = \sum\nolimits_{i=1,\ldots,c/\Delta\tau} q_{Mj}(i)/(c/\Delta\tau) = f \cdot \Delta\tau \;\; \forall i$$

this case occurs for long streets, usually longer than 600 m.

The dispersion is described by:

$$q_{D3}(i + \Delta i) = \eta q_M(i) + (1 - \eta) q_M(i + \Delta i - 1) \tag{8.6}$$

The parameter η is usually assumed equal to $1/(1 + 0.4\ t_{jk})$, a coefficient different from 0.40 may be used. Thus as said above the longer the travel time is the closer to uniform profile the dispersed flow pattern is. This consideration is the main element in distinguishing isolated (see Chapter 7) vs. interacting approaches.

The total flow f is unchanged by dispersion. No delay occurs during dispersion; anyhow users experience the travel time.

Other traffic models available in TRANSYT are:

- **Congested Platoon Dispersion Model (CPDM)**, not described for brevity' sake.
  This model is an enhancement of the PDM that takes into account congestion along the street between two junctions, that is the travel time along a street depends on the flow.
- **Cell Transmission Model (CTM)**, described in Chapter 3.
  This model is well suited for small junction networks with short stretches of streets, such as downtown urban networks. This model well describes the backward effects on the exit capacity of the upstream junctions due to the downstream queuing phenomena. Thus the entering capacity of a street as well as the travel time along it depend on the flow and the length of the queue. This model does not describe dispersion.

An example of the application of Equation (8.5) is shown in Figure 8.4, assuming that there is still a queue when the green period ends, and

the length of each sub-interval is $\Delta\tau = 5$ s (for simplicity's sake, usually values less than 1 s are commonly used),
the cycle length is $c = 120$ s,
the effective green time is $g = 25$ s,
the saturation flow is $s = 0.48$ vehicles/s (1,780 vehicles/h), therefore the maximum number of vehicles over 5 s is $5 \cdot 0.48 = 2.4$,
the total flow is 0.12 vehicles/s = 432 vehicles/h.

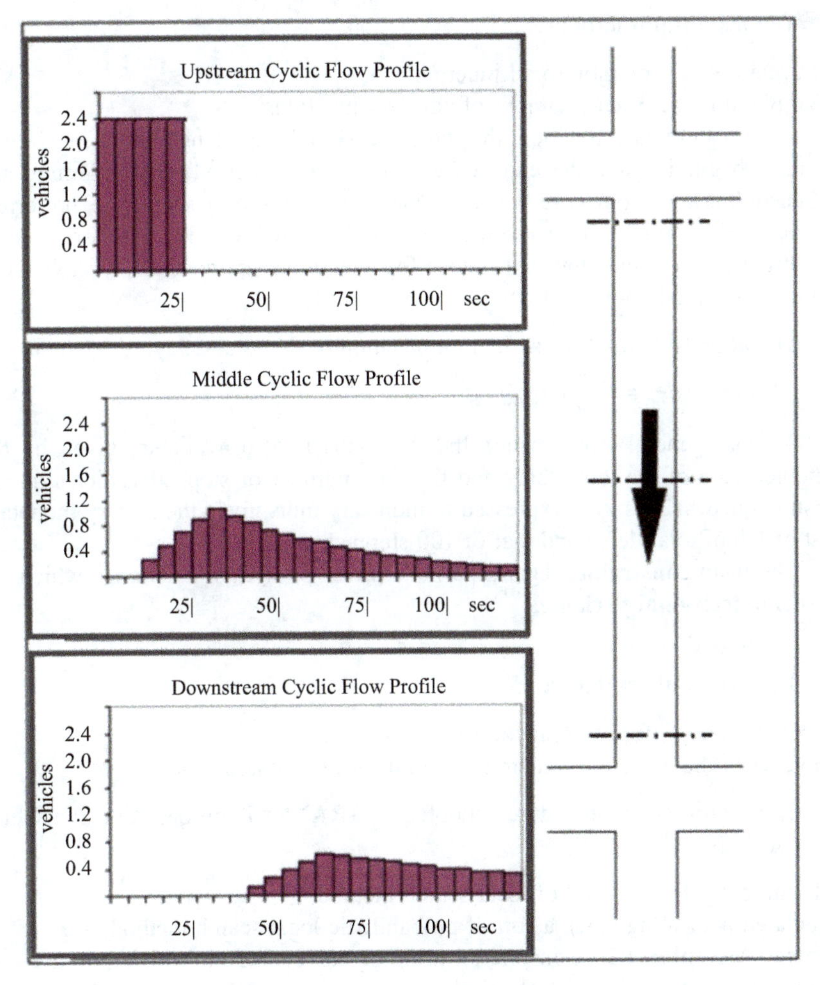

Figure 8.4 Example of platoon dispersion along a street $\Delta\tau = 5$ s, $c = 120$ s, $g = 25$ s, $s = 0.48$ vehicles/s (1,780 vehicles/hour) the maximum number of vehicles over 5 s is $5 \cdot 0.48 = 2.4$

8.4.2 *TRANSYT optimisation method – coordination*

Knowing the signal plan of each junction

8.4.2.1 Optimisation of the node offset of each junction

After the signal plan of each junction has been specified, TRANSYT – **coordination** aims at designing the node offsets of the signal plan of a network, or an arterial, with respect to the minimisation of a Level-of-Service indicator computed through the Traffic Model described above. The whole optimisation model is described below.

The main **input data** are:

the cycle time c, common to all junctions,
the travel time between each pair of adjacent junctions,
the signal plan of each junction, the stage matrix and stage lengths (Green Timing) or the beginning and the end of each green-amber period (Green Timing and Scheduling); in both case, capacity factor maximisation should be preferred, since delay minimisation would be based on delay functions for isolated approaches not consistent with those for interacting approaches [TRANSYT is fully integrated with Junctions.].

The **objective function** is the minimisation of the total delay:

$$Minimise\ d_{T*} = \sum\nolimits_i f_i d_{U*}(f_i, g_i) \tag{8.7}$$

A more general Performance Indicator (PI) is also available, given by the weighted sum of the total delay and the total number of stopped vehicles due to existing queues. This PI is expressed in monetary units given the unitary monetary cost of 1 h of total delay and that of 100 stopped vehicles.

The main **constraints** regard the node offset ϕ_j of each junction j, which are the main **decisional variables**:

$$\phi_j \leq c \ \forall j$$

The main **output data** are:

the node offset ϕ_j for each junction j,
the value of the objective function is useful for plan assessment.

The optimisation procedures adopted in TRANSYT are based on metaheuristics such as

Hill-climbing (HC), a kind of local search method,
Simulated Annealing (SA), a sort of probabilistic local search method,
Genetic Algorithm (GA), inspired by biological evolution, which produces an evolving set (populations) of solutions.

As for all metaheuristics generally, some parameters must be designed to get an effective solution algorithm. A few trials for application at the hands might be useful.

8.4.3 TRANSYT optimisation method – synchronisation

Knowing the stage matrix of each junction

8.4.3.1 Optimisation of the offsets and of the stage durations at each junction

After the stage matrix of each junction has been specified, TRANSYT – **synchronisation** aims at designing both the stage lengths of each junction signal plan and the node offsets of the network signal, with respect to the minimisation of a Level-of-Service indicator. The whole optimisation model is very similar to that for coordination.

The stage matrix for each junction as an input data can be obtained as described in Section 3 of Chapter 7, or by applying a method for Green Timing and Scheduling, such as SICCO or OSCADY [TRANSYT is fully integrated with Junctions].

8.5 Summary

This chapter introduces basic variables and methods for the analysis and control of a system of signalised junctions, including methods for arterials, such as Green-Wave, or MaxBand and its extension MultiBand, as well as for both arterials and networks, such as the well-known and widely used **TRANSYT** method, implemented in a commercial software fully integrated with methods in **Junctions** software, see below.

Worth of further comments are:

Scheduled synchronisation. The values of the offsets, the length of each stage t_{ki}, and the stage matrix Δ_k of each junction k are determined at the same time after a set of feasible possible stages is defined for each junction k (a method addressing this problem has recently been presented in Cantarella *et al.*, 2015);

Complete synchronisation. The values of the offsets and the signal plan of each junction k are determined at the same time given the compatibilities among streams for each junction, this methodology has never been discussed in the literature to the authors' knowledge.

Optimal control of oversaturated signals is usually based on queue allocation strategies, such as the store and forward method introduced by Michalopoulos P. G. and Stephanopoulos G. for single junctions (1977a) and for systems of jucntions (1977b).

Junctions is a software package by Transport Research Laboratory, see Chapter 7.

TRANSYT (TRAffic Network StudY Tool) is a traffic engineering software developed by the Transport Research Laboratory (Robertson, 1969). It is used to model signalised highway networks and has the ability to model platooning. The current release is version 16 (edited from Wikipedia 2024, 10, 31).

https://trlsoftware.com/news/junctions-signal-design/transyt/transyt-16-now-available/.

TRANSYT 7F is a traffic simulation and signal timing optimisation program. The original TRANSYT version 7, developed by the Transport Research Laboratory in the United Kingdom (see above) was 'Americanised' for the Federal Highway Administration (FHWA); thus the '7F'. TRANSYT-7F continues to undergo further development and is currently maintained by the University of Florida's McTrans Center (edited from Wikipedia 2024, 10, 31).

https://mctrans-wordpress-prd-app.azurewebsites.net/highway-capacity-software-hcs/transyt-7f/.

References

A huge literature is available on these topics, some relevant papers are in the list of references below, besides those quoted in the main text.

Cantarella G.E., and Vitetta A. (2010). *La regolazione semaforica di intersezioni stradali: metodi ed applicazioni*. Collana Trasporti – Franco Angeli Editore.

Cantarella, G.E., de Luca, S., Di Pace, R., and Memoli, S. (2015). Network Signal Setting Design: Meta-heuristic optimisation methods. In *Transportation Research Part C: Emerging Technologies*, 55, pp. 24–45.

Gartner N.H. (1976). Area traffic control and network equilibrium, *in Traffic Equilibrium Methods, Lecture Notes in Economics and Mathematical Systems*, Vol. 118, M. Florian, ed., Springer-Verlag, Berlin, 1976, 274–297.

Gartner N.H., Assmann S.F., Lasaga F., and Hous D.L. (1990). MULTIBAND—A Variable-Bandwidth Arterial Progression Scheme. In *Transportation Research Record Journal of the Transportation Research Board No*. 1287, p. 212–222.

Gartner N.H., Assmann S.F., Lasaga F., and Hom D.L. (1991). A multiband approach to arterial traffic signal optimization. *Transportation Research*, 25B, 55–74.

Gartner, N.H., Stamatiadis, C., and Tarno, P.J. (1995). Development of advanced traffic signal control strategies for intelligent transportation systems: Multilevel design. *Transportation Research Record*, 1494, 98–105.

HCM. (2022). *Highway Capacity Manual 7th Edition: A Guide for Multimodal Mobility Analysis*. National Academies of Sciences Engineering and Medicine, Transportation Research Board.

Little J.D.C. (1966). The synchronisation of traffic signals by mixed-integer-linear-programming. *Operations Research*, 14, 568–594.

Little J. D. C., M. D. Kelson, and N. H. Gartner. (1981). MAXBAND: A Program for Setting Signals on Arteries and Triangular Networks. In *Transportation Research Record*, 795, pp. 40–46.

Michalopoulos P. G. and Stephanopoulos G. (1977a). Oversaturated signal systems with queue length constraints-I. Single Intersection. *Transportation Research*, 11(6), 413–421.

Michalopoulos P. G. and Stephanopoulos G. (1977b). Oversaturated signal systems with queue length constraints-II. Systems of intersections. *Transportation Research*, 11(6), 423–428.

Robertson D.I. (1969). TRANSYT method for area traffic control. *Traffic Engineering & Control*, 10, 276–281.

Roess R. P., Prassas E. S., and McShane W.R. (2010). *Traffic Engineering*. Pearson College Div.

Sun W., Mouskos K.C. (2002). Network-Wide Traffic Responsive Signal Control in Urban Environments. NCTIP Research Report, New Jersey Institute of Technology (http://www.transportation.njit.edu/nctip/).

TRANSYT-7F (1987). User's Manual, Release 5.0. Federal Highway Administration.

TRANSYT-7F (1991). User Guide, Methodology for Optimizing Signal Timing, vol. 4. Transportation Research Center, University of Florida.

TRL (2021). *Junctions is a software package by Transport Research Laboratory*, see Section 7.5.3.

TRL (2024). *TRANSYT (TRAffic Network StudY Tool) is traffic engineering software developed by the Transport Research Laboratory*, see above.

Warberg Andreas, Jesper Larsen, and Rene Munk Jørgensen. (2008). Green Wave Traffic Optimization – A Survey. D T U Compute. Technical Report, n. 2008-01. *Informatics and Mathematical Modelling*.

Wu, Xiaoping, Deng, Shuai, Du, Xiaohong and Ma, Jing. (2014). Green-Wave Traffic Theory Optimization and Analysis. *World Journal of Engineering and Technology*, 2, 14–19.

Chapter 9
Dynamic analysis and control

Roberta Di Pace[1]

> The enjoyment of scientific research also means coming up against obstacles to overcome, coming up with even better investigation tools and even more complex theories while endeavouring to always move forward despite knowing that we will likely get closer to comprehending reality, without ever fully being able to understand it.
>
> Margherita Hack

Outline. *This chapter reviews the basic definitions and notations for signalised junction analysis and control. It introduces and reviews some common optimisation methods for dynamic control with reference to single junctions and networks of signalised junctions.*

The primary limitation of fixed-time strategies (described in detail in Chapters 4 and 5) is that their settings are based on historical data rather than real-time information. This can lead to oversimplifications because: traffic demand is not constant, even within a single day; demand patterns can vary from day to day, such as during special events; long-term changes in demand may cause the optimised settings to become outdated; turning movements also fluctuate in the same way as demand and can be influenced by how drivers adjust their behaviour in response to new signal settings, aiming to minimise their personal travel time; incidents and other disruptions can unpredictably affect traffic conditions.

For these reasons, traffic-responsive strategies if designed correctly, could be more efficient, though they come at a higher cost due to the need for a real-time control system, including sensors, communication networks, a central control room, and local controllers. Relevant methods are based on dynamic control through the application of traffic-responsive strategies activated by real-time flows/arrival detection.

One of the most used flow-based single junction strategies are SIGCAP (Allsop 1976), SIGSET (Allsop 1972), and SICCO (Improta and Cantarella, 1984; Cantarella and Improta, 1988) (discussed in section 2.4) whilst for network control strategies is the TRANSYT method, first developed by Robertson in 1969 and then enhanced in successive releases (Vincent *et al.*, 1980; Chard and Lines, 1987; Binning *et al.*, 2010) (see section 2.5).

[1]Department of Civil Engineering, University of Salerno, Italy

Fixed timing strategies may perform poorly when actual flows are greatly different from those used for optimisation due to within-day fluctuations, as well as day-to-day variations. To overcome such a limitation, some authors have developed strategies based on real-time observed flows, despite them generating high operational costs (in terms of sensors, communications, local controllers, etc.). Methods that fall within such a group are SCOOT (Split Cycle Offset Optimisation Technique; Hunt *et al.*, 1981; Bretherton *et al.*, 1998; Stevanovic *et al.*, 2009) and SCATS (Luk, 1984; Stevanovic *et al.*, 2008) for network junctions. These require traffic data to be updated online to get input flow for the optimiser (such as TRANSYT, Binning *et al.*, 2010) and arrange green timings, offsets, and cycle time duration. Whichever optimisation method is employed, in general, Network flow-based Traffic Control strategies require within-day-dynamic traffic flow modelling. Several approaches can be adopted for within-day dynamics in a transportation network as discussed in Chapter 1.

Concerning the strategies that require knowledge of vehicle arrivals, MOVA (Vincent and Pierce, 1988) has been developed for single junctions whilst for network traffic responsive control strategies a brief review has been hereunder summarised.

Gartner (1983) gives a detailed description of the rolling horizon approach in OPAC. Rather than aiming at dynamic programming, OPAC uses a technique named Optimal Sequential Constrained Search to plan for the entire horizon, penalising queues left after the horizon. This strategy does not consider explicitly splits, offsets, or cycles. Based on prespecified staging, they calculate the optimal values for the next few switching times in real time, over a future time horizon H, starting from the current time t and the currently applied stage. The rolling horizon procedure is employed for real-time application of the results. Hereby, the optimisation problem is solved in real time over a time horizon H (e.g., 60 s) using measurement-based initial traffic conditions and demand predictions over H, but results are applied only for a much shorter roll period h (e.g., 4 s), after which new measurements are collected and a new optimisation problem is solved over an equally long time horizon H, and so forth. The rolling horizon procedure avoids myopic control actions while embedding a dynamic optimisation problem in a traffic-responsive (real-time) environment.

UTOPIA (Urban Traffic Optimisation by Integrated Automation, Mauro and Di Taranto, 1989; Mauro, 2002) is a hybrid control system that combines online dynamic optimisation and offline optimisation. This is achieved by adopting a hierarchy structure with a wide-area level and a local level. An area controller continuously generates and provides a reference plan for local controllers. The local controllers then adapt the reference plan and coordinate signals dynamically in adjacent junctions. The rolling horizon approach is then used again by the local controller to optimise single junction design variables. To automate the process of updating reference plans that are generated by TRANSYT, an AUT (Automatic Updating of TRANSYT) module is developed. These traffic data are continuously updated by evaluating the mean of the flows collected by some of the detectors in the network. The data are processed to predict traffic flow profiles for different parts of the day to be used when calculating new reference plans. AUT generates the data to be used in the TRANSYT computation.

9.1 Single junction

Strategies for signalised single junction are classified as:

- *Fixed-time* [pre-timed] signal control (FT) involves preset, fixed signal timings (known as time plans); a detailed presentation of these methods is provided in Chapter 5. Separate time plans can be established for various times of day, such as morning peak, mid-day, afternoon peak, and night, each tailored to suit the typical traffic conditions of that period.
- *Vehicle-actuated* control (VA/semi and fully actuated) adjusts green time and cycle length dynamically, based on real-time traffic detection in the signalised approaches or lane groups (signal groups). The decision to extend or terminate the green light is based solely on the traffic conditions for the approach or group currently receiving the green light.

The difference between semi-actuated and fully actuated traffic signal control lies in how each approach responds to vehicle detection:

- *Semi-actuated control*: Vehicle detectors are placed only on the minor or side approaches, not on the major road. This means the signal timing for the main approach is typically fixed, while the green time for the minor approaches adjusts based on detected traffic. Semi-actuated control is often used when there is a significant difference in traffic volume between the main and side streets, as it reduces delays on the main road while responding to demand on the side streets.
- *Fully-actuated control*: Detectors are installed on all approaches, allowing the signal timing to adjust based on real-time traffic demand for each direction. Fully-actuated signals respond dynamically to traffic conditions on all approaches, making them suitable for intersections where traffic volume fluctuates significantly across all directions, enhancing overall intersection efficiency.

In summary, semi-actuated control adjusts only the timing for the side streets based on detected demand, while fully-actuated control adapts the signal timing on all approaches.

Self-optimised real-time control uses adaptive green time allocation and cycle length based on continuous real-time optimisation to enhance traffic flow for all signalised approaches at the intersection. This approach takes a comprehensive view of conditions across all directions, enabling more balanced and efficient signal timing adjustments.

9.1.1 Actuated signals

In Vehicle Actuated Control (VA) systems, the control system adjusts traffic signal timings based on the presence and flow of vehicles. This is typically done using vehicle detectors installed in all signalised approaches to detect vehicle passages and/or vehicle presence. The vehicle detection function represents whether a vehicle is detected at a given time at a specific detector location. This information is used for the following purposes, to register:

- demand for green light for vehicles arriving during red signal in the approach;
- demand for green time extension for vehicles arriving during green light in the approach;
- presence of vehicles within the detection area in the approach after the termination of green, i.e., overflow of a queue to the next signal cycle.

LHOVRA (Brude and Larsson, 1988) is an example of signal group control and other vehicle-actuated control systems, the green duration is dynamically adjusted based on real-time demand. The signal timing for a green phase g_i in each direction (for each stream *i*) can be updated as a function of the detected flow $f_i(t)$, traffic demand, and historical data:

$$g_i(t) = g_{mini} + \alpha \cdot f_i(t) + \beta \cdot d_i(t)$$

where

- g_{mini} is the minimum green time for stream *i*;
- $f_i(t)$ is the traffic flow function at time *t* for stream *i*;
- $d_i\,(t)$ represents the detection of vehicles at time *t* for stream *i*;
- α is a constant that adjusts the green time relative to the flow rate;
- β is a constant that adjusts the green time based on the presence of vehicles.

In a system like LHOVRA, overlapping green phases are used to optimise the use of green time for vehicles travelling in different directions (through lanes and turning lanes). The effective green time g_i in an overlapping phase is determined by the junction's capacity and vehicle demand:

$$g_i = \min(g_{maini}, g_{turningi})$$

where

- g_{maini} is the green time for the main direction for stream *i*;
- $g_{turningi}$ is the green time for the turning lanes for stream *i*;
- the *overlap* ensures that both directions can get green at the same time if traffic conditions allow.

The total cycle time *c* can be dynamically adjusted based on real-time traffic conditions, particularly if vehicle demand fluctuates significantly.

9.1.2 Self-organising control

Vehicle Actuated (VA) control offers adaptability to random short-term fluctuations in traffic, which can substantially lower average delays compared to Fixed-Time control. However, since the initiation and extension of a green phase in VA control are solely based on detected traffic demand on a particular approach, it lacks a mechanism for optimal, system-wide traffic control, therefore self-organising control strategies are introduced. These strategies are based on the local vehicle delay considering the impacts of traffic in all approaches.

In the Miller algorithm control approach, the control function is calculated as the difference in delay between the benefits gained by allowing additional vehicles to pass through the junction during a green extension and the increased delay experienced by queued vehicles on the secondary street due to this extension. Based on this control function, the decision to extend the current green phase is evaluated at specific, regular intervals.

The method operates on a straightforward principle: it calculates the impact of ending the green phase immediately versus ending it after a delay of t seconds, testing various values of t (e.g., 0.5 s, 1.0 s, 1.5 s, up to around 20 s). If, for every value of t, it is determined that ending green now is more beneficial, then the green phase ends. Otherwise, the green phase continues. This calculation is performed multiple times per second, and each result prompts a fresh decision based on updated traffic data at the junction.

The advantage of the Miller algorithm is that it is simple to implement as the total consequences of a decision do not have to be calculated. The risk with the Miller algorithm is that it under certain conditions might tend to postpone problems as the calculations are only done for one cycle.

Between 1982 and 1988, the British Transport Research Laboratory (TRL) continued developing self-optimising traffic strategies, resulting in the MOVA (Microprocessor Optimised Vehicle Actuation) strategy for single junctions (Vincent and Pierce, 1988). A major difference between MOVA and LHOVRA strategy lies in the type of traffic control each employs: MOVA uses stage control, while LHOVRA relies on signal group control. This gives LHOVRA an edge in flexibility, as signal group control allows for more responsive real-time traffic adjustments. MOVA's control strategy blends mathematical optimisation with heuristic algorithms. When a stage turns green, it follows four sequential steps:

1. *absolute minimum green time*: Each stage has a predefined minimum green time (typically 7 seconds in the UK), and each link or signal group has its own minimum as well.
2. *variable minimum green time*: This step adjusts the green time to account for vehicles between the exit detector and stop line.
3. *saturation flow monitoring*: The stage remains green as long as at least one relevant link maintains discharge at saturation flow.
4. *optimisation start*: Once saturation flow ends on all links, the optimisation process begins.

MOVA's optimisation algorithm, more complex than Miller's original method, sets high maximum green times for each stage, which are rarely reached. Additionally, there is a cap on cycle length to prevent excessive pedestrian delays, with MOVA determining green splits within this maximum cycle. Using a microscopic traffic model, MOVA predicts the position of each vehicle between the IN detector and the stop line.

Every half-second, MOVA evaluates whether the total delay will be minimised if the current stage remains green for intervals of 0.5 s, 1.0 s, 1.5 s, and so forth, or if it should switch to red. This involves comparing areas on a delay graph (e.g., area 1A with areas 1B and 2 in Figure 9.1). If area 1A is the largest, green continues; otherwise, it goes to red. In cases of oversaturation, where significant queues remain on one or more approaches at the end of the green, MOVA automatically switches to a different approach. Instead of the Miller algorithm, it applies a heuristic strategy to maximise capacity, prioritising queue reduction.

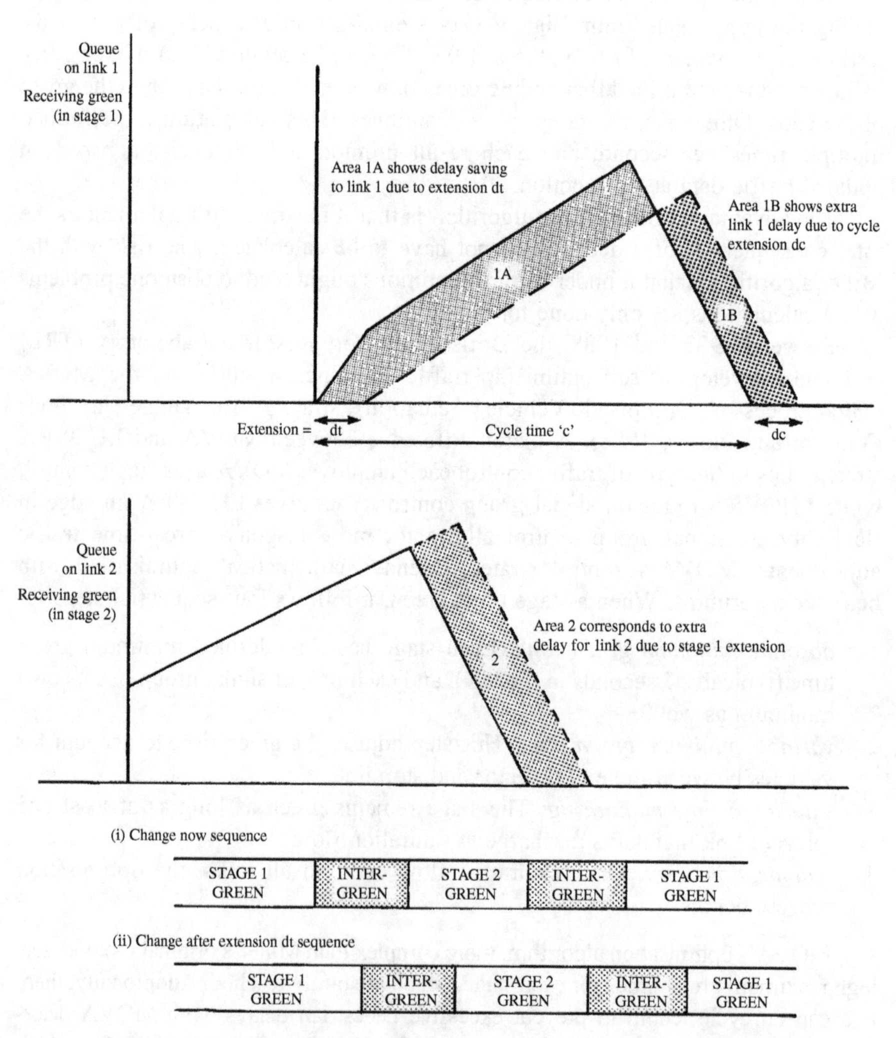

Figure 9.1 Principles of self-organising signal control (Vincent and Pierce, 1988)

9.2 Networks

Strategies for network signalised junctions are classified as:

- *Fixed time*: Operation of all signalised intersections within the coordinated system uses pre-determined, fixed-time parameters (cycle time, green times, offsets) based on multiple time plans that remain constant over periods of the day, depending on flow evaluation derived from historical data;
- *Timing plan selection*, when the plan is periodically updated in real-time by choosing a plan from among a library of pre-timed signal plans depending on detected flows;
- *Timing plan computation*, when the plan is periodically updated in real-time by computing a new plan depending on detected flows.

This section will focus on the third category which can be further classified as described in the following:

- *Centralised*: which simultaneously achieves the optimal values of all decision variables (e.g., by optimising the green timings and the offset of all junctions in the case of network optimisation); in the centralised system, such as SCATS (Sims and Dobinson, 1979, Lowrie, 1982) and SCOOT (Hunt *et al.*, 1981, Robertson and Bretherton, 1991), a master computer is used to adjust the cycle lengths, offsets and splits of all signals on a cycle-by-cycle basis, such that a 'best' timing plan can be found for the operation of the entire network.
- *Distributed*: which has a sequential procedure to obtain the optimal values of the decision variables (e.g., by first optimising the green timings at each junction within a network, and then the offsets for the whole network); in the distributed system, such as OPAC (Gartner, 1983) separate calculations are taken to determine the phase sequences and durations of each signal, with an attempt to optimise the performance of each local junction.
- *Decentralised:* which requires only local information to optimise the decision variables of each junction independently of others on the same network.

9.2.1 Centralised strategies

9.2.1.1 SCOOT

The Split Cycle Offset Optimisation Technique (SCOOT), developed by the UK's Transport and Road Research Laboratory (TRRL) and others, is a fully adaptive traffic signal control system that adjusts signal timings based on real-time data from detectors installed at junctions (see Figure 9.2). Designed for quick and stable responses to traffic fluctuations, SCOOT seeks to minimise an objective function that includes terms for vehicle stops, delays, and congestion across all signalised junctions within a coordinated network. Weights can be assigned to prioritise specific components or specific links within the network.

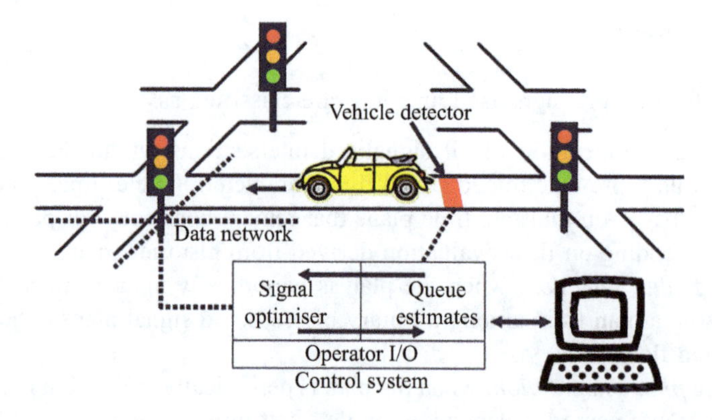

Figure 9.2 SCOOT detectors layout

SCOOT operates on three main principles:

1. *Cycle Flow Profile* (CFP): This is an adaptive measure updated every few seconds, like the approach used in the TRANSYT system but automated for real-time updates in SCOOT. TRANSYT depends on average flow data, saturation flows, and travel time estimates, while SCOOT's adjustments are continuously refreshed with live data.
2. *Online Model Update and Queue Estimation*: Queue lengths are estimated in real time, using detector data collected every second from points upstream and downstream of each junction. This process is also inspired by TRANSYT's methodology but operates at a higher frequency in SCOOT, with queue calculations updated every few seconds.
3. *Incremental Optimisation*: SCOOT uses a flexible coordination approach that allows cycle lengths to expand, or contract based on the real-time CFP. This process involves small, frequent adjustments to cycle lengths to maintain alignment with detected flow patterns. While some adjustments may not be ideal for specific junctions, the overall effect across the network compensates by achieving smoother traffic flow.

Together, these principles allow SCOOT to manage traffic adaptively, optimising signal timings in response to current conditions for reduced congestion and delays across the network (Figure 9.3).

In more detail during the operation of a signalised network, the system checks whether it is appropriate to make small adjustments to: cycle duration, phase durations, and offsets. This check is performed at each cycle with the goal of reducing the highest saturation levels. The cycle duration check is carried out by dividing the junction network into multiple sub-areas, and the same cycle duration is applied to all junctions within each sub-area. Similarly, the offset check is carried out at each cycle, aiming to improve the outflow conditions between junctions.

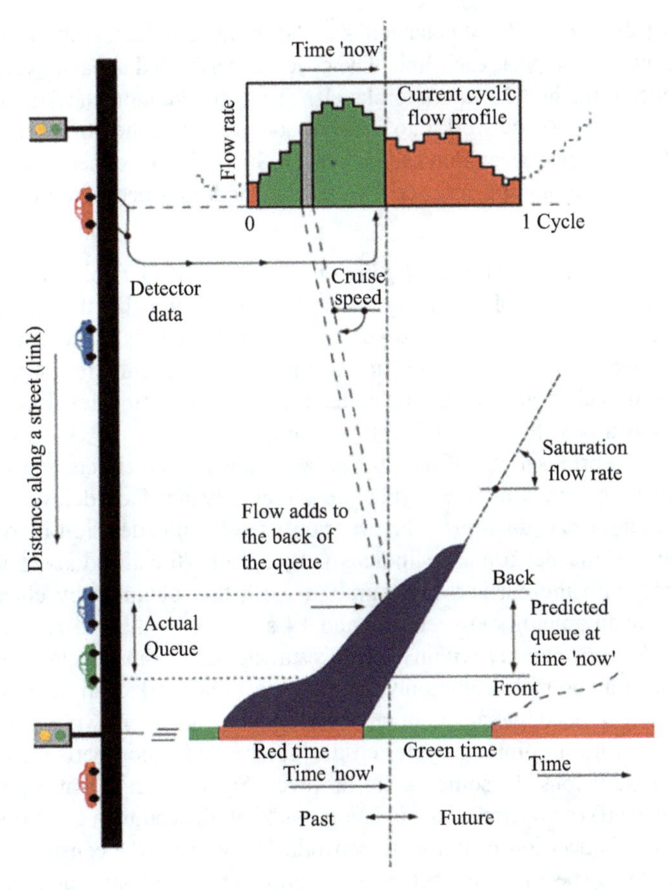

Figure 9.3 SCOOT traffic flow model (Hunt et al., 1982)

To manage congestion conditions, SCOOT uses two strategies:

Virtual barriers: These represent capacity reductions at certain accesses to prevent traffic from increasing in sensitive areas. This strategy is implemented by increasing the green time for exits from the sensitive area and reducing the green time for entries into the same area.

Weight assignment to congested links: This strategy reduces traffic on links with higher saturation by artificially increasing the flow on the most congested links, thereby extending the green time for these branches.

During the operation of a junction network, the system checks whether it is appropriate to make small adjustments to:

- Cycle duration (through the Cycle Time Optimiser procedure);
- Phase durations (through the Split Optimiser procedure);
- Offsets (through the Offset Optimiser procedure).

Each of these procedures determines a performance index based on the predictions of delays and stops at each link. The check is performed at each cycle with the goal of reducing the highest saturation levels. The cycle duration check is carried out by dividing the junction network into multiple sub-areas, and the same cycle duration is imposed on all junctions within each sub-area. Similarly, the offset check is carried out at each cycle, aiming to improve the outflow conditions between junctions.

In more detail:

- The green time optimisation (*splits optimiser*) in SCOOT is activated a few seconds before each phase change (stage), assessing whether the change should be delayed or advanced by a small amount (1-4 seconds). The criterion used for this decision is the minimisation of the maximum saturation degree for the various branches accessing that junction. The congestion level can be considered as a possible criterion in this case.
- The offset optimisation (*offsets optimiser*) evaluates the effects generated by a change in offset, which can also be either advanced or delayed by ±4 s between adjacent junctions. The criterion used for this decision is to compare the sum of the performance indices (PI), derived for all adjacent junctions, obtained with the current configuration, with that obtained by changing the offsets by an amount between -4 s and +4 s.
- The *cycle time optimiser* within SCOOT estimates the optimal cycle for the region every 2.5 min or, more commonly, every 5 min. A network controlled by SCOOT is typically divided into a certain number of regions; junctions within each region, operating with a common cycle, ensure good coordination between the traffic signal installations. In some cases, however, SCOOT (in a manner similar to Transyt itself) can operate some junctions with half the common cycle (i.e., double cycle) if this decision benefits the network. Cycle times are constrained to vary only within a specific range between a minimum practical value and a maximum value, determined based on safety considerations. The criterion used in cycle time optimisation is to allow the regions with the most heavily loaded junctions to operate, if possible, with a maximum saturation degree of approximately 90%.

SCOOT also allows the three optimisation processes described above to be used individually (for example, in the case of a simple coordination need, the splits optimiser is not activated, but only the offset optimiser is used).

9.2.1.2 SCATS

The Sydney Coordinated Adaptive Traffic System (SCATS) was developed by the Department of Main Roads of New South Wales, Australia and operates at two levels: strategic and tactical: at the strategic level, a regional computer calculates appropriate signal timings and offsets for all junctions within sub-areas (each containing up to ten junctions) based on average traffic conditions; at the tactical level, each local controller is empowered to adjust green time dynamically to accommodate the varying demand at each junction (Sims and Dobinson, 1980).

SCATS integrates theoretical models with library-based plans. It calculates cycle times and splits for critical junctions based on actual traffic demand. Timing plans for sub-areas, which include a critical junction and its neighbouring signalised junctions, are selected to align with the signal timing needed for the critical junction. The optimisation goal of SCATS is to minimise vehicle stops and delays, with a particular focus on reducing stops on radial roads. In some cases, the system can be tailored to achieve specific objectives, such as maximising the throughput of the street network.

Overall, SCATS is a hybrid system that combines plan selection and generation with local adaptation. The network architecture consists of groups of signals, each forming a 'subsystem' that includes a critical junction and its neighbouring junctions. These subsystems are controlled by regional computers, and the system can be expanded by adding more regional computers. A central computer typically oversees the regional computers, which can dynamically adjust the subsystems, although each junction can only belong to one subsystem at a time.

In more detail, SCATS operates at two control levels: 'strategic' and 'tactical' (Lowrie, 1982).

Strategic control calculates signal timings based on typical traffic conditions. It determines the common cycle for a subsystem using flow and occupancy data from detectors. The subsystem can consist of up to ten junctions, prioritising a predefined critical junction. The strategic level also uses specific detectors placed at key points within the network.

Tactical control focuses on optimising individual junctions within the constraints defined by strategic control. Local controllers have some flexibility, allowing them to skip stages with no demand, though some stages (such as those on main roads) cannot be skipped. The tactical level uses similar detectors to the strategic level but places them at different locations.

The information from the strategic detectors is used by algorithms to calculate, on a cycle-by-cycle basis, the phase split plan, internal and external offset plans, and cycle length for the subsystem, along with any incremental adjustments to splits and offsets. SCATS employs reactive/case-based control logic to calculate multiple split plans for each junction. It identifies the phase with the highest traffic demand as a 'stretch phase'. The algorithm computes phase splits based on the green time required for each phase during peak periods, and it adjusts the cycle length based on the green time of the selected stretch phase.

The system's response to events is gradual to prevent oscillating offsets. Offsets are recalculated each cycle but only implemented when at least three out of the previous five cycles have suggested a change. When a detector is triggered for a certain duration, an alarm is raised in the control room, and operators can verify the issue using video cameras.

9.2.2 Distributed strategies

9.2.2.1 Optimised Policies for Adaptive Control (OPAC)

The Optimised Policies for Adaptive Control (OPAC) strategy, designed for traffic signal control, enables both individual control at junctions and coordinated control across a network. It was the first US developed strategy for real-time, traffic-adaptive signal control and operates as a distributed control approach using a dynamic optimisation algorithm that calculates signal timings to reduce total junction delays and stops. This algorithm factors in both real-time and projected demand to set phase durations, limited only by minimum and maximum green times and, in coordinated mode, by an adaptable cycle length and offset, updated continuously based on current data.

As a distributed strategy, OPAC eliminates the need for fixed cycle times, instead adjusting signal timings to minimise traffic performance metrics like vehicle delay and stops within a set minimum and maximum phase durations.

The strategy has evolved through four versions, each outlined below.

OPAC-1

The first version, known as OPAC-1, applies Dynamic Programming (DP) techniques to solve the traffic control problem. DP is a global optimisation approach for multistage decision-making, making it a useful benchmark for comparing other strategies. Although it guarantees globally optimal solutions, OPAC-1 requires complete data on vehicle arrivals throughout the entire control period, making it unsuitable for real-time application due to its intensive processing demands and the lack of real-time data for the entire period.

OPAC-2

The second algorithm, OPAC-2, serves as a foundational component for developing an online, distributed control strategy. OPAC-2 operates by dividing the control period into stages, each lasting T seconds, roughly equivalent to a typical traffic cycle, though potentially longer. Each stage is further split into intervals, typically 2–5 s long. For each stage, enough phases must be available to ensure that no optimal solution is overlooked. Phase switching times are measured from the start of the stage in intervals.

In each stage, for a given switching sequence, the performance function for each approach is defined by summing the initial queue length plus new arrivals minus departures for each interval. The optimisation problem aims to minimise vehicle delays by determining the best switching times for each stage, given the initial queues and arrivals for each interval. This is achieved using the Optimal Sequential Constrained Search (OSCO) method, which exhaustively searches all valid combinations of switching times within each stage to identify the optimal set. Finally, the switching times are subject to constraints on minimum and maximum phase lengths.

OPAC-3

OPAC-2 could theoretically use a traffic prediction model to estimate traffic over each stage, but such predictors often prove less reliable than historical data. To address this limitation and make better use of available flow data, OPAC-3 introduced a ‘rolling horizon’ concept to OPAC-2.

In OPAC-3, each stage is divided into n intervals and is referred to as the Projection Horizon (or Horizon) – the period for projecting traffic patterns and calculating optimal phase changes. The horizon length typically aligns with an average cycle duration. Detectors positioned upstream collect real-time arrival data for the initial k intervals (the head of the horizon), while a simple moving average model estimates flow for the remaining n-k intervals (the tail of the horizon). The optimal switching policy is computed for the entire horizon, but only the changes

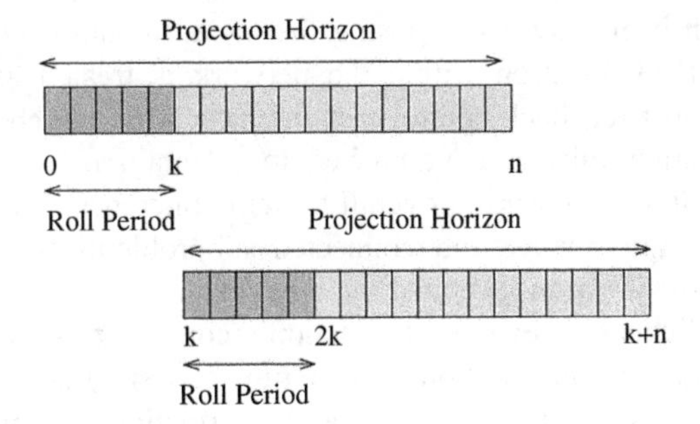

Figure 9.4 OPAC – Implementation of the rolling horizon approach (Gartner et al., 2001)

within the head portion are implemented. This approach allows adjustments when updated real-time data become available. After the head period ends, a new horizon is defined, beginning where the previous head period left off, and the process repeats (see Figure 9.4).

OPAC-4

In the FHWA's Real-Time Traffic-Adaptive Signal Control System (RT-TRACS) project, the University of Massachusetts, Lowell (UML) enhanced the OPAC control logic by adding an optional coordination and synchronisation feature for arterial and network-wide implementations. This advanced version, called Virtual-Fixed-Cycle OPAC (VFC-OPAC), allows the cycle reference point, or yield point, to vary around a virtual fixed cycle length and offset, enabling synchronisation phases to end sooner or extend as needed to respond to changing traffic conditions. VFC-OPAC uses a three-layer control structure:

Layer 1: Local Control Layer – This layer performs the OPAC-3 rolling horizon procedure, continually calculating optimal switching sequences for the Projection Horizon while respecting the virtual-fixed-cycle constraint set by Layer 3.
Layer 2: Coordination Layer – This layer optimises junction offsets, updating them once per cycle.
Layer 3: Synchronisation Layer – This layer sets the network-wide virtual-fixed-cycle length, recalculating it every few minutes or as specified by the user. The cycle length can be separately determined for groups of junctions as needed.

The cycle length and offsets are dynamically adjusted over time to adapt to fluctuating traffic conditions, enabling a more responsive and efficient traffic management system.

9.2.2.2 Urban Traffic Optimisation by Integrated Automation (UTOPIA)

The Urban Traffic Optimisation by Integrated Automation (UTOPIA) system is a closed-loop traffic control strategy that operates on two levels: area-level and local (junction) level. UTOPIA applies large-scale systems control theory to tackle the comprehensive traffic control challenge, utilising a hierarchical and decentralised structure.

The approach begins by decomposing the overall control problem in a topological manner. Each junction within the network is treated as a subproblem, managed by a robust feedback control that interacts with neighbouring junctions. Consistent interaction rules are established to maintain network stability, with higher-level coordination ensuring overall system coherence. This division of the control challenge into smaller, interconnected subproblems is grouped into two main categories: the junction level and the area level.

At the local (junction) level, each junction's controller calculates the optimal signal phase sequence and duration in real time, considering local traffic data, coordination criteria set by the area level, and information from nearby junctions.

At the area level, a central system oversees medium- and long-term traffic forecasting and control for the entire area. It dynamically optimises signal plans for adaptive coordination, allowing the system to respond to evolving traffic patterns.

The system's architecture emphasises the roles of the junction and area levels, which together manage the overall control strategy; each level includes an observer for analysing real-time traffic data and a controller for determining optimal traffic flow conditions. Both area and local levels use a rolling horizon technique to optimise traffic control, allowing continuous adjustment based on the latest data (Mauro and Di Taranto, 1990).

The junction level

The junction level forms the lower level of the UTOPIA control system, where a localised control operates for each traffic light junction or zone (a zone may contain several connected junctions). This local control interacts with adjacent local controls and the area-level control to optimise traffic flow.

Each local control has two key components: an observer and a controller.

- Observer: The observer continuously updates the estimated state of the junction, using available data on traffic counts and traffic light states. It uses a microscopic model that tracks the vehicles on each incoming link, grouping them by their predicted arrival times at the stop line (for example, 3-s intervals). The state vector of the junction includes the state vectors of all incoming links, organised by these arrival steps. Additionally, the observer estimates slow-changing parameters for each link, such as travel times, turn percentages, and saturation flows.
- Controller: The controller decides the optimal signal settings for the traffic lights by minimising a function that adapts to the current traffic conditions at the junction. This optimisation is performed over a 'time horizon' (for example, 120 s) and is recalculated at regular intervals (such as every 6 s), which is also the duration over which the selected settings are applied. This rolling horizon approach enables the controller to adapt continuously to real-time changes.

To ensure that control remains optimal at the network level, the controller's function incorporates the strong interaction concept, which means that the function considers the states of neighbouring junctions and applies constraints set by the area-level control. The function is composed of several weighted costs: time lost by vehicles, vehicle stops, maximum queue length for each link, time lost by public transport vehicles at the junction, deviation from previous signal settings, and conditions at adjacent junctions. Weights for these costs are periodically updated by the area-level control.

An additional feature of the local control is its ability to manage local over-saturation problems, which arise from unpredictable traffic surges or changes in network conditions. The observer detects critical conditions at specific links, and the controller can respond with two types of actions:

1. Increasing Throughput: This action focuses on the junction receiving incoming vehicles. It relaxes green time constraints both at the junction and at down-stream receivers, increasing queue weights for critical links to prioritise flow.
2. Decreasing Demand: If the increasing throughput action fails, this action reduces demand at the sending junction by adding a new cost element to the control function, representing the cost of sending vehicles to nearby saturated junctions. This action helps alleviate congestion at critical points by temporarily reducing the flow of vehicles sent to already saturated areas (Figure 9.5).

The area level

To achieve smooth traffic flow under any demand condition across the entire controlled area, the area-level control sets the appropriate functional objectives for junction controllers, applies suitable constraints, and provides local observers with target data.

At the area level, control is handled by two primary components: the observer and the controller.

- Observer: The area-level observer analyses overall traffic conditions using real-time traffic counts from the network, combined with statistical traffic patterns, to predict main routes and traffic volume at the origin points of these routes. Based on a time-discrete model updated every 3 min, the observer models major routes using a macroscopic network representation of 'storage

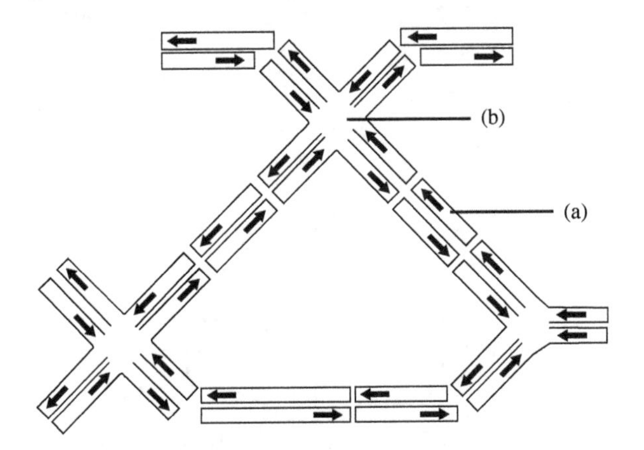

Figure 9.5 UTOPIA – storage units representing main routes; a 'links' and b 'nodes' (Mauro and Di Taranto, 1989)

units'. Each storage unit corresponds to a segment of an arterial road or a junction node (which may include multiple real junctions). The model overlays fixed routes onto this network, with each storage unit tracking the number of vehicles for each route at each time step.

- Controller: The controller optimises the area-level network function by adjusting 'fictitious' controls, such as average speed and saturation flows within each storage unit, subject to constraints that mirror real-world traffic limitations. The main objective of the network function is to minimise the total travel time for private vehicles moving through the area. Optimisation uses a rolling horizon technique, recalculating every 30 min. Whenever the observer updates its traffic predictions, the area-level controller recalculates the optimal controls. These optimised controls are then translated into commands and information for the junction level, ensuring that both levels work in harmony for effective traffic management across the network (Figure 9.6).

9.2.3 Decentralised strategies

The concept of back pressure was first introduced by Tassiulas and Ephremides (1992) in the field of telecommunications. The theoretical approach is like how water flows through a network of pipes via pressure gradients when flow tends away from high-pressure regions and toward low-pressure regions.

The authors proposed a strategy aimed at maximising the capacity of a wireless router network. The network is seen as a directed graph where queues are stored in nodes.

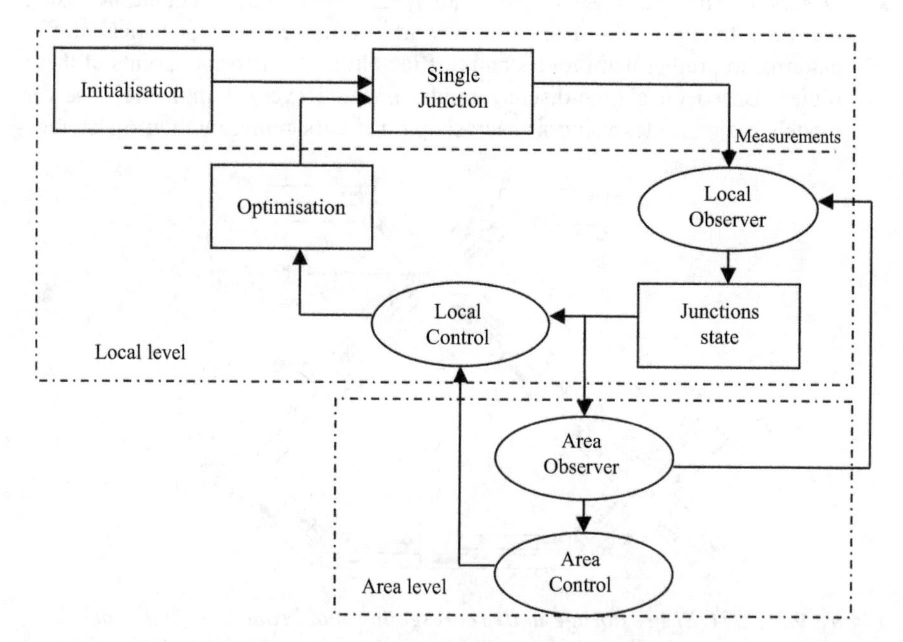

Figure 9.6 UTOPIA – framework description

Clients can enter at any node. The arcs represent servers that transmit data. Clients move between an origin and a destination and are divided into classes. Since the servers are interdependent, the goal is to find a scheduling of the arcs to activate to serve clients while maximising the number of clients served. The philosophy of this strategy is essentially based on calculating the difference in queue lengths between two connected nodes and activating the arc with the highest weight. Time is divided into a set of uniform intervals (time slots).

The approach has been successively applied to the traffic control problem (Varaiya, 2009; Wongpiromsarn *et al.*, 2012; Gregoire *et al.*, 2014a, b). Further enhancements to the method were introduced by Taale *et al.* (2015) to carry out the pressure of the links through the integration with the route guidance.

In traffic management, back pressure refers to how congestion in specific parts of a road network impacts traffic flow in upstream areas. This approach helps optimise traffic distribution across a network by adjusting upstream signal timings based on downstream congestion levels. Essentially, when a downstream road segment or intersection becomes congested, the system reduces green time at upstream signals, thereby limiting the flow of additional vehicles into the congested area. This approach acts like a 'release valve', preventing further build-up and stabilising traffic flow. Overall, back pressure in traffic control is a method to ensure that the flow of traffic is balanced throughout the network, especially in situations where congestion is building up, helping to reduce overall travel time and improve traffic management in urban areas.

In more detail:

At each junction, the number of queued vehicles is monitored. This provides information on the 'pressure' or demand at each approach within the intersection.

Then is computed the pressure by considering the difference between the queue at each approach and the capacity of the downstream intersection to accept more vehicles. If an intersection has a high queue and the downstream intersection has available capacity, this creates a high-pressure difference.

Consequently, the control strategy prioritises giving the green light to the approach with the highest positive pressure difference. This approach allows vehicles to move from high-pressure intersections toward lower-pressure intersections, aiming to balance vehicle flow across the network; the back-pressure control continuously recalculates and adjusts signal timings to adapt to real-time demand across the network, helping prevent localised congestion from spreading upstream.

In essence, the back-pressure strategy minimises network-wide congestion by dynamically adjusting signal timings based on where traffic pressure is highest, improving overall flow. This approach is particularly effective in complex urban networks with high variability in traffic demand.

Consider a network composed of N nodes that operate in time slots of duration t. An overview of the algorithm is provided below:

let

a and b be the nodes,

Q_a and Q_b be the number of vehicles respectively at nodes a and b (i.e., the current number of vehicles at each node also called queue backlog)

p be the feasible phase

$\mu_{ab,p}$ be the service rate (i.e., transmission rate) of the phase p at which the flows of vehicles cross the link and represents the number of vehicles it can transfer from node a to node b

then
the weight of each turning movement can be defined as:

$$w_{ab}(t) = Q_a(t) - Q_b(t)$$

and the pressure γ_p for each feasible phase p is given by

$$\gamma_p = \Sigma_a \Sigma_b \mu_{ab,p} w_{ab}(t)$$

then the considered approach aims to activate the phase, p^*, with the highest value of pressure:

$$p * (t) = \text{argmax} \{\gamma_p(t)/p \in P\}$$

However, Varaiya (2009) proposed an alternative approach in which the green timing is allocated by considering the application of the logit function to calculate the phase pressure. In particular,
let

$r_{a,b}$ be the turning probability at the end node b,
η, be a model parameter tuning the sensitivity of green proportions with respect to the pressures (Chow and Sha, 2016, Gregoire *et al.*, 2014a, b)

the weight of each movement can be defined as:

$$w_{ab}(t) = Q_a(t) - \Sigma a r_{a,b} Q_b(t)$$

thus, the back pressure term is computed by considering the difference between the incoming links and the outgoing ones and the pressure for each feasible phase, it is calculated as in the previous case.

Finally, the green phase for phase p is computed with reference to the following logit function on the base of the pressure term:

$$V_p = e^{\eta \gamma p(t)} / \Sigma_i e^{\eta \gamma i(t)}$$

9.3 Summary

This chapter reviews the basic definitions and notations for signalised junction analysis and control. It introduces and reviews some common optimisation methods for dynamic control with reference to single junctions and networks of signalised junctions. The chapter discusses the details of MOVA for single junction, SCOOT, SCATS, OPAC, and UTOPIA for flow/arrival-based strategies.

Further comments are warranted regarding the store-and-forward strategy, which is not discussed in the chapter:

STORE-AND-FORWARD MODELLING, introduced by Gazis and Potts in 1963, has been widely applied in traffic network control. The primary purpose of using this model for road traffic control is to simplify the traffic flow representation by avoiding the use of discrete variables. This simplification is crucial because it enables the use of efficient optimisation techniques, including linear, quadratic, and nonlinear programming. Alternatively, a multivariable regulator approach can be used – such as the TUC (Traffic-Responsive Urban Control) strategy developed by Diakaki and colleagues in 2002 – which calculates network split adjustments in real-time, while separate algorithms determine cycle time and offsets in parallel.

Signal group; Stage control

Signal Group Control is a traffic management method that assigns specific signal phases to groups of vehicle movements at an intersection. A signal group typically controls one or more movements governed by the same signal phase. For instance, a signal group may manage all vehicles moving straight in one direction or may oversee both left-turn and straight-through movements from a single approach at the intersection. Each signal group is allocated a green light period during which vehicles in that group can proceed. The duration of this green light is determined by traffic demand, sensor data, and the chosen traffic control strategy. Consider a junction with four approaches: north, south, east, and west. Each approach might have two signal groups:

- Northbound Approach: *Signal Group 1* (through traffic), *Signal Group 2* (left turn).
- Southbound Approach: *Signal Group 3* (through traffic), *Signal Group 4* (left turn).
- Eastbound Approach: *Signal Group 5* (through traffic), *Signal Group 6* (left turn).
- Westbound Approach: *Signal Group 7* (through traffic), *Signal Group 8* (left turn).

For example, if there is heavy traffic in the northbound direction, the system may allocate additional green time to Signal Group 1, which controls through traffic, while decreasing the green time for Signal Group 2, which manages left-turning vehicles. This approach enables the system to adapt dynamically to real-time traffic flow, providing more efficient movement for the high-demand direction.

The stage control manages a set of signal groups that operate together during a particular signal stage within the junction's cycle. A stage consists of one or more signal groups activated simultaneously for a designated period within the signal cycle, allowing movements that do not conflict to proceed together. For instance, a stage may include straight-through traffic from different directions or a combination of straight-through and right-turn movements if they can safely occur at the same time.

- *Stage 1*: All through movements (northbound and southbound) get green (This might include Signal Groups 1 and 3).
- *Stage 2*: Left-turn movements (northbound and southbound) get green (This might include Signal Groups 2 and 4).

References

A huge literature is available on these topics, some relevant papers are in the list of references below, besides those quoted in the main text.

Allsop, R. E. (1972). SIGSET: A computer program for calculating traffic capacity of signal-controlled road junctions. *Traffic Engineering & Control* 12, 58–60.

Allsop, R. E. (1976). SIGCAP: A computer program for assessing the traffic capacity of signal-controlled road junctions. *Traffic Engineering & Control* 17, 338–341.

Binning, J. C., Crabtree, M. R., and Burtenshaw, G. L. (2010). TRANSYT 14 user guide. Transport Road Laboratory Report nr AG48. APPLICATION GUIDE 65 (Issue F).

Bretherton, R. D., Wood, K., and Bowen, G. T. (1998). SCOOT Version 4. In: *Proceedings IEE 9th International Conference on Traffic monitoring and Control*. London.

Brude, U., and Larsson, J. (1988). Traffic safety effects of lhovra signals: Analysis and results (No. 5).

Cantarella G.E. and Improta G. (1988). Capacity factor optimization for signalized junctions: a graph theory approach. *Transportation Research Part B* 22, 1–23.

Chard, B. M., and Lines, C. J. (1987). Transyt—the latest developments. *Traffic Engineering & Control* 28, 387–390.

Chow, A. H., and Sha, R. (2016). Performance analysis of centralized and distributed systems for urban traffic control. *Transportation Research Record*, 2557(1), 66–76.

Diakaki, C., Papageorgiou, M., and Aboudolas, K. (2002). A multivariable regulator approach to traffic-responsive network-wide signal control. *Control Engineering Practice*, 10(2), 183–195.

Gartner, N. H. (1983). OPAC: A demand-responsive strategy for traffic signal control. *Transportation Research Record*, No. 906, pp. 75–84.

Gartner, N. H., Pooran, F. J. and Andrews, C. M. (2001). Implementation of the OPAC adaptive control strategy in a traffic signal network. *Proc. 4th IEEE Conference on Intelligent Transportation Systems*, pp. 197–202.

Gazis, D. C., and Potts, R. B. (1963). The oversaturated intersection. In: *Proceedings of the 2nd International Symposium on Traffic Theory*, London, U.K., pp. 221– 237.

Gregoire, J., Frazzoli, E., de La Fortelle, A., and Wongpiromsarn, T. (2014a). Back-pressure traffic signal control with unknown routing rates. *IFAC Proceedings Volumes*, 47(3), 11332–11337.

Gregoire, J., Qian, X., Frazzoli, E., De La Fortelle, A., and Wongpiromsarn, T. (2014b). Capacity-aware backpressure traffic signal control. *IEEE Transactions on Control of Network Systems*, 2(2), 164–173.

Hunt, P. B., Robertson, D. I., Bretherton, R. D., and Royle, M. C. (1982). The SCOOT on-line traffic signal optimisation technique. *Traffic Engineering and Control*, 23(4), 190–192.

Hunt, P. B., Robertson, D. I., Bretherton, R. D., and Winton, R. I. (1981). SCOOT – a traffic responsive method of coordinating signals. RRL Report LR 1041, Road Research Laboratory, U.K.

Improta G. and Cantarella G. E. (1984). Control System Design for an Individual Signalized Junction. *Transportation Research part B* 18, 147–167.

Lowrie, P. R. (1982). The Sydney Coordinated Adaptive Traffic system Principles, Methodology, Algorithms. *IEE Conference*, 207, 67–70. Department of Main Roads, N.S.W., Australia.

Luk, J. Y. K. (1984). Two traffic-responsive area traffic control methods: SCAT and SCOOT. *Traffic Eng. Control* 25, 14–22.

Mauro, V. (2002). UTOPIA—Urban Traffic Control—Main Concepts, *Presented at the EU–China ITS Workshop*, Beijing, China.

Mauro, V., and Di Taranto, C. (1990). UTOPIA. *IFAC Proceedings Volumes*, 23(2), 245–252.

Mauro, V., and Di Taranto C. (1989). Utopia. In: *Proc. of the 2nd IFAC-IFIP-IFORS Symposium on Traffic Control and Transportation Systems*, pp. 575–597.

Robertson, D. I. (1969). *TRANSYT: a traffic network study tool. RRL Report LR 253, Road Research Laboratory*, England.

Robertson, D. I., and Bretherton, R. D. (1991). Optimizing networks of traffic signals in real time-the SCOOT method. *IEEE Transactions on Vehicular Technology*, 40(1), 11–15.

Sims, A. G., and Dobinson, K. W. (1980). The Sydney coordinated adaptive traffic (SCAT) system philosophy and benefits. *IEEE Transactions on Vehicular Technology*, 29(2), 130–137.

Sims, A. G., and Dobinson, K. W. (1979). Department of Main Roads Sydney, Australia. *In Proceedings of the International Symposium on Traffic Control Systems*, Held at Berkeley, California, August 6–9, 1979 (Vol. 2, p. 19). University of California.

Stevanovic, A., Kergaye, C., and Martin, P. T. (2008). Field evaluation of SCATS traffic control in park city, UT, *Presented at 15th World Congress on ITS*, New York City.

Stevanovic, A., Kergaye, C., and Martin, P. T. (2009). SCOOT and SCATS: a closer look into their operations, 09-1672. In: *Proceedings of the 88th Annual Meeting of the Transportation Research Board*, Washington, D.C.

Taale, H., van Kampen, J., and Hoogendoorn, S. (2015). Integrated signal control and route guidance based on back-pressure principles. *Transportation Research Procedia*, 10, 226–235.

Tassiulas, L., and Ephremides, A. (1992). Jointly optimal routing and scheduling in packet ratio networks. *IEEE Transactions on Information Theory*, 38(1), 165–168

Varaiya, P. (2013). Max pressure control of a network of signalized intersections. *Transportation Research Part C: Emerging Technologies*, 36, 177–195.

Varaiya, P. (2009). A universal feedback control policy for arbitrary networks of signalized intersections. Available from: https://people.eecs.berkeley.edu/~varaiya/papers_ps.dir/090801-Intersectionsv5.pdf.

Vincent, R. A., Mitchell, A. I., and Robertson, D. I. (1980). *User guide to TRANSYT version 8*. Transport and Road Research Laboratory Report, LR888, Crowthorne, Berkshire, UK.

Vincent, R., and Pierce, J. (1988). 'MOVA': Traffic responsive, self-optimizing signal control for isolated intersections. Research Report 170, Transport and Road Research Laboratory.

Wongpiromsarn, T., Uthaicharoenpong, T., Wang, Y., Frazzoli, E., and Wang, D. (2012). Distributed traffic signal control for maximum network throughput. In *2012 15th International IEEE Conference on Intelligent Transportation Systems* (pp. 588–595). IEEE.

Part III

Intelligent transportation systems

Chapter 10
Advanced traffic management strategies

Roberta Di Pace[1] *and Franco Filippi*[1]

Many study how to lengthen life, when instead, we should be focusing on broadening it.

Luciano De Crescenzo (from the movie '*32 Dicembre*')

Outline. *This chapter provides an overview of some approaches adopted for advanced traffic management. In particular, the focus is on ramp metering and variable speed limits. For each one of them, some main strategies are reviewed.*

10.1 Ramp metering

10.1.1 Introduction

One of the earliest forms of ramp metering (RM) involved a police officer manually directing traffic flow from a ramp. For example, this strategy was implemented on a stretch of the Eisenhower Expressway in Chicago in the 1960s. The ramp metering system was then experimentally introduced in other major cities in the United States and, over time, it was analysed and updated with advancements in technology. Eventually, it began to be adopted in Europe and Oceania as well. In the 1980s, specifically in Europe, the United Kingdom installed ramp metering for the first time on the M6 near Walsall (later updated) in 1986. The Netherlands introduced ramp metering in Amsterdam in 1989, later expanding to other cities such as Rotterdam, Utrecht, and Paris. In Germany, ramp metering is present on highways around Hamburg, Munich, and many other areas of the Rhine-Ruhr region.

Ramp metering (RM) is described as a system composed of traffic lights, detectors, and signs aimed at optimising the entry time of a vehicle from the access ramp to the highway. Traffic lights may have two or three phases, depending on the presence of the yellow light. Detectors are instruments, usually sensors, designed to monitor the positions and movements of vehicles, detect the presence of a vehicle positioned at the stop line, and monitor the length of the queue on a ramp or the traffic flow on the mainline. In this way, road signs assist drivers in further operations.

[1]Department of Civil Engineering, University of Salerno, Italy

10.1.2 Fixed time strategies

Without using real-time data, fixed-time ramp metering techniques are developed offline for specific periods of the day based on consistent historical needs. Their foundations are basic static models. Sections of a highway with many on- and off-ramps are separated by a single on-ramp. Thus:

$$q_j = \sum_{i=1}^{j} \alpha_{ij} r_i \tag{10.1}$$

where,

- q_j is defined as the mainline flow of section j,
- r_i, on-ramp volume of section i in veh/h,
- $\alpha_{ij} \in [0, 1]$, vehicle portion that enters the freeway in a section i and does not exit upstream of j.

Some constraints are defined, for instance, to avoid congestion:

$$q_j \leq q_{cap,j} \quad \forall j$$

where, $q_{cap,j}$ is the capacity of section j. On the other hand, further constraints are:

$$r_{j,\min} \leq r_j \leq \min\{r_{j,\max}, d_j\}$$

where,

- d_j is the demand,
- $r_{j,\max}$, ramp capacity at on-ramp j,
- $r_{j,\min}$, minimum flow allowed to enter the highway constantly.

Wattleworth (1965) was the first to propose this strategy. There are more comparable methods in Chen *et al.* (1974), Schwartz and Tan (1977), Wang (1972), and Yuan and Kreer (1971). As an objective criterion, one could want to balance the ramp queues (i.e., minimising $\sum_j (d_j - r_j)^2$), maximise the overall travelled distance ($\sum_j \Delta_j q_j$, where Δ_j is the length of the section), or maximise the number of served cars ($\sum_j r_j$, which is equal to minimising the total time spent). These formulations result in problems that can be easily addressed with accessible computer programs, such as linear or quadratic programming.

Fixed-time techniques have one major disadvantage: their parameters are dependent on historical data instead of current data. The following might make this an oversimplified statement.

- Even throughout the same time of day, demands change.
- Certain days may have differing demands because of things like special events.
- Long-term demand changes cause the optimised settings to 'aging'.
- Similar to how demands are changing, the proportions α_{ij} may also fluctuate as a result of drivers attempting to reduce their individual journey times in response to the newly optimised signal settings.
- Traffic conditions might be unpredictably disrupted by incidents and further disruptions.

Therefore, fixed-time ramp metering strategies might result in either underuse of the highway or overflow of the mainstream flow (congestion) since real-time measurements are not available. Ramp metering is a sensitive but effective control mechanism. Insufficient accuracy in ramp metering schemes may result in underutilisation of mainstream capacity or failure to prevent congestion from occurring (e.g., due to groundlessly strong metering).

10.1.3 Reactive strategies

10.1.3.1 Local ramp metering

Demand-capacity strategy

This strategy (Masher *et al.*, 1975) consists of the following:

$$r(k) = \{(q_{cap} - q_{in}(k-1), \quad if\ o_{out}(k) \leq o_{cr}; r_{min}, else\} \tag{10.2}$$

where,

- q_{cap} is the highway capacity downstream of the ramp;
- q_{in}, highway flow measure upstream of the ramp;
- o_{out}, highway capacity measure downstream of the ramp;
- o_{cr}, critical occupancy (when the freeway flow becomes maximum);
- $r_{\min}$, predefined minimum ramp flow value.

To achieve the downstream motorway capacity q_{cap}, the demand-capacity strategy aims to increase the measured upstream flow $q_{in}(k-1)$ as much ramp flow $r(k)$ as necessary. Although the ramp flow is reduced to the lowest flow $r_{\min}$ in order to prevent or relieve the congestion, if for any reason, the downstream measured occupancy $o_{out}(k)$ becomes overcritical (i.e., a congestion may arise). It is evident that this is more of an open-loop disturbance-rejection policy than a closed-loop approach (see Figure 10.1), and open-loop strategies are typically known to be highly sensitive to a variety of non-measurable disturbances.

Occupancy strategy

Based on the same principles as the demand-capacity approach, the occupancy strategy (Masher *et al.*, 1975) uses occupancy-based estimation which, in some

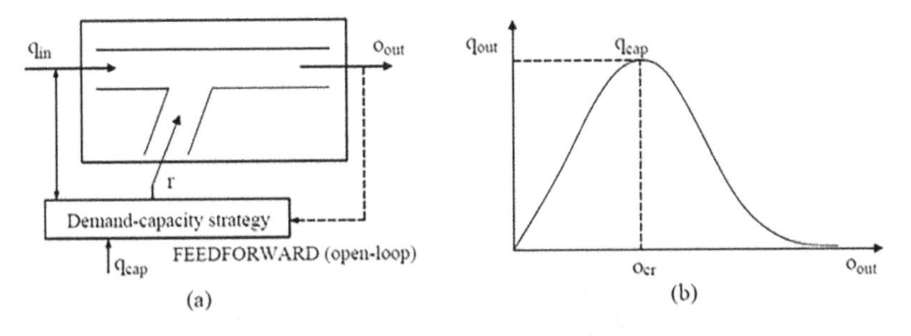

Figure 10.1 (a) Feed-forward (open loop) – Demand-Capacity strategy; (b) fundamental diagram

cases, may reduce the associated implementation cost. If the left-hand side of the fundamental diagram in Figure 10.1 is approximated with a straight line, one has:

$$q_{in} = (v_f \cdots o_{in})/g \tag{10.3}$$

where v_f is the free speed of the highway; and g is the g-factor (Jia *et al.*, 2001). Replacing this equation with the upper part of the Demand-Capacity Strategy we will have the following:

$$r(k) = K_1 - K_2 \cdot o_{in} \cdot (k-1) \tag{10.4}$$

where,

- $K_1 = q_{cap}$
- $K_2 = v_f/g$
- $r(k)$ is truncated if it exceeds a range $[r_{\min}, r_{\max}]$, defining $r_{\max}$ is the ramp's estimated flow capacity.

Thus, regarding local metering, the strategy adopted is based on the percentage of occupancy. This step of the algorithm relies on the assumption of using a linear interpolation diagram as depicted in Figure 10.2: within this diagram, we have a representation of metering rates as a function of the upstream occupancy coefficient of the controlled road section. Based on historical data, the goal is to determine the interpolation diagram whose extremes are represented by the minimum metering rate and the maximum metering rate as functions of respectively maximum and minimum levels of occupancy coefficients (see Figure 10.2). Observing the diagram, we deduce that the higher the occupancy rate along the mainline, the more restrictive the metering rate value should be, meaning a lower volume of traffic will be allowed to flow. Conversely, for low occupancy rates, the metering rate will be higher as it will be possible to allow the free flow of vehicles.

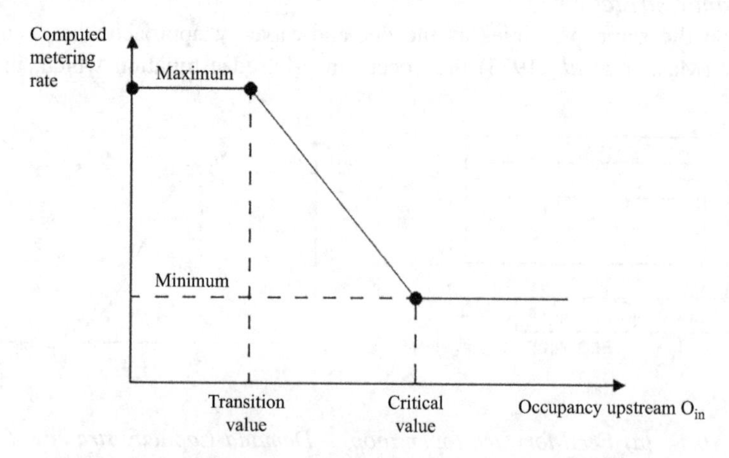

Figure 10.2 Occupancy coefficient strategy diagram (Masher)

At the tactical level, reactive ramp metering tactics are used with the aim of maintaining the highway traffic conditions around predefined set values, depending on real-time readings.

ALINEA

ALINEA (Papageorgiou *et al.*, 1991.) is a local control strategy based on a closed-loop operating algorithm. The reason why this algorithm is considered a closed-loop strategy is because the metering rate at a certain time instant is a function of the metering rate at the previous time instant. At each feedback, the implementation of the strategy explicitly depends on what happens at the previous time instant.

By measuring the occupancy coefficient, detected by a loop detector located downstream of the ALINEA ramp, the strategy evaluates, in case of exceeding a certain threshold value, the difference between the measured occupancy percentage and the desired reference value (typically, but not necessarily, this reference value coincides with the critical threshold value above which the control strategy is activated). The goal is to bring this occupancy below the critical value to ensure the correct functioning of the highway below its capacity value.

The used equation with the aim of determining the metering rate at each discrete* interval k is the following:

$$r(k) = r(k-1) + K_R\left[\widehat{O} - O_{out}(k-1)\right] \tag{10.5}$$

where,

- $r(k)$ and $r(k-1)$ current metering rate and metering rate from the previous time interval, respectively.
- K_R regulation parameter (which is greater than 0).
- $\widehat{O}$ set (desired) value for the downstream occupancy.
- O_{out} occupancy on the mainline downstream of the entrance ramp to be regulated.

Regarding the regulation parameter K_R, Papageorgiou *et al.* (1991) provide several reference values depending on the experimental case to which it refers. Specifically, the choice of this value depends on a series of experimental considerations as it allows for the adjustment of the oscillations existing between one iteration and the next of the metering rate.

The operating logic of ramp metering is that when the sensor on the mainline of an entrance ramp downstream detects an occupancy value lower than the critical one, then the metering rate is increased, allowing a greater number of vehicles to access the highway. Conversely, if the value detected by the measurement controller is higher than the reference value, then the metering rate is reduced, allowing fewer vehicles to access the highway to reduce congestion levels by smoothing the inflow of vehicles. When the calculated metering rate is much higher than the

*ALINEA is a control strategy whose implementation occurs through successive time steps defined by a temporal horizon over which the implementation takes place. This horizon is divided into further sub-intervals, and for each of them, feedback is carried out.

maximum set value, then the ramp meters are not activated (meaning there will always be a green light), forcing the vehicles entering from the ramp to experience random and unnecessary waits. When the metering rate is lower than the minimum reference value, then the metering rate is set equal to the minimum value to ensure a minimum outflow of vehicles entering from the ramp.

Regarding queues on the ramp, they are considered in the evaluation of the flow to which the metering rate obtained through the control strategy is applied.

ALINEA, like many control strategies, operates by establishing two priority levels. The first focuses on the degree of congestion on the mainstream, with the consequent need to reduce the inflow to support decongestion. The second priority is related to the simultaneous formation of queues at the entrance ramps. Since the primary aim of these strategies is to reduce congestion along the mainstream, it is not acceptable (in terms of fairness) to excessively increase waiting times at the entrance ramps. This is equivalent to setting a critical threshold for the occupancy percentage above which traffic flow to the entrance ramps is not allowed. If this critical value is exceeded, the queue override is activated, allowing the free flow of vehicles from the ramps, indicated by the ramp meters always showing green.

In ALINEA, therefore, two extreme configurations of the system are distinguished, as a result of which the strategy is not activated; in other words, the metering rate varies within an interval $[r_{\min}; 1]$. The metering rate will be equal to its upper bound when the strategy is not activated because it is 'useless' and when the extent of queues on the ramps is so high that the only way to eliminate the phenomenon is to allow the free flow of traffic. The value of rmin, on the other hand, represents the metering rate to be applied, under normal conditions (not in queue override mode) of the strategy's operation, to allow a minimum outflow from the ramp, regardless of the occupancy conditions of the mainline, and to limit the waiting time of vehicles at the ramp-metering device to acceptable values.

Comparing ALINEA with the already treated strategies, it is important to note that the demand-capacity strategy reacts to excessive occupancies O_{out} only after a threshold value O_{cr} is exceeded, and in a rather crude way, while ALINEA reacts smoothly even to slight differences $\widehat{O} - O_{out}(k)$, and thus it may prevent congestion by stabilising the traffic flow at a high throughput level. It is easily seen that in a stationary state (i.e., if q_{in} is constant), $O_{out}(k) = \widehat{O}$ results automatically from the above equation, although no measurements of the inflow are explicitly used in the strategy.

All control schemes determine appropriate ramp flows. When ramp metering is implemented in a traffic cycle, the duration of the green phase is changed by using the following formulation:

$$g = (r/r_{sat}) \cdot c \qquad (10.6)$$

where,

- c is the fixed cycle time, and
- r_{sat}, the ramp saturation flow.

The duration of the green phase g is limited by $g \in [g_{min}, g_{max}]$, with $g_{min} > 0$ avoiding the closure of the ramp, and $g_{max} \leq c$. A constant-duration green phase allows exactly one vehicle to pass in the case of a one-car-per-green realisation. Therefore, by adjusting the red-phase time between a minimum (zero) and a maximum value, the ramp volume can be regulated. Thus, ALINEA may be written in terms of green- or red-phase duration, i.e.,:

$$g(k) = g(k-1) + K_R{}' \left[\widehat{O} - O_{out}(k) \right] \tag{10.7}$$

where,

- $K_R{}' = K_R \cdot \frac{c}{r_{sat}}$
- $g(k-1)$ which might be a value of g between g_{min} and g_{max} applied in the previous time step to avoid the wind-up phenomenon in the regulator.

10.1.3.2 Multivariable regulator strategies

The objectives of multivariable regulators for ramp metering are identical to those of local ramp metering techniques: they try to run highway traffic conditions close to a set of predetermined values. Multivariable regulators use all mainstream measurements $o_i(k), i = 1, \ldots, n$ that are available on a stretch of highway to simultaneously calculate the ramp flow values $r_i(k), i = 1, \ldots, m$ for all controllable ramps included in the stretch (Papageorgiou *et al.*, 1990). This is in contrast to local ramp metering, which is carried out independently for each ramp based on local measurements. Due to coordinated control actions and more complete information provision, this offers possible gains over local ramp metering.

Some multivariable control strategies are analysed in the following sections such as METALINE, ALINEA/Q, RMPS, Linked, and FLOW.

ALINEA/Q

This strategy, developed by Smaragdis and Papageorgiou (2003), represents an advancement over the traditional local strategy algorithm, ALINEA. It introduces a more sophisticated control strategy for managing queues induced on ramps by using the ramp-metering device. In particular, this strategy utilises special detectors such as video detectors, which, unlike the loop detectors used by ALINEA, allow for measuring the queue length at ramps. While loop detectors have a local application, video detectors enable measuring the queue length across the entire area as they are area detectors.

This algorithm calculates two metering rates: one is the same as calculated in ALINEA, while the other provides the minimum metering rate needed to keep the queue length at the ramps below the maximum allowed value. This rate is calculated based on the queues using the following formula:

$$r'(k) = -1/T \quad [\widehat{w} - w(k)] + d(k-1) \tag{10.8}$$

where:

- $r'(k)$ is the minimum metering rate applied in time interval k, which prevents excessive queue buildup on the ramp;

- $\widehat{w}$ is the maximum allowed queue length;
- $w(k)$ is the number of vehicles in the queue at the ramp in time interval k;
- T is the observation period;
- $d(k-1)$ is the number of vehicles entering the ramp in time interval k-1.

The strategy involves identifying the higher metering rate obtained through ALINEA and the rate obtained through queue control:

$$r(k) = \max\{r(k);r'(k)\} \tag{10.9}$$

The use of this algorithm offers a range of benefits, notably enabling adjustments related to queues with greater stability, thereby avoiding system oscillations seen with the straightforward application of ALINEA. By calculating the queue length in each interval, the algorithm effectively manages to keep the queue length value in check while maintaining a high metering rate. This prevents significant reductions or excessive variations between iterations, which could otherwise lead to pronounced oscillations in strategy implementation.

METALINEA

As a generalisation and extension of ALINEA, the multivariable regulator approach METALINEA computes the metered on-ramp volumes using (in the following equation bold variables denote vectors and matrices).

$$\mathbf{r}(k) = \mathbf{r}(k-1) - \mathbf{K_1}[\mathbf{o}(k) - \mathbf{o}(k-1)] + \mathbf{K_2}\left[\widehat{\mathrm{O}} - \mathbf{O}(k)\right] \tag{10.10}$$

where,

- $\boldsymbol{r} = [r_i, \ldots, r_m]^T$ is the vector of m controllable on-ramp volumes,
- $\boldsymbol{o} = [o_1, \ldots, o_n]^T$, vector of n measured occupancies on the highway stretch,
- $\boldsymbol{O} = [O_1, \ldots, O_m]^T$, subset of $\boldsymbol{o}$ that includes m occupancy locations for which pre-specified set values $\widehat{O} = \left[\widehat{O}_1, \ldots, \widehat{O}_m\right]^T$ may be given.

It is important to understand that there can never be more set-valued occupancies than managed on-ramps due to control theory constraints. For inclusion in the vector $\boldsymbol{O}$, one bottleneck point is usually chosen downstream of each controlled on-ramp. Lastly, the constant gain matrices for the regulator are $\boldsymbol{K}_1$ and $\boldsymbol{K}_2$, which need to be appropriately built (Papageorgiou *et al.*, 1990).

Based on modelling data and field studies, the following conclusions have been drawn on the relative efficacy of METALINEA and ALINEA:

- ALINEA takes barely any design work, while the METALINEA application requires a very involved process based on cutting-edge control-theoretic techniques (LQ optimum control).
- Under recurring congestion, METALINEA was found to offer no advantages over ALINEA (the latter operated individually at each controlled on-ramp) for urban motorways with a high density of on-ramps.
- METALINEA outperforms ALINEA in the event of nonrecurrent congestion (caused by an incident, for example) because it has access to more extensive measurement data.

Ramp Metering Pilot Scheme (RMPS)

This algorithm was developed based on ALINEA and applied with a series of modifications introduced by Gould *et al.* (2002). The steps followed by the algorithm, summarised briefly, are as follows:

- Calculation of correct values for flow, speed, and occupancy coefficient (see the following equation).
- Activation/deactivation of the algorithm operation cycle.
- Calculation of the cycle duration of the ramp-metering device.
- Queue control on the ramp.
- Verification of queue admissibility on the ramp and possible queue override.

Now let us describe each step of the algorithm individually. The first step involves 'flattening' the flows, speeds, and occupancy using the following formula (here specifically written for flows, although applicable to speeds and occupancy with the same criteria):[†]

$$F_S(t) = [1 - K_F]F_S(t-1) + K_F F(t) \tag{10.11}$$

This formula allows for smoothing out the values of flows, speeds, and occupancies. Furthermore, whenever the flows obtained through the corrective formula exceed a certain maximum value and, at the same time, the speeds fall below the minimum allowed value, the strategy is activated.

The activation of the ramp meters is related to the activation of a pre-timed phase plan, meaning the cycle time and the green and red (and possibly yellow) times are predefined.

The interruption of ramp metering occurs when the flows drop below the minimum allowable value, while the speeds exceed the threshold value. In such cases, the metering policy becomes ineffective and loses its significance. In reality, the timing plan of the ramp meters is parameterised according to minimum and maximum occupancy values. In other words, one of the predefined timing plans is chosen depending on whether the estimated occupancy value, obtained from a formula like the above equation, falls within the specified interval for that plan.

In extreme cases where the occupancy value exceeds the maximum threshold value, the cycle with maximum duration is selected, whereas the cycle with minimum duration is activated in the opposite scenario.

The reference values of the ramp metering timing plan are calculated by simulating the operation of an ALINEA-type strategy.

The strategy as described above reflects how it is implemented in America; in England, it is implemented using the same criteria but without following a predefined table, rather than following an activated strategy that prescribes the extension of green times. In this case, with a predetermined and fixed cycle duration, the minimum red time is used to calculate the maximum green time; the

[†]This form allows for a flattening of the flows similar to what the coefficient K achieves in ALINEA to reduce oscillations between one iteration and the next.

duration of the green time will be constrained to the maximum green time, so whenever this value is reached, the maximum green time has been exhausted. Alternatively, if the number of vehicles that have passed the detector exceeds a certain threshold value, and even if the duration of the green time has not reached the maximum green, it will be interrupted to prevent the ramp flow from interfering with the flow along the mainstream, adopting a priority policy for the mainstream.

The next step after determining the cycle duration is related to adjusting based on the queues at the ramps. The occupancy coefficient on the ramps is detected through detectors and weighted to define 'queue adjustment scores'. Based on the adjustment score, the cycle length and how much it should be increased are determined.

Linked

This algorithm (Taylor *et al.*, 1998) is based on the development of a Proportional-Integral-Plus (PIP) controller. It is a feedback control with state variables.

The basis of the control strategy design prescribes using, in addition to the input and output variables, the integral value of the error to describe the system state. There is also a design of the control strategy in terms of *non-minimal state space* (NMSS), within which a *local linear model* (LLM) is formulated. Through this strategy, both current measurements and those made in previous time intervals are considered at each time interval of control implementation, both upstream and downstream of the access ramp. Additional variables are also calculated for each entrance and exit ramp. The LLM for each point within the entrance ramp will be described by the following equations:

$$o_{j,k} = ao_{j,k-1} + bo_{j-1,k-1} + co_{j+1,k-1} + dq_{on,k-1} \tag{10.12}$$

where,

- $o_{j,k}$ is the value of the occupancy coefficient measured at point j downstream of the ramp (through sensor localisation) at time k;
- $o_{j,k-1}$ is the value of the occupancy coefficient measured at the same point j and at the previous time;
- $o_{j-1,k-1}$ is the value of the occupancy coefficient measured at the previous time and upstream of point j;
- $o_{j+1,k-1}$ is the value of the occupancy coefficient measured in the previous interval and upstream of point j;
- $q_{on,k-1}$ is the flow at the entrance ramp in the previous time interval;
- a, b, c, d are parameters that need to be properly estimated.

$$\begin{aligned} \boldsymbol{x}_k &= \boldsymbol{F}\boldsymbol{x}_{k-1} + \boldsymbol{G}\boldsymbol{u}_{(k-1)} + \boldsymbol{D}\boldsymbol{y}_{d,k-1} + \boldsymbol{g}_2 o_{up,k-1} + \boldsymbol{g}_3 o_{down,k-1} \\ \boldsymbol{y}_k &= \boldsymbol{H}\boldsymbol{x}_k \end{aligned} \tag{10.13}$$

where,

- x_k represents the state vector at time instant k;
- $\boldsymbol{u}_{(k-1)}$ is the control vector of input variables, namely the flows at the entrance ramps, at time k-1;

- $y_{d,k-1}$ is the reference vector of occupancy;
- $o_{up,k-1}$ is the upper limit value of the occupancy upstream of the ramp;
- $o_{down,k-1}$ is the lower limit value of the occupancy downstream of the ramp;
- y_k is the control vector of output variables;
- The matrices F, G, D, and H are determined through the vectors g_2, g_3.

Now, let us consider an example where we assume that along the entire mainstream, there are 11 sensors located, and in addition, we have four sensors located along the entrance ramp at positions 3, 5, 7, and 9. Occupancy coefficients are detected upstream and downstream along the mainstream, $o_{up,k} = o_{o,k}$ and $o_{down,k} = o_{10,k}$, respectively. In addition, $y_{dk}^T=[o_{d3,k,}\ o_{d5,k,}\ o_{d7,k}\ o_{d9,k}]$; $u_k^T=[q_{on3,k,}\ q_{on5,k,}\ q_{on7,k,}\ q_{on9,k}]$, while the state vector will be $x_k^T=[o_{1,k,}\ o_{2,k}\ o_{3,k}\ o_{4,k}\ o_{5,k,}\ o_{6,k}\ o_{7,k,}\ o_{8,k}\ o_{9,k}\ z_{3,k}\ z_{5,k,}\ z_{7,k}\ z_{9,k}]$. 'The integral of the error' is computed for the last four elements of the state vector using the following formula:

$$z_{j,i} = z_{j,k-1} + \left(y_{dj,k} - o_{j,k}\right) \tag{10.14}$$

Regarding the matrices, they will be computed using the following expressions:

While concerning the vector $\mathbf{u}_k$, it will be computed according to the control law, namely $\boldsymbol{u}_k = -\boldsymbol{K}\boldsymbol{x}_k$, where the matrix K represents the *Gain Matrix* discussed in Chapter 1. Within this discussion, we have observed that among the various methods for determining the gain matrix, the preferable approach is the *Linear Quadratic Control (LQ)* technique, where the matrix is determined through the minimisation of the following cost function:

$$J = \sum_{k=1}^{\infty} \left\{\boldsymbol{x}_k^T \boldsymbol{Q} \boldsymbol{x}_k + \boldsymbol{u}_k^T \boldsymbol{R} \boldsymbol{u}_k\right\} \tag{10.15}$$

where **Q** and **R** are matrices defined as semipositive and positive, respectively.

This algorithm is actually considered merely a highly theoretical formalisation as it does not explicitly relate to the length of queues on the ramps.

FLOW

FLOW (Jacobson *et al.*, 1989) is the most commonly used among a series of heuristic-based strategies, which, as such, do not provide closed-form solutions. FLOW is an area-wide (coordinated) algorithm aimed at regulating the outflow at one or more bottleneck locations identified on the mainline through observation by the analyst. The algorithm aims to return the bottlenecks to non-critical operating conditions by acting on one (or generally, multiple) ramps upstream of the bottleneck and limiting controls.

The purpose of this algorithm is to calculate, for each access ramp, both a local metering rate that does not consider the possibility of cooperation among all controlled ramps to achieve desired outflow conditions at the bottleneck and a coordinated metering rate that takes into account the fact that the regulation effect of the various ramps may differently and synergistically affect the bottleneck. The choice of the final metering rate will be directed towards the value that provides the greatest restriction, i.e., the smaller of the two.

In particular, this metering control algorithm is characterised by three components: the calculation of local metering rates, the calculation of coordinated metering rates, and finally an adjustment of the metering values obtained based on the queue lengths at the access ramps.

In practice, the operation of this metering control algorithm can then be articulated into three successive levels of implementation:

- Determination of metering rates using a local strategy (the local strategies known to us are of the open-loop or closed-loop type; in this case, the adopted strategy will be an open-loop strategy, namely the occupancy capacity strategy).
- Determination of metering rates using the coordinated strategy.
- Adjustment of metering rates based on the queue lengths at the ramps.

The overall operation of the FLOW algorithm can thus be represented as shown in Figure 10.3.

In the following paragraphs, the main steps of the algorithm are analysed in greater detail.

The next step is to determine specific indices that formalise the effect of regulating the inflows from a certain number of ramps on the outflow conditions of the section of the mainline to be controlled (bottleneck). For each bottleneck identified along the mainline, it is necessary to define a certain influence area (see Figure 10.4), which is a

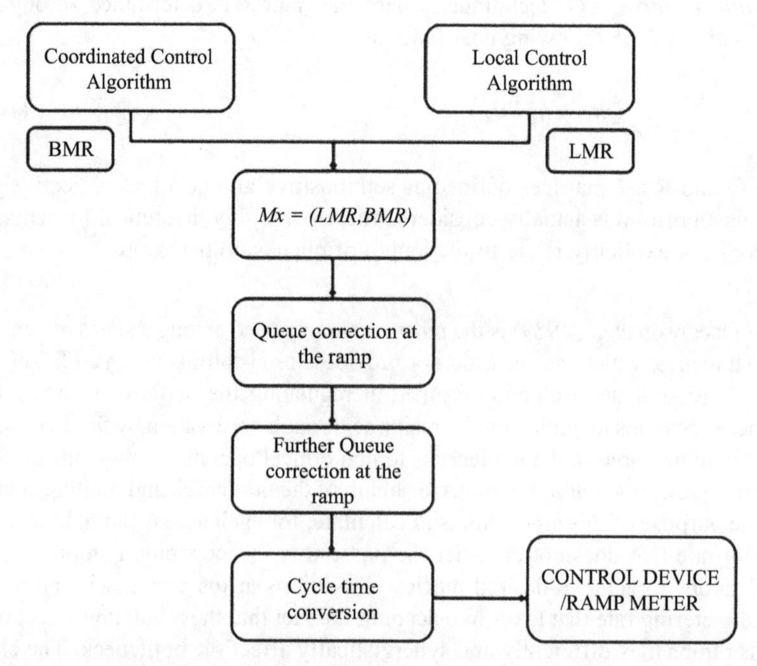

Figure 10.3 FLOW general function scheme (BMR – Bottleneck Metering Rate/ LMR – Local Metering Rate)

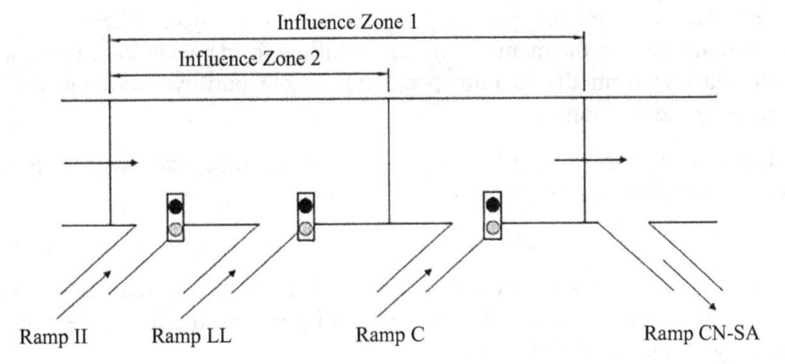

Figure 10.4 Example of a diagram with influence areas

set of ramps whose regulation is assumed to have a non-negligible effect on the outflow conditions at the bottleneck. For a given highway infrastructure, multiple bottlenecks may have been identified; each of them has its own influence area.

Within each influence area, there are several entrance ramps, and it is delimited by a first entrance ramp upstream and the bottleneck section downstream. For each entrance ramp (j), a weighting factor (w_{Fj}) is determined based on the distance of the ramp from the bottleneck and the historical value of the traffic volume entering the ramp and desired to continue at least to the bottleneck.

In other words, the goal is to construct an estimate (approximate and based on historical observations or offline simulations) of the effect that the propagation of incoming flow from the ramp would have on the outflow conditions at the bottleneck.

For example, concerning the case study depicted in Figure 10.4, for influence area 1, the weighting factors assigned to Ramps C, LL, and II are 0.5, 0.3, and 0.2, respectively. Similarly, for influence area 2, the assigned weights for Ramps II and LL are 0.6 and 0.4, respectively. It should be noted that the weighting factors of all ramps within the same influence area must be normalised to have a unit sum.

The sensors required to implement the strategy must be positioned at least at the extremes of the influence area. Additionally, queue sensors need to be placed on the controlled ramps.

The calculation of the BMR (Bottleneck Metering Rate) is triggered by the completion of two conditions introduced below. The first condition relates to the occupancy in the section downstream of the influence area, which must exceed the threshold occupancy value. The second condition pertains to the fact that the sum of vehicles entering the section and the entrance ramp must exceed the sum of vehicles exiting the section and leaving the mainline through the exit ramp (indicating an accumulation of vehicles at the bottleneck). The two preceding conditions can be analytically interpreted using the following relationships:

- *Condition on capacity*: $o \geq o_{th}$, where o is the average occupancy value detected by the sensor located downstream, and o_{th} represents the preset occupancy threshold value.

- *Condition on vehicle queueing*: $q_{in} + q_{on} \geq q_{out} + q_{off}$, where q_{in} is the incoming flow on the mainline, q_{on} is the inflow from the entrance ramp, q_{out} is the outflow from the mainline, and q_{off} is the outflow from the mainline entering the exit ramp.

The reduction in volume for the i-th section and time interval $t+1$ is determined by the following relationship:

$$U_{i,(t+1)} = (q_{IN_{it}} + q_{ON_{it}}) - (q_{OUT_{it}} + q_{OFF_{it}}) \tag{10.16}$$

This formula allows for the determination of the necessary reduction in traffic volume. This reduction is distributed among all ramps in the influence zone proportionally to the calculated weights:

$$BMRR_{ji(t+1)} = U\left(i,_{(t+1)}\right) \cdot (WF_j) / \left(\sum_{j}^{n} (WF_j)_i \right) \tag{10.17}$$

The final metering rate for the coordinated control strategy is then obtained by subtracting the reduction rate for each bottleneck from the measured flow for each entrance ramp in the previous time interval, i.e.,

$$BMR_{ji(t+1)} = q_{ON_{it}} - BMRR_{ji(t+1)} \tag{10.18}$$

The metering rate associated with the coordinated control strategy will be differentiated based on the different influence zones to which a ramp belongs. Ultimately, the assigned value will be the most restrictive. Furthermore, the choice of the most restrictive value is not only relative to the values obtained through the coordinated strategy but also among these and those obtained through the local strategy.

The most restrictive value between the metering rate associated with the coordinated strategy and the local strategy is referred to as SMR (*Synthetic Metering Rate*). This index undergoes further adjustment based on the formation of the queue on the ramp. The adjustment is calculated under two different conditions that are activated when the queue on the ramp backs up to two appropriately predefined positions. If the queue reaches the first sensor positioned further downstream of the ramp, the SMR value is slightly relaxed in an attempt to stop the queue from backing up. If this measure is not sufficient and the queue backs up to a second sensor positioned further upstream of the ramp beyond which there are risks of backup along the ordinary roadway, the relaxation of the metering rate is more radical.

In practice, the first relaxation of the metering rate is generally sufficient to ensure an additional outflow from the ramp of about three vehicles per minute (180 vehicles/hour). The more radical relaxation consists, instead, of suspending the ramp metering policy (queue override).

10.2 Variable speed limits: model specification, performance analysis, and impacts

Variable Speed Limit (VSL) systems are Intelligent Transportation System (ITS) solutions that allow for dynamic changes in the posted speed limits in response to prevailing traffic, incidents, and/or weather conditions. VSL systems use traffic travel time information, volume detection, and road weather conditions to determine appropriate speeds at which drivers should travel, given current traffic and road conditions. Changes in posted speed limits are indicated by displays on overhead variable signs or along the roadway. VSL systems have significant potential to be used as an incident management tool and have a significant impact on traffic operations, congestion management, safety, and environmental sustainability on major roadways. The key benefits of implementing VSLs can be summarised as follows:

- Safety improvements, by reducing speed differences between vehicles travelling in the same lane and/or adjacent lanes. This synchronises driver behaviour and discourages lane-changing, thus reducing the likelihood of collisions.
- Resolution of traffic jams: when traffic is near capacity, any disruption in traffic flow can lead to a traffic jam. VSLs can restore highway capacity by slowing down traffic, thereby delaying or in some cases preventing the occurrence of traffic jams.
- Capacity enhancement and environmental benefits: since congestion is also associated with increased fuel consumption and emissions, the capacity of VSLs to improve traffic flow also leads to environmental benefits.

The VSL control strategies developed so far can be divided into:

- Reactive approaches, which rely on simple rule-based logic. In this type of approach, decisions change in real time based on pre-selected thresholds of traffic flow, occupancy, or average speed; considering threshold values, changes in posted speeds are then indicated on variable sign displays.
- Proactive approaches, where control strategies have the intrinsic ability to act coordinated across the entire network proactively, anticipating the complex behaviour of dynamic systems.

10.2.1 Reactive approach

Reactive control strategies primarily aim at harmonising speed differences and stabilising traffic flow. Examples of such systems have been developed by Zackor (1979), Smulders (1990), Smulders and Helleman (1998), Rama (1999), and Piao and McDonald (2008). According to Zackor (1979), the driver adjusts the speed based on weather conditions, traffic situation, and individual driver-vehicle characteristics. Driver behaviour can then be improved through variable information reflecting the current situation regarding road section capacity and traffic safety.

When traffic volume is low, the desired speed is determined by road geometry, road surface conditions, specific vehicle characteristics, and individual driver characteristics. As traffic volume (vehicles per unit of time) and traffic density (vehicles per unit distance) increase, the probability that a vehicle approaches another vehicle travelling at a slower speed and cannot be overtaken also increases. Consequently, the approaching vehicle must reduce its speed, at least temporarily. The degree of deceleration (and acceleration) and, therefore, the distance from a preceding vehicle, as well as lane-changing based on the position and speed of nearby vehicles, influence speed distribution.

By indicating a variable speed adapted to the real-time situation (such as maximum or recommended speed), drivers modify their behaviour in two ways: they roughly adjust their desired speed to the indicated speed and relax their aggressive driving behaviour by adjusting their following distance. This results in a lower average speed difference between following vehicles and fewer accelerations and decelerations. The informational content of the indication fundamentally suggests that overtaking is unnecessary under any circumstance because, further down the road sections, traffic density does not allow for higher speeds. A necessary condition for this type of indication is trust in the road sign, which must be continuously confirmed by experience.

Literature on the effectiveness of VSL systems in simultaneously improving both mobility and safety has produced conflicting results (Lee *et al.*, 2006; Abdel-Aty *et al.*, 2006, 2007; Allaby *et al.*, 2006). Findings varied from location to location based on congestion level and road network topology. Lee *et al.* (2006) demonstrated that real-time VSL systems can reduce accident potential but at the expense of longer travel times. On the other hand, Abdel-Aty *et al.* (2006) indicated that VSL systems provide a significant reduction in accident probability only under uncongested conditions. However, no substantial safety benefit associated with VSL application in congested conditions was found. In addition to safety improvement, Park and Yadlepati (2003), Lavansiri (2003), Pei-Wei *et a*l. (2004), and Lyles *et al.* (2004) demonstrated the effectiveness of some VSL systems in improving traffic flow and reducing travel time for vehicles passing through work zones. In a recent study, Talebpour *et al.* (2013) studied the impact of early detection of shock waves on congestion formation and safety using speed harmonisation as a control strategy in a connected vehicle environment. A reactive algorithm based on drivers' cognitive risk showed a significant improvement in traffic flow characteristics under congested conditions.

Limitations of rule-based strategies can mainly be attributed to the reactive rather than proactive nature of control; when VSL actions are implemented, traffic conditions may already have reached collapse, and VSL control may do little to resolve the situation.

10.2.2 Proactive approach

However, most proactive VSL strategies developed are based on second-order macroscopic traffic flow models and utilise aggregated data (such as average speed, flow, and density) from point detection technology. The deployment of such technologies corresponds to high installation, maintenance, and communication costs, as well as high failure rates (Herrera *et al.*, 2010). Additionally, this relatively coarse data aggregation obscures many features of interest, such as any changes in traffic state within the aggregation interval (Wu and Liu, 2014). Moreover, these macroscopic models used for VSL design do not fully reflect individual driver behaviour in traffic flow. When traffic is in a congested state, any disturbance in the flow can create shock waves that can lead to traffic collapse. Such shock waves stem from microscopic driver behaviour, such as sudden decelerations, mergers, or lane changes, resulting in non-uniform distances. The use of a macroscopic traffic model cannot fully reflect the occurrence of such disturbances (Khondaker and Kattan, 2015).

To address the limitations of proactive approaches, the Model Predictive Control (MPC) approach has been developed, where future traffic is predicted before congestion occurs, and corrective VSL strategies are implemented in the system to reduce incoming flow and resolve shock waves before traffic reaches collapse. The objective function used is described below.

10.2.2.1 Objective function

Let,

- m be a highway section divided into N_m segments of length L_m and λ_m lanes;
- $\rho_{m,i}(k)$ be the traffic density;
- T be the time interval used for traffic flow prediction;
- $w_o(k)$ be the queue length;
- $v_{ctrl,m,i}$ be the speed limit imposed on segment m, link I.

$$J(k_c) = T \sum_{k=Mkc}^{M(kc+Np)-1} \left\{ \sum_{(m,i)\in I\_links} \rho_{(m,i)}(k) L_m \lambda_m + \sum_{o\in Iorig} w_o(k) \right\} + \alpha_{speed} \sum_{l=k\_c}^{k\_c+N\_c-1} \sum_{(m,i)\in Ictrl} \left(v_{ctrl,m,i}(l) - v_{ctrl,m,i}(l-1) \right) / \left(v_{free,m} \right) \Big)^2 \tag{10.19}$$

where,

- I_{links} is the set of indices of all pairs of segments and links;
- I_{orig} is the set of all origins;
- I_{ctrl} is the set of pairs of indices of links and segments where speed control is applied.

This objective function contains a term for Total Time Spent (TTS) and a term that penalises abrupt variations in the speed limit control signal. The variation term is weighted by the non-negative parameter α_{speed}. Thus, it consists of two terms: one term for Total Time Spent (TTS) and one term that penalises large variations in the control signal. In the TTS term, the total number of vehicle hours spent on the highway section and in the originating queue is summed. The penalty term for variations in the control signal is included to express a preference for more uniform signals. The compromise between these two terms is expressed by the relative weight of these terms in the objective function. Since the main objective of the controller is to minimise TTS, the weights are chosen such that the TTS term is more important.

10.2.2.2 Constraints

In general, for the safe operation of a speed control system, it is required that the maximum decrease in speed limits that a driver can encounter $v_{maxdiff}$ be limited.

The maximum constraints on speed difference for the three situations are formulated as follows:

$$v_{ctrl,m,i}(l-1) - v_{ctrl,m,i}(l) \leq v_{maxdiff} \tag{10.20}$$

For all m, i, l such that $(m,i) \in I_{ctrl}$ and $l \in [k,\ldots,k+N_c-1]$.

$$v_{ctrl,m,i}(l) - v_{ctrl,m,(i+1)}(l) \leq v_{maxdiff} \tag{10.21}$$

For all m, i, l such that $(m,i) \in Ictrl$, $(m,i+1) \in I_{ctrl}$ and $l \in [k,\ldots,k+N_c-1]$.

$$v_{ctrl,m,i}(l-1) - v_{ctrl,m,(i+1)}(l) \leq v_{maxdiff} \tag{10.22}$$

For all m, i, l such that $(m,i) \in I_{ctrl}$, $(m,i+1) \in I_{ctrl}$ and $l \in [k,\ldots,k+N_c-1]$. In addition to safety constraints, a minimum value $v_{ctrlmin}$ may be set for speed limits:

$$v_{ctrl,m,i}(l) \leq v_{ctrlmin} \tag{10.23}$$

For all $(m,i) \in I_{ctrl}$ and $l \in [k,\ldots,k+N_c-1]$

The current VSL design strategy can be improved in a Connected Vehicle environment; specifically, Vehicle-to-Vehicle (V2V) and Vehicle-to-Infrastructure (V2I) communication initiatives provide a basis for detecting individual vehicle trajectories. These microscopic or single-vehicle level data can be used as more accurate inputs for designing advanced traffic control devices aimed at reducing congestion and improving road safety. The main advantage of using microscopic data is that driver behaviour can be described in detail. For example, analysing individual trajectory data is important to identify the position and magnitude of shock wave formation that can be created at the single vehicle level, such as a vehicle changing lanes or abruptly stopping. This step is crucial for promptly activating advanced traffic control devices. Consequently, studies focusing on individual driver behaviour (e.g., acceleration/deceleration, lane changes, overtaking, etc.) rather than aggregated behaviour are needed to develop advanced and robust traffic control devices.

10.2.3 Improvement of traffic operation through VSL

10.2.3.1 Impacts of VSL on traffic flow

The impact of VSL on the traffic flow diagram simply consisted of replacing the left portion of the flow-occupancy curve with a straight line, while Cremer (1979) developed a quantitative model for the flow-occupancy diagram (Figure 10.5) modified by VSL.

To develop an integrated VSL system for safety, it is necessary to consider both the effects of VSL on traffic flow and safety during operation. The proposed VSL control algorithm contains two main modules: (1) traffic flow analysis module and (2) accident risk assessment module. The traffic flow analysis module uses an extension of the METANET traffic flow model to analyse the potential effects of variable speed limits on traffic flow. Additionally, as an integrated VSL system for

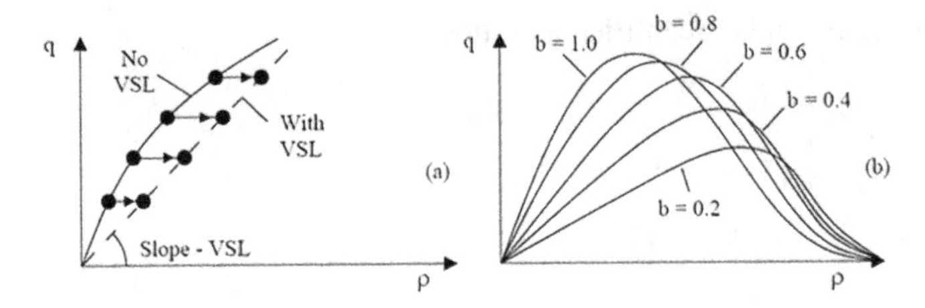

Figure 10.5 Modification of the flow-occupancy diagram by VSL

safety, one of the main objectives of the system is to reduce the incidence of accidents. Finally, the best VSL control strategies are obtained by solving optimisation problems to minimise the risk of accidents for the controlled highway section.

10.2.3.2 Traffic flow analysis module

The traffic flow analysis module adopts an extension of the METANET model to evaluate the effects of VSL on traffic flow. In the METANET model, highway sections are divided into segments where each segment has uniform characteristics in terms of geometry. A highway section m is divided into N_m segments of length L_m and λ_m lanes. The traffic flow in each segment at time instant $t = kT$ is represented by traffic density variables $\rho_{m,i}(k)$ [vehicles/lane/mile], average speed $v_{m,i}(k)$ [miles per hour], and traffic volume $q_{m,i}(k)$ [vehicles/hour], where T is the time interval used for traffic flow prediction.

The traffic variables for each segment of the highway section are calculated using the following equations:

$$\rho_{m,i}(k+1) = \rho_{m,i}(k) + T/(L_m\lambda_m)\left[q_{m,i-1}(k) - q_{m,i}(k)\right] \tag{10.24}$$

$$q_{m,i}(k) = \rho_{m,i}(k) \cdot v_{m,i}(k) \cdot \lambda_m \tag{10.25}$$

$$\begin{aligned} v_{m,i}(k+1) = {} & v_{m,i}(k) + T/\tau\left(V\left(\rho_{m,i}(k)\right) - v_{m,i}(k)\right) \\ & + T/L_m v_{m,i}(k)\left(v_{m,i-1}(k) - v_{m,i}(k)\right) \\ & - +\eta T/(\tau L_m)\left(\rho_{m,i+1}(k) - \rho_{m,i}(k)\right)/\left(\rho_{m,i}(k) + \kappa\right) \end{aligned} \tag{10.26}$$

$$V\left(\rho_{m,i}(k)\right) = v_{f,m} \cdot exp[-1/a_m \cdot \left(\rho_{m,i}(k)/\rho_{cr,m}\right)^a_m \tag{10.27}$$

where,

- $v_{f,m}$ represents the free flow speed at link m;
- $\rho_{cr,m}$, the lane critical density at section m;
- a_m, τ, η, κ are the parameters which must be calculated.

A quantified model was used to illustrate the modified flow-density diagram by VSL. The VSL rates $b_m(k)$ represent the ratios between the applied variable speed limits and the original constant speed limits. The effects of VSL on the flow-density diagram can be quantified as follows:

$$v'_{f,m} = v_{f,m} b_m(k) \tag{10.28}$$

$$\rho'_{cr,m} = \rho_{cr,m}\{1 + A_m[1 - b_i(k)]\} \tag{10.29}$$

$$a'_m = a_m[E_m - (E_m - 1)b_i(k)] \tag{10.30}$$

where,

- $v_{f,m}$, $\rho_{cr,m}$, α_m represent the non-VSL values for the three parameters;
- A_m and E_m are constant parameters representing the impacts of VSL on the fundamental diagram and must be estimated based on real data.

10.2.3.3 Improvement of safety through VSL

From a traffic safety perspective, VSL systems are designed to enhance traffic safety and reduce the occurrence of accidents. VSL systems focused on traffic safety typically include a function to quantify the risk of accidents, known as an accident prediction model. These accident prediction models have been developed to quantify the likelihood of accidents occurring and are used to decide when to activate VSL control and evaluate VSL performance on traffic safety.

For logistic regression models, let us assume that the occurrence of an accident has outcomes $y = 1$ or $y = 0$ with respective probabilities p and 1-p; the real-time accident prediction model can be represented as:

$$\text{logit}(p) = \log(p/(1-p)) = \beta 0 + \mathbf{X\beta} \tag{10.31}$$

where,

- β_0 is the intercept;
- $\overline{X}$ is the vector of explanatory variables;
- $\overline{\overline{\beta}}$ is the vector of coefficients for the explanatory variables.

VSL systems are set to be activated when predetermined levels of accident risk are reached; Lee *et al.* (2006) suggested four threshold levels of potential accident values for merging/diverging road sections and straight road sections separately. As the threshold values increase, the intervention of the VSL system is undertaken less frequently; Abdel-Aty *et al.* (2009) used the speed difference between the upstream average speed and the average speed of the VSL station of interest as a measure to determine whether VSL implementation is necessary or not. After the VSL system takes control of traffic flow, accident risks are monitored in real-time. If the risks of high-probability accidents have been reduced with lower speed limits, the speed limits will gradually increase and return to the baseline condition.

The common procedures of VSL control strategies oriented towards traffic safety approaches are shown in the Figure 10.6:

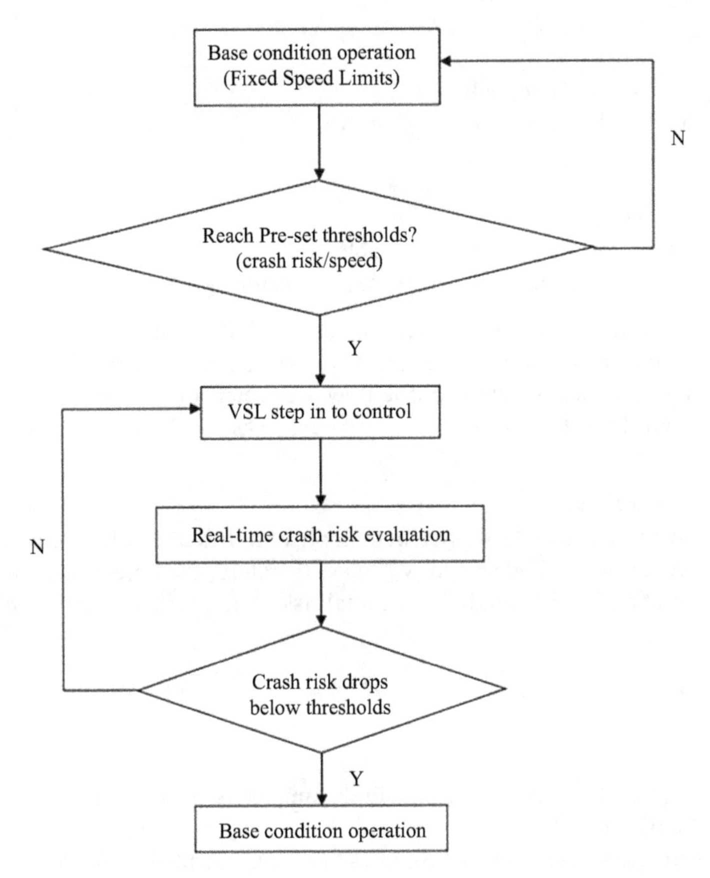

Figure 10.6 VSL control strategies common logical diagram

Risk assessment module

In this module, an accident risk assessment model is incorporated to evaluate the danger of accidents in real time. Road accidents are complex events involving hazardous human and environmental factors, road geometric characteristics, and traffic flow conditions. Since microsimulation software cannot directly replicate accidents, a surrogate measure of traffic safety needs to be proposed to assess the safety improvements brought by VSL systems. The development of real-time accident risk assessment models is the frequently adopted approach to quantify the risk of accident occurrences in VSL simulation studies.

The accident risk assessment model utilises a logistic regression model to measure the risk of accidents with historical accident data and real-time traffic data matched to each accident case. Assuming that the occurrence of an accident has outcomes $y = 1$ (accident cases) or $y = 0$ (non-accident cases) with respective probabilities p and 1-p, the logistic regression can be explained as follows:

$$\begin{aligned} y &\sim \text{Binomial}(p) \\ \text{logit}(p) &= \log(p/(1-p)) = \beta_0 + \mathbf{X\beta} \end{aligned} \tag{10.32}$$

where,

- β_0 is the intercept;
- $\overline{X}$ is the vector of explanatory variables;
- $\overline{\beta}$ is the vector of coefficients for the explanatory variables.

The three traffic flow parameters of the METANET model (average speed, density, and traffic volume) are the candidate explanatory variables in the accident risk assessment model. Since the traffic flow parameters of the time interval can be obtained from the METANET model, accident risks for the time interval can be calculated.

Optimisation of VSL

The objective is to use the VSL system to improve traffic safety. Let us assume that the road section m is divided into N_m links; therefore, the objective function for VSL optimisation is set to minimise the total risk of accidents for section m at time step k+1:

$$\sum i = 1^{Nm} CR_{i(k+1)} = e^{\beta}0^{+X(k+1)\beta} / (1 + e^{\beta}0^{+X(k+1)\beta}) \tag{10.33}$$

where,

- $\overline{X}(k+1)$ is the vector of traffic flow parameters provided by the extended METANET model;
- $b_i(k)$ represents the optimal VSL rates to be implemented for segment i at time step k, with $b_i(k) \in [b_{\min}, 1]$ and $b_{\min} \in (0, 1)$ being the allowable lower limit for VSL rates.
- Combining all the equations described above, the optimisation problem can be represented as:

$$b_i(k+1) = f[b_i(k), \boldsymbol{u}(k)], b_i(0) = 1 \tag{10.34}$$

where,

- b_i is the VSL rate of segment i in link m;
- $\overline{u}$ is the traffic flow state vector containing input data for the METANET model (information on speed, density, and volume).

From the equation, it can be seen that the input data of this model are the traffic flow parameters and the current VSL rates, and the optimal VSL rates for the next time step are the only outputs. Furthermore, considering traffic operability and safety, constraints are set for:

(a) the maximum increase in average travel time for the VSL control area compared to non-VSL control cases which is 5%;

(b) the maximum difference between two adjacent indicated speed limits is ten miles per hour (spatial constraint);
(c) the maximum difference between two consecutive VSL control time steps is ten miles per hour (temporal constraint).

The constraint for controlling the increase in average travel time can be formulated as:

$$\sum_{i=1}^{Nm} L_m/v_i(k+1) \leq (1+t_m)\sum_{i=1}^{Nm} L_m/v_i'(k+1) \tag{10.35}$$

where,

- $v_i'(k+1)$ is the average speed under non-VSL control $(b_i'(k) = 1)$;
- t_m is the rate of increase in average travel time.

10.3 Summary

This chapter provides an overview of some approaches adopted for advanced traffic management. In particular, the main focus is on ramp metering and variable speed limits. For each one of them, some main strategies are reviewed. Furthermore, the main tools of the control theory required are discussed in the appendices.

Traffic control strategies can be implemented in an integrated way. Concerning the strategies discussed in the chapter, it has been demonstrated in the literature (see Hegyi *et al.*, 2005) that speed limits can complement ramp metering, when the traffic demand is so high that ramp metering is not efficient anymore.

More recently the literature on traffic management strategies has been focused on mixed traffic flow (Shang *et al.*, 2023; Arora and Kattan, 2023).

References

Some suggestions for further readings are reported below among the huge literature available on these topics.

Abdel-Aty, M., and Dhindsa, A. (2007). Coordinated use of variable speed limits and ramp metering for improving safety on congested freeways. In: *86th TRB Annual Meeting*, Washington, DC.

Abdel-Aty, M., Dilmore, J., and Dhindsa, A. (2006). Evaluation of variable speed limits for real-time freeway safety improvement. *Accident Analysis & Prevention*, 38(2), 335–345

Abdel-Aty, M., Haleem, K., Cunningham, R., and Gayah, V. (2009). Application of variable speed limits and ramp metering to improve safety and efficiency of freeways. In *2nd International Symposium on Freeway and Tollway Operations* (pp. 1–13).

Allaby, P., Hellinga, B., and Bullock, M. (2006). Variable speed limits: safety and operational impacts of a candidate control strategy for an urban freeway. *IEEE Transactions on Intelligent Transportation Systems,* 8(4), 671–680.

Arora, K., and Kattan, L. (2023). Operational and safety impacts of integrated variable speed limit with dynamic hard shoulder running. *Journal of Intelligent Transportation Systems*, 27(6), 769–798.

Chen, C.I., Cruz Jr, J.B., and Paquet, J.G. (1974). Entrance ramp control for travel-rate maximization in expressways. *Transportation Research*, 8(6), 503–508.

Cremer, M. (1979). *Der Verkehrsfluss aud Schnellstrassen*. Springer-Verlag, Berlin.

Gould, C., Munro, P., and Hardman, E. (2002). M3/M27 Ramp Metering Pilot Scheme (RMPS)-implementation and assessment. *Eleventh International Conference on Road Transport Information and Control*, pp. 161–167.

Hegyi, A., De Schutter, B., and Hellendoorn, H. (2005). Model predictive control for optimal coordination of ramp metering and variable speed limits. *Transportation Research Part C: Emerging Technologies*, 13(3), 185–209.

Herrera, J.C., Work, D.B., Herring, R., Ban, X., Jacobson, Q., and Bayen, A.M. (2010). Evaluation of traffic data obtained via GPS-enabled mobile phones: the mobile century field experiment. *Transportation Research Part C*, 568–583.

Jacobson, L.N., Henry, K.C., and Mehyar, O. (1989). Real-time metering algorithm for centralized control. *Transportation Research Board*, 1232, 17–26.

Jia, Z., Chen, C., Coifman, B., and Varaiya, P. (2001). The PeMS algorithms for accurate, real-time estimates of g-factors and speeds from single-loop detectors. In *ITSC 2001. 2001 IEEE Intelligent Transportation Systems. Proceedings* (Cat. No. 01TH8585) (pp. 536–541). IEEE.

Khondaker, B., and Kattan, L. (2015). Variable speed limit: an overview. *Journal of Transportation Letters*, 7(5), 264–278. http://dx.doi.org/10.1179/1942787514Y.0000000053.

Lavansiri, D. (2003). Evaluation of Variable Speed Limits in Work Zones. PhD Dissertation, Michigan State University.

Lee, C., Hellinga, B., and Saccomanno, F. (2006). Evaluation of variable speed limits to improve traffic safety. *Transportation Research Part C: Emerging Technologies*, 14(3), 213–228.

Lyles, R.W., Taylor, W.C., Lavansiri, D., and Grossklaus, J. (2004). A field test and evaluation of variable speed limits in work zones. In *Transportation Research Board Annual Meeting* (CD-ROM), Washington, DC.

Masher, D.P., Ross, D.W., Wong, P.J., Tuan, P.L., Zeidler, H.M., and Petracek, S. (1975). Guidelines for design and operation of ramp control systems. *Transportation Research Board*, 537 p.

Papageorgiou, M., Blosseville, J.M., and Haj-Salem, H. (1990). Modelling and real-time control of traffic flow on the southern part of Boulevard Périphérique in Paris: Part II: Coordinated on-ramp metering. *Transportation Research Part A: General*, 24(5), 361–370.

Papageorgiou, M., Hadj-Salem, H., and Blosseville, J.M. (1991). ALINEA: A local feedback control law for on-ramp metering. *Transportation Research Record*, 1320(1), 58–67.

Park, B.K., and Yadlepati, S.S. (2003). Development and testing of variable speed limit logics at work zones using simulation. In: *Proceedings of 82nd TRB Annual Meeting*, Washington, DC.

Pei-Wei, L., Kyeong-Pyo, K., and Gang-Len, G. (2004). Exploring the effectiveness of variable speed limit controls on highway work-zone operations. *Journal of Intelligent Transportation Systems,* 8(3), 155–168.

Piao, J., and McDonald, M. (2008). Safety impacts of variable speed limits – a simulation study. In: *Proceedings of the 11th International IEEE Conference on Intelligent Transportation Systems* (pp. 833–837).

Rama, P. (1999). Effects of weather-controlled variable speed limits and warning signs on driver behavior. *Transportation Research Record,* 1689, 53–59.

Schwartz, S.C., and Tan, H.H. (1977). Integrated control of freeway entrance ramps by threshold regulation. In *1977 IEEE Conference on Decision and Control including the 16th Symposium on Adaptive Processes and A Special Symposium on Fuzzy Set Theory and Applications* (pp. 984–986). IEEE.

Shang, M., Wang, S., and Stern, R.E. (2023). Extending ramp metering control to mixed autonomy traffic flow with varying degrees of automation. *Transportation Research Part C: Emerging Technologies*, 151, 104119.

Smaragdis, E., and Papageorgiou, M. (2003). Series of new local ramp metering strategies: Emmanouil smaragdis and markos papageorgiou. *Transportation Research Record*, 1856(1), 74–86.

Smulders, S. (1990). Control of freeway traffic flow by variable speed signs. *Transportation Research Part B: Methodology*, 24(2), 111–132.

Smulders, S., and Helleman, D.E. (1998). Variable speed control: state-of-the-art and synthesis. In: *Road Transport Information and Control*, No. 454 in Conference Publication (pp. 155–159), IEEE.

Talebpour, A., Mahmassani, H.S., and Hamdar, S.H. (2013). Speed harmonization: Evaluation of effectiveness under congested conditions. *Transportation Research Record*, 2391(1), 69–79.

Taylor, C., Meldrum, D., and Jacobson, L. (1998). Fuzzy ramp metering: Design overview and simulation results. *Transportation Research Record*, 1634(1), 10–18.

Wang, C.F. (1972). On a ramp-flow assignment problem. *Transportation Science*, 6(2), 114–130.

Wattleworth, J.A. (1965). *Peak-Period Analysis and Control of a Freeway System* (No. Hpr 1/4/). San Antonio, TX, USA: Texas Transportation Institute.

Wu, X., and Liu, H.X. (2014). Using high resolution event-based data for traffic modeling and control: an overview. *Transportation Research Part C,* 42, 28–43.

Yuan, L.S., and Kreer, J.B. (1971). Adjustment of freeway ramp metering rates to balance entrance ramp queues. *Transportation Research*, 5(2), 127–133.

Zackor, H. (1979). Self-sufficient control of speed on freeways. In: *Proceedings of the International Symposium on Traffic Control Systems*, vol. 2A (pp. 226–249). Berkeley, California.

Chapter 11
Advanced traveller information systems

Stefano de Luca[1] *and Roberta Di Pace*[1]

The only thing that will redeem mankind is cooperation.

Bertrand Russell

Outline. *This chapter aims to discuss the main elements concerning the Advanced Traveller Information Systems. In particular, it introduces basic notions, approaches, and solutions to: learning process modelling, pre-trip route choice modelling, and en-route choice modelling.*

Advanced Traveller Information Systems (ATIS) dispatch traffic information to travellers to assist them in making several different travel choices. ATIS may affect pre-trip and en-route choices, but they may also affect departure time choices and, in the same case, pre-time mode choices. The role of ATIS is more than ever fundamental for a sustainable and effective transportation system due to the technological revolution currently underway in the automotive industry (connected, automated or autonomous vehicles), but also in terms of communications, the presence of alternative fuel vehicle powertrains, and/or new protocols. Nowadays, it is not only important to minimise travel time but also to give information which is consistent with the specific vehicle characteristics (e.g., battery autonomy) and with the driver's expectations, which are increasing as rapidly as the overall social and technological ecosystem is expanding. Indeed, ongoing technological advances require personalised and dynamic travel information applications. In this context, the simulation of a transportation system requires, more than ever, to model drivers' behaviour coherently with the complex ecosystem in which it happens. The choice under ATIS may be schematised as a process mainly affected by users'/travellers' beliefs, preferences, habits, inertia, and perceived payoff (utility) of the available alternatives. At the same time, attitudes and perceptions may concur to define travellers' preferences, as well as travellers' socioeconomic and trip characteristics. In this classical framework (Ben-Akiva and Lerman, 1985), we may distinguish several different behaviours that may require specific interpretative paradigms and specific modelling solutions. To this aim, this chapter aims to introduce basic notions, approaches and solutions to: (i) learning process modelling

[1]Department of Civil Engineering, University of Salerno, Italy

(Section 11.2), (ii) pre-trip route choice modelling (Section 11.3), and (iii) en-route choice modelling (Section 11.4).

11.1 Problem statement

11.1.1 Introduction

ATIS may directly influence travel choices, hence flow propagation and network performance. In the ATIS research area, several issues may be identified: the design and implementation of the technology, the definition of the goals to be pursued, the interpretation and modelling of driver behaviour, and the design of the information to be dispatched to achieve the predefined goals. Research into ATIS dates to the 1950s and has evolved considerably over time. We may distinguish different kinds of ATIS: ATIS to make travellers aware of non-recurring congestion when special events or accidents occur; ATIS to prevent or mitigate the disruption of traffic flow propagation; ATIS to improve personalisation/customisation of the information with the aim of supporting routing, wayfinding, and so forth; ATIS to improve network (or subnetwork) performance through control or travel demand management strategies.

Irrespective of the goals to achieve and the methodological framework in which ATIS will be implemented, a key issue concerns the need to interpret and model the driver's reactions to information. This requires (i) a definition of the behavioural paradigm underlying the choice phenomenon, (ii) formalisation of the most effective theoretical solutions, and (iii) identification of the variables that may affect the choice process. All the above requisites have become ever more complex and challenging since, as with any kind of information system, the accuracy and reliability of the information dispatched can determine the success or failure of the ATIS.

Overall, we may distinguish three different types of information, namely descriptive information, prescriptive information, and mixed information. Descriptive information (whether quantitative or qualitative) may concern the working conditions of the whole network or of specific infrastructures/nodes of the same network, or supply information on trip characteristics. Such information does not indicate alternative solutions but leaves the burden of choice on the user. The user is called upon to interpret the information and undertake pre-trip or en-route choices by forming their own expectations of travel times. The information conveyed may be instantaneous or predictive. If reliable, the system nevertheless allows user trip comfort to be increased while the potential improvement in general network working conditions remains to be appraised. In terms of interpretation, the decisional process is very similar to the decisional process without information.

Descriptive information also helps non-systematic users distinguish alternatives better (or differently in the case of systematic users) and arrive at a more precise perception of costs; moreover, information can reduce the number of alternatives available.

Information of a prescriptive nature supplies guidelines on what path/itinerary to follow and can be pre-trip or en-route. Undoubtedly, such information is the most useful for travellers although it requires knowledge of trip origins and destinations, that is to say, the path chosen in advance. Prescriptive information further reduces choice possibilities but, although instantaneous and predictive information can be distinguished, prescriptive information must be much more reliable and must thus predict system evolution over time. From the modelling standpoint, prescriptive information boils everything down to a problem of choice between suggested alternatives, or a problem of acceptance of the only alternative proposed.

Mixed information stands for all information of a prescriptive nature to which, however, descriptive details are associated. Its usefulness lies in making the user aware of what is occurring on the network and facilitating the acceptance of prescriptive information.

The dispatched information can be further classified according to its dynamics. Indeed, the information can be static or dynamic with reference to traffic conditions. In the static case, the dispatched information mainly relates to the network structure. In other words, it typically refers to the network topology and to free-flow (or static average) travel times (or distances). Static ATIS dispatches static information, which is known far in advance and changes infrequently (for instance when the network structure changes).

In the dynamic case the information somehow seeks to represent the traffic conditions, and, given the network structure, it changes according to these conditions (or some estimates of these conditions). Dynamic ATIS dispatch (real-time) information changes frequently as the network conditions change. Even if it is sometimes maintained (e.g., Avineri and Prashker, 2006) that static (but appropriate) information can lead to good results (even better than some kinds of dynamic information), dynamic information is potentially more effective (albeit much more unlikely to be well designed).

Within the dynamic case, two more kinds can be distinguished. In the first sub-case, the information could be based on the instantaneous knowledge (or estimate) of traffic conditions, for instance on instantaneous travel times. In the second sub-case, it is based on predictive estimates of traffic conditions, such as actual travel times, where actual travel times are those that the travellers will have experienced at the end of their trips, generally differing greatly from instantaneous travel times.

For uncongested networks (or those far removed from saturation conditions), static and dynamic information practically coincide. Similarly, the instantaneous and actual information is practically coincident not only in the uncongested case but also if the network can be considered in within-day static conditions. However static information is inherently incorrect in the case of congested networks, as is instantaneous information in the case of within-day dynamic traffic conditions. Of course, nothing ensures that the actual (predictive) information is correct in every case, this depends on the quality of the traffic predictions that depend, in turn, on the anticipatory route guidance problem (see for instance Bottom *et al.*, 1998; Bierlaire and Crittin, 2001; Dong *et al.*, 2006) which should guarantee that the dispatched information must be consistent with the reactions induced on the travellers and the effects on the network.

The type of information greatly influences the process of accepting the supplied information, defining choice alternatives, and forming expectations about the variables in question. Moreover, the process of choice learning differs according to the variation in the user class in question and, especially, the variation in familiarity with the network and the systematic nature of the trip undertaken.

Overall, the process underlying route choice may be structured into the following phases:

(a) acquisition and elaboration of information,
(b) use of information,
(c) choice.

As with consumer choices prior to purchasing a good, it may be very important to represent the user's cognitive process and his/her subsequent capacity to elaborate on the information supplied. In this scenario, two latent variables can be distinguished and defined, one representing cognitive involvement, the other the capacity to process information. By cognitive involvement, we mean user engagement towards the problem of route choice. In particular, the user's decision-making process may thus be schematised by a hierarchical structure in which the user first chooses whether to acquire the information then decides whether to use it, and then select the route. The underlying hypothesis is that the information is used if its utility exceeds a certain threshold. Modelling the phenomenon may lead to first estimating the propensity of acquiring information for its use, then estimating the propensity to use it upon having and, finally, estimating the propensity to act according to the information supplied or the propensity to choose the alternative proposed (or built) by the information system (or the information process).

Within the proposed general framework, the type of information varies, different choice behaviours can be observed and therefore different behavioural paradigms, modelling solutions and/or variables may be adopted. Several challenging tasks and issues may be identified.

First, how to represent and model the compliance and concordance issues with information. The matter of the traveller's compliance is evident in the case of descriptive/prescriptive information, and it may be defined as the evidence that the traveller follows the recommendation of the descriptive/prescriptive information. By contrast, a concordant-with-ATIS traveller chooses the suggested route either because he/she trusts in the system (is compliant) or because he/she would have chosen the route in any case, independently on ATIS indications. Compliant-with-ATIS traveller is always concordant-with-ATIS, thus the set of concordant travellers contains the set of compliant travellers, and the probability of being concordant is greater than the probability of being compliant.

Second, the effect of the ATIS (in)accuracy and/or the effect of the ATIS reliability.

The information is accurate if the dispatched travel times are those that the travellers will have experienced by the end of their trips. Similarly, now in the prescriptive case, the system is said to be reliable (Dhivyabharathi *et al.*, 2018) if the suggested route is found by the traveller to have been the minimum-time one. Similarly, reliability is also defined for the descriptive case; in this case, the system is said to be reliable if the route that exhibits the minimum dispatched travel time is found by the traveller to have been the actual minimum-time route (even if the dispatched travel time may be wrong).

11.1.2 A general choice process paradigm under ATIS

The drivers' choice process under ATIS may be schematised as follows.

Let,

- i be the generic user (class of users),
- s the trip purpose,
- od the origin-destination pair,
- m the transport mode used,
- g the day of the trip,
- t the start of the trip.

The generic trip may be schematised as a sequence of nodes that the user crosses and at which choices may be made. In such a context, simulation of path choice behaviour may be structured into the following two main problems:

1. characterisation of behaviour at the trip origin,
2. characterisation of en-route behaviour.

11.1.2.1 Pre-trip choice behaviour

At the trip origin, the dynamics of the decisional process of each user may be schematised in a succession of interdependent decisions and reasonably structured according to the following order:

(I) the trip behaviour,
(II) the choice alternatives;
(III) the choice itself.

By trip behaviour (decisional dimension I) we mean the approach adopted by the user when undertaking the trip. To this end, we may distinguish:

(orgin-a1) completely preventive behaviour
(orgin-a1) mixed preventive/adaptive behaviour

If *completely preventive* it is hypothesised that there is no within-period dynamic of the choice process, path choice is exhausted at the trip origin* and the sequence of nodes that the user will cross is known and cannot be changed by events or information encountered along the path. Such an approach is reasonable if the available alternatives do not allow efficient deviations from the path (very different paths), or the user believes he/she is unable to modify his/her own choice during the trip due to little experience with the network characteristics (user unfamiliar with the network, non-systematic trip), due to absence of information on the functional characteristics of the network itself or due to a fairly passive attitude to searching for more efficient paths. If *preventive/adaptive mixed*, it is supposed that users extend their own decisional process to the duration of the trip and the chosen path cannot be defined *a priori*.

*Or if the system working conditions (e.g., stationary or in dynamic equilibrium) are such that one can imagine simulating the whole choice process at the trip origin.

Having decided the type of behaviour, the user is called upon to define the choice alternatives (decisional dimension II). In a completely preventive context, the choice alternatives consist of identifiable routes on the transport supply system which connect the origin-destination pair in question. If it is *mixed preventive/ adaptive*, it is hypothesised that the user can make path diversions. To this end, two possible behaviours can be distinguished:

(origin-b1) *'reference path'*
(origin-b1) *'strategy'*

In the reference path approach, the user chooses a path in advance and, at specific nodes (diversion nodes), evaluates the possibility of changing their choice, analysing alternative solutions. Each solution will originate in the diversion node and will be taken into consideration on the basis of the working conditions of the network which the user has experienced or was supplied exogenously (information).

In the '*strategy*' approach the user chooses a path set and, within the strategy, evaluates which sub-path (itinerary) to use on the basis of the functional characteristics encountered and/or the information supplied.

Irrespective of the approach adopted and the type of behaviour within a specific approach, the choice alternatives may be:

(origin-c1) *combined with other choice dimensions*
(origin-c2) *non-combined*

As regards the estimate of path/itinerary flows, path choice may be interpreted irrespective of other trip choices (rigid demand vs. elastic demand). The choice of path/itinerary is strongly interrelated with other trip characteristics and it is therefore reasonable to imagine that it may be simulated contemporaneously with the departure time interval, with the choice of destination and/or transportation mode[†]. In a dynamic simulation approach and in the presence of a short-term time horizon, the problem of path/route choice is studied separately or, at most, combined with the choice of departure time. This applies so much more if there is a pre-trip system of information for the user.

Lastly, downstream of the previous decisions, the choice problem may be introduced (decisional dimension III). In this case, we mean the set of behaviours that lead to the choice of the final path/route.

To this end, it is worth distinguishing whether the choices are influenced by choices already made in previous periods, i.e., by experience accumulated during previous trips. Two hypotheses may be formulated:

(origin-d2) *choice is not tied to choices made in the same context (purpose, od, time interval) but on previous days*
(origin-d2) *choice is tied*

[†]Not to speak of more complex approaches that interpret path choice within simulation of user (household) activity.

The first hypothesis involves simulation of the choice process alone; in the second case, rather than choice, one may speak of the process of updating the choice, or the process by which users update their own choices from one day to the next on the basis of their experience or information supplied to them.

In general, user choice behaviour may be interpreted in light of a decisional process in which the user processes the information which is formed endogenously or which is supplied exogenously, and chooses by adapting to the context in question. In general, we may distinguish between a cognitive process, a learning process and a decisional process.

The cognitive process and the learning process envisage the acquisition by the user of information from the outside world. Such information may be supplied *ad hoc* by an information system, may be part of the user's experience, or may be acquired during the trip through direct observation of successive phenomena. The information supplied exogenously undergoes a process of pre-trip evaluation to which the user associates a degree of reliability and filters the information deemed important.

Under the learning process, the user receives information and learns from the information received, although it may not necessarily apply to every user. If, for users who make systematic trips, a learning process can be imagined, for those who make non-systematic trips it is hard to imagine learning from the information supplied. At the same time, there are users who, albeit systematic, may not be subject to a learning process as they are already consolidated in following choices resulting from their own experience. Cognitive and learning processes allow users to continuously update their perceptions and formulate new expectations on the variables that affect the choice process[‡].

Lastly, in the choice process, the user defines the set of alternatives from which to choose, associates with each alternative some characteristics deemed important, and defines rules and/or criteria according to which to choose. On comparing the individual alternatives, the user then makes his/her own choice.

11.1.2.2 En-route choice behaviour

Downstream of the choice process and upon variation in the possible choice strategies, the user begins his/her trip and begins to acquire information from direct experience or from the information system(s) present. The user compares what he/she knows from experience with what he/she receives, and begins to form expectations about the path characteristics of reference (strategy) and the possible alternative paths (itineraries within the pre-selected strategy[||]). En-route choices are therefore tied to possible trip behaviour (decisional dimension I) that the user has decided to adopt (e.g., choice of a path or a strategy) and to choices made at the origin (e.g., the path or strategy chosen).

Also in this case, it is possible to distinguish a cognitive process, a process of learning and of forming/updating attributes regarding the remaining part of the

[‡]Clearly, the same considerations may be extended to the process of updating choices introduced above.
[||]It is hypothesised that the user does not question the strategy chosen in advance.

path. Upstream of this process and once a possible diversion node has been reached, the user decides whether to change their own path of reference (if that is the approach) or to choose how to continue the trip within the pre-trip chosen strategy.

In a *reference path* approach the user reconsiders their pre-trip choice, assesses whether it is worth changing path and, if so, it is reasonable to imagine that the path is chosen following the same decisional rules used at the trip origin: definition of the choice set and choice among alternatives.

In a *strategy* approach, the user, at each diversion node belonging to the strategy, decides on the sequence of links up to the next diversion node. Clearly, the user's decisional behaviour is different and maybe schematised as a choice among a set of alternatives known *a priori* whose costs may be different from those perceived at the beginning of the trip.

Such behaviour may be reproduced at each possible diversion node and, also in this case, various hypotheses may be formulated in which behaviour may be interpreted:

(node j-a1) *the choice of diversion is not tied to diversion choices made in the same context (purpose, od, time interval) but in previous days*
(node j-a2) *the choice is tied*

but also:

(node j-a1) *the choice of diversion is not tied to diversion choices made at previous nodes*
(node j-a2) *the choice is tied*

In the first hypothesis (**a1**) it is necessary to study the process of choice updating at the node in question between one period of reference and another. In the second hypothesis (**b1**) choice behaviour at each diversion node is conditioned by what happened at the previous nodes.

11.2 Modelling the learning process in route choice

Learning involves identifying the true values of variables that influence the outcome of a process, such as travel time or the reliability of ATIS. Route choice, in general, can be seen as a dynamic and adaptive learning process shaped by three main factors: (i) external information sources (e.g., travel time predictions, route guidance data), (ii) users' past experiences (e.g., perceived travel times and reliability), and (iii) psychological factors (e.g., risk attitude, habits, and preferences). The dynamic decision-making process in daily route choice can be explained by models that update perceptions and expectations (e.g., Van der Mede and Van Berkum, 1996; Mahmassani and Liu, 1999) or by examining observed preferences and their link to perceived payoffs (e.g., Arentze and Timmermans, 2003; Ben-Elia and Shiftan, 2010). Factors like memory decay, reinforcement, and inertia also play a role (e.g., Bogers *et al.*, 2006).

Several learning paradigms have been applied to model these processes. Under uncertain and risky conditions, approaches such as Bayesian learning (Chorus *et al.*, 2009), random utility theory, cumulative prospect theory, reinforcement learning (RL), or belief-based learning (e.g., Avineri and Prashker, 2005) have been used. Among these, RL is particularly popular. In RL, the likelihood of repeating an action increases with its associated payoff. This paradigm captures the adaptive behaviour where individuals strive to maximise expected payoffs based on their experiences and observations. Variants of RL incorporate cognitive and psychological factors, such as bounded rationality (e.g., Arthur, 1994; Cominetti *et al.*, 2010) or forgetting behaviour (e.g., Hurkens, 1995).

Another widely used learning framework is belief-based learning, where individuals form and update beliefs about the expected outcomes of actions based on new information or experiences. A prominent model in this category is the fictitious play (FP) process (e.g., Fudenberg and Levine, 1995), which assumes that players predict others' behaviours based on observed choice frequencies and respond with a best-response strategy. Unlike RL, FP is predictive, relying on assumptions about full observability, which are often unrealistic. Variants of FP, such as sampled FP (Lambert *et al.*, 2005) and joint strategy FP (JSFP, Marden *et al.*, 2009), address these limitations.

Bayesian learning (e.g., Jordan, 1995; Chorus *et al.*, 2009) offers another belief-based approach, using Bayes' rule to update beliefs based on observed outcomes.

In the following sections five learning models are introduced: RL, extended RL (ERL), JSFP, Bayesian learning (BL), and a variant of BL. These models account for the variability in travellers' payoff perceptions and ATIS information, while also examining factors like memory decay and inertia in route learning.

11.2.1 Reinforcement learning and extended reinforcement learning

Reinforcement learning (RL) can be interpreted as a stimulus-response mechanism oriented to obtain higher rewards over time; it is based on the following behaviour principles (Erev and Roth, 1998):

(a) The higher the received payoffs for a realised action, the more likely it will be chosen in the future;
(b) The learning effect seems high initially and then attenuates over time;
(c) The choice behaviour is probabilistic in response to uncertainty.

Several RL models have been proposed for the modelling strategy learning process of human decision-making (Erev and Roth, 1998; Feltovich, 2000). The general RL framework can be described as follows:

1. Initially, each traveller has no prior information regarding travel times of routes; each route of any given o/d pair (r) has the same propensity (q):

$$q_{ai}^{0} = c_i, \quad \forall r, \forall i \in I_r, \forall a_i \in A_i \tag{11.1}$$

where c_i is a constant (in our case, the negative average travel time of the routes of the o/d pair);

2. The reinforcement of an action (route choice) for traveller i is a function of the difference between the payoff associated with the route and the minimum payoff of actions in the traveller's choice set on day t:

$$\Delta^t_{ai} = u^t_{ai} - \min_{a'_i \in A_i} \left[u^t_i(a'_i, a'_{-i}) \right], \quad \forall r, \ \forall i \in I_r, \forall a_i \in A_i \tag{11.2}$$

where a'_i is any admissible choice within set A_i and a'_{-i} is the set of all choices of all travellers except i for all o/d pairs in the network (not restricted to I_r);

3. The propensities are updated over all actions on day t+1 as

$$q^{t+1}_{a_i} = q^t_{a_i} + \Delta^t_{a_i}, \quad \forall r, \forall i \in I_r, \forall a_i \in A_i \tag{11.3}$$

We can consider the propensity as a cumulative payoff or a payoff sum.

4. The choice probability of a_i is determined by a proportional rule related to the cumulative propensities as

$$p^t_i(a_i) = q^t_{a_i} / \sum_{a'_i \in A_i} q^t_{a'_i}, \quad \forall r, \forall i \in I_r, \ \forall a_i \in A_i \tag{11.4}$$

Equation (11.2) states the reinforcement of actions with higher payoff (behaviour principle (a). Equations (11.3) and (11.4) are related to the behaviour principles (b) and (c) because the denominator increases over time (resulting in a stabilised learning curve) and the choice is determined by a probabilistic rule (any route has a probability of being chosen). Note that the choice probability function in Eq. (11.4) is linear. One can replace it with a more generalised non-linear or power function (Laslier *et al.*, 2001).

The previous model can be adapted in order to relax some assumptions by introducing two psychological considerations: 1) experimentation and 2) recency (forgettable or bounded memory of travellers). The first states that travellers will not be stuck on choices with higher propensities and always have the possibility of choosing other actions. This effect can explain travellers' exploratory behaviour to gain more information about the payoffs of other routes as well as novelty-seeking behaviour (Arentze and Timmermans, 2005). Moreover, this random term can capture travellers' random choices when payoff variability is significant. The second reflects the fading-memory effect of travellers over time, which has been revealed in many studies as a relevant variable to explain travellers' route choice behaviour (Avineri and Prashker 2005, 2006).

For the first expression, the experimentation probability ε in Eq. (11.4) has been introduced (Ma and Di Pace, 2017) and obtain the ERL model as

$$\begin{cases} a^t_i \in |\mathrm{R} \text{ and } (A_i) & \text{with prob.}\varepsilon \\ p^t_i(a_i) = q^t_{a_i} / \sum_{a'_i \in A_i} q^t_{a'_i} & \text{with prob.} 1 - \varepsilon \end{cases} \tag{11.5}$$

where R and (A_i) is a random action in A_i. Hence Eq. (11.4) is a special case of Eq. (11.5) when $\varepsilon = 0$. Note that we can specify the experimentation probability as a function of payoff variability (Erev and Barron, 2005) to model random choice behaviour when the payoff variability effect is significant in the case of low information accuracy.

The second relaxation considers a recency effect $\alpha \in [0, 1]$ in Eq. (11.3) which attenuates the effect of past experiences on the current best choice (Erev and Roth, 1998):

$$q_{a_i}^{t+1} = (1 - \alpha)q_{a_i}^t + \Delta_{a_i}^t, \quad \forall r, \forall i \in I_r, \forall a_i \in A_i \tag{11.6}$$

Note that if $\alpha = 0$ all day-to-day propensities have the same relevance. By contrast, when $\alpha = 1$, only current experience is taken into account for updating the player's propensity. Hereafter, we call the extended RL model based on Eqs. (11.5) and (11.7) ERL.

Another remark is that the reinforcement-based model follows a propensity-proportional probabilistic choice rule, contrary to best-replay strategies whereby players make their best choice with the highest expected payoff.

The RL/ERL models stated above take into account only the revealed payoffs of travellers' choices. When the information system becomes available for travellers from day k, he/she can learn the reliability of the information system over time and combine the received information with his/her past experience. For this issue, two alternatives are proposed. The first is based on a *weighted updating of propensity perception* (Horowitz, 1984); the second alternative is based on the *probability matching assumption* (Shanks *et al.* 2002), drawn from past experimental studies of human decision-making, which states decision-makers match their choice probability with rates of positive payoff.

As regards the first approach, past studies show the influence of information on travellers' route choice decisions is based on the updating process of travellers' payoff perceptions (Horowitz, 1984; Mahmassani and Chang, 1986; Van der Mede and Van Berkum, 1996; Mahmassani and Liu, 1999; Selten *et al.* 2007; Chorus and Dellaert, 2012). The principle asserts a weighted combination of an individual's historical experiences and newly obtained information. We adopt this principle for cumulative propensity updating by assuming that travellers use the information provided to pre-update the cumulative propensity of each alternative in the choice set, make their choice decision accordingly, and then use revealed payoffs to correct their cumulative propensity *a posteriori*. The temporally cumulative propensity $\widehat{q}_{a_i}^t$ of route a_i at the beginning of day t for traveller i is temporally updated as

$$\widehat{q}_{a_i}^t = q_{a_i}^t + \tilde{\Delta}_{a_i}^t g_i^{t-1}, \quad \forall r, \ \forall i \in I_r, \forall a_i \in A_i \tag{11.7}$$

where $\tilde{\Delta}_{a_i}^t$ is the difference of expected payoff (provided estimated travel time by ATIS) and the minimum payoff of actions for traveller i on day t-1. The propensity increment is weighted by travellers' experienced information reliability g_i^{t-1}, reflecting the confidence of a traveller in his received information. The information reliability (concordance of recommended shortest route and *a posteriori* revealed the shortest route) perceived for traveller i up to the end of day t is defined as:

$$g_i^t = \frac{1}{t-k+1}\sum_{w=k}^{t}\delta_{\{y_i^w = a_i^{w*}\}} \tag{11.8}$$

where,

$t \geq k$

y_i^w, y, and a_i^{w*} respectively are the system-recommended route and the revealed best route for traveller i on day w,

$\delta_{\{y_i^w = a_i^{w*}\}}$ is an indicator function being 1 if both the recommended route and revealed best route are the same for traveller i on day w, and 0 otherwise.

Travellers use the weighted cumulative propensity to determine their route choice by Eq. (11.4) for the RL model and Eqs. (11.5) and (11.6) for the ERL model. When the payoffs are revealed at the end of each day, the cumulated propensity is first restored to $q_{a_i}^t, \forall a_i \in A_i$ and then Eqs. (11.2)–(11.4) are applied based on the revealed payoffs. Equation (11.7) integrates the traveller's perceived information reliability on his reinforcement updating in a day-to-day route choice learning context.

With respect to the second approach, empirical studies have shown such a simple strategy can give a good prediction in repeated choices (Erev and Barron, 2005). As a result, we assumed that travellers are compliant with the received information, proportional to their perceived information accuracy. This assumption states that the higher the reliability of the information system, the more frequently travellers would like to follow the system's route recommendation (compliance). This assumption has been confirmed in the literature (Ma and Di Pace, 2017) In case of non-compliance, travellers choose their routes by the probabilistic rules of Eqs. (11.4) to (11.6). It is a generalisation of RL in the sense that travellers respond to the behaviour of the information system based on their experienced information reliability. We argue that such a hybrid scheme in which travellers are assumed to learn the reliability of pre-trip information and respond in accordance with their experience is plausible.

11.2.2 Belief model based on the joint strategy fictitious play

The reinforcement-based learning paradigm states that travellers base their decisions on their past experiences. Good/bad outcomes in the past will be reinforced or avoided in the future. However, it has been revealed that humans may forecast future situations or the reactions of other players and act in accordance with uncertainty or limited/imprecise information (Fudenberg and Levine, 1998a, 1998b; Young, 2004). A plausible learning model is the FP model (Brown, 1951) in which a traveller assumes that all others make their respective decisions independently and make a choice which best responds to the empirical frequencies of choices made by all other travellers (Monderer and Shapley, 1996). Since the FP model assumes that each traveller has complete information about both the actions of their competitors and the associated payoffs, it is unrealistic for general n-player repeated games with numerous players and limited/incomplete information. To relax this strict assumption, the JSFP process (Marden *et al.* 2009) assumes that only the payoffs of joint actions are available for travellers, and their choices are based on the empirical frequencies of outcomes of all travellers' joint actions. In this context, a belief-based player would observe the outcome (ATIS reliability and experienced travel time) and update their future beliefs based on the received outcome and information. He/she then makes a choice given their beliefs in order to gain better future payoffs.

Let the expected payoff of a choice a_i at the beginning of day t be computed as traveller i's empirical average payoff up to day t-1:

$$\bar{u}^t_{a_i} = \frac{1}{t-1}\sum_{w=1}^{t-1} u^w_{a_i}, \quad \forall r, \forall i \in I_r, \forall a_i \in A_i \tag{11.9}$$

In the JSFP process, each traveller constructs his best-response choice set on day t as:

$$B^t_i = \left\{ a_i \in A_i : a_i = \arg\max_{a'_i \in A_i} \left[\bar{u}^t_{a'_i} \right] \right\} \tag{11.10}$$

The set of best-response choices represents the routes with maximum expected payoffs up to day t. We generalise the classical FP model by introducing an inertia term to capture travellers' unwillingness-to-change behaviour. A traveller then chooses an action based on the following rules (Marden *et al.* 2009):

$$\text{If } a^{t-1}_i \in B^t_i \Rightarrow a^t_i = a^{t-1}_i \tag{11.11}$$

$$\text{If } a^{t-1}_i \notin B^t_i \text{ then } \begin{cases} a^t_i \in \mathrm{R} \text{ and } (B^t_i) & \text{with prob. } 1-\sigma \\ a^t_i = a^{t-1}_i & \text{with prob. } \sigma(\text{inertia}) \end{cases}$$

where R and (B^t_i) is the result of a random choice in B^t_i. Equation (11) states that if a previous route choice remains in the set of best responses on a current day, this action is retained. Otherwise, the traveller randomly chooses a route among the ones in the best action set with probability (1–σ) or repeats his previous route choice with inertia probability σ (Srinivasan and Mahmassani, 2002). It is interesting to note that the JSFP model supposes individuals construct their beliefs based on the empirical average payoffs of actions and use best response strategies to make their choices. Moreover, traveller's 'habit' or 'inertia' behaviour is modelled by the inertia probability σ, which takes into account the traveller's risk aversion behaviour as reported in psychology and transportation research literature.

In order to take into account travel time information provided by the system, let us assume travellers may adjust their expected payoff estimations of available routes according to the predicted travel time given by the information system. Similar to the RL/ERL models, we assume the weighted updating of payoff perception where travellers use a linear model to combine past experience and travel time information. The perception of an expected payoff a_i on day t is updated, according to the following Eq. (11.12), instead of Eq. (11.9), as a weighted average of the empirical average payoff to day t-1 and the payoff $v^t_{a_i}$ supplied by the information system. The weight is determined by travellers' experienced system-information reliability (Eq. 11.8) and thus the update expected payoff is defined as:

$$\bar{u}^{t,updated}_{a_i} = \frac{1-g^t_i}{t-1}\sum_{w=1}^{t-1} u^w_{a_i} + g^t_i v^t_{a_i}, \quad \forall r, \forall i \in I_r, \forall a_i \in A_i \tag{11.12}$$

Hence, the more reliable the travel information system is, the higher its influence on the traveller's expected payoff. The above updating rule captures the impact of information accuracy on travellers' expected payoff, influencing their choice decisions. By following the same best-response strategy, the JSFP model with information reliability can be formulated to model travellers' strategy learning process under travel information.

11.2.3 Bayesian learning model

The Bayesian learning process is a belief-based strategy learning process which assumes each traveller updates his beliefs based on revealed outcomes and chooses actions which maximise his/her expected payoffs in the future. In the past, different learning models based on BL have been proposed (Jha *et al.*, 1998; Avineri and Prashker, 2005; Chorus *et al*, 2009). In our route choice experiment, each traveller has a prior belief regarding his route choice set to make a shorter time route choice. At the end of each day, the travel times and the correctness of the recommended route are revealed to each traveller. The Bayesian learning process can be described as follows (Young, 2004):

(a) Each traveller has a prior belief about his route choice set;
(b) At the end of each day, the payoffs and reliability of provided information are revealed. Each traveller updates his prior belief accordingly, based on the Bayesian update rule (described later);
(c) Each traveller chooses a response consistent with their belief.

It is worth noting that it has been proven that if the individual beliefs are compatible with the outcome of actual play, the Bayesian update and the expected payoff-maximisation principle lead to correct predictions in future plays and converge to Nash equilibria (Kalai and Lehrer, 1993). The term 'compatible' means the players' prior belief is positively related to the payoff being received in their next play.

Based on the expected payoff up to day t (Eq. 11.9), we assume the traveller i's belief $p^t_i(a_i)$ for the route a_i at the beginning of day t is determined by a logit model (exponential response rule) (McFadden, 1974; Avineri and Prashker 2006):

$$p^t_i(a_i) = \frac{e^{\theta \bar{u}^t_{a_i}}}{\sum_{a'_i \in A_i} e^{\theta \bar{u}^t_{a_i}}} \quad \forall r, \forall i \in I_r, \forall a_i \in A_i \tag{11.13}$$

where $\theta \geq 0$ is a sensitivity parameter, assumed to be identical across individuals. The higher the value of θ is, the more probable travellers would choose a route with higher expected payoffs. When $\theta = 0$ the travellers' route choice decision is not sensitive to the payoffs of routes. The behaviour principle is based on rational expectations behaviour, assuming that the higher the expected payoff of an action is, the more likely that it will be chosen.

Two Bayesian updating processes may be proposed. The first Bayesian updating model (BL) assumes travellers use prescriptive information to update their perception of the reliability of recommended routes provided by the system. Based on the perceived reliability variation, travellers comply with their route recommendations accordingly. The second Bayesian updating model is based on the model of Jha *et al.* (1998) (BL_{Jha}), which uses descriptive information to update travel time perception distributions in two stages: at the pre-trip stage, it is updated with their received information; at the post-trip stage, it is updated with their revealed experienced travel time. The disadvantage of this model is that it may violate bounded rationality behaviour (Simon, 1955) since this model needs complicated computations of means- and variances-updating of travel time perceptions.

11.2.3.1 BL based on prescriptive information

As the information system provides route recommendations to travellers, each traveller can decide to comply or not with his/her recommended routes based on his/her perceived reliability of the system, revealed *a posteriori*. The proposed BL model consists of updating the travellers' beliefs regarding the prescriptive information based on Bayes' rule. Thus, each traveller's belief is updated by combining their prior ones and their perception of the reliability of the information system. Each traveller's perception of reliability can be stated as '*given a recommended route h on day t, what is the probability that this route is the best route*?' As travellers have no prior information about how routes are recommended, direct computation of this conditional probability is infeasible. An alternative way is using Bayes' rule based on the travellers' empirical frequencies of the best routes up to day t. Then the belief based *on the reliability of the information system* at the beginning of day t+1 for any route $h \in A_i$, which can be updated by the Bayes' rule as

$$p_i^{t+1}(a_i^{t*} = h|y_i^t = h) = \frac{p_i^t(a_i^{t*} = h)\, p_i^t(y_i^t = h|a_i^{t*} = h)}{\sum_{h' \in A_i} p_i^t(a_i^{t*} = h')p_i^t(y_i^t = h'|a_i^{t*} = h')}, \forall h \in A_i \qquad (11.14)$$

where a_i^{t*} is the best action (with minimum travel time) for traveller i on day $t.y_i^t$ is the recommended route for traveller i on day t.

In case the denominator is zero,

$$p_i^{t+1}(a_i^{t*} = h|y_i^t = h) = 1/|A_i|, \quad \forall r, \forall i \in I_r, \forall h \in A_i \qquad (11.15)$$

where $|A_i|$ denotes the number of routes in traveller i's choice set A_i.

Note that the travellers' belief-updating regarding the reliability of the information system based on Bayes' rule makes the travellers predict the behaviour of the system with increasing accuracy as new outcomes are revealed.

The above Bayesian learning model for taking into account a traveller's belief regarding reliability

1. Computes empirical probability distribution of the best action at the beginning of day t+1 as

$$p_i^{t+1}(a_i^{t*} = h = \frac{1}{t}\sum_{w=1}^{t}\delta_{\{h=a_i^{w*}\}}, \quad \forall r, \forall i \in I_r, \forall h \in A_i \tag{11.16}$$

where $\delta_{\{h=a_i^{t*}\}}$ is an indicator function being 1 if the best route for traveller i on day t is h, and 0 otherwise.

2. Computes the conditional probability $p_i^t(y_i^t = h|a_i^{t*} = h)$ as the empirical frequency of the corrected recommendation up to day t:

$$p_i^t(y_i^t = h|a_i^{t*} = h) = \frac{\sum_{w=1}^{t}\delta_{\{y_i^w=a_i^{w*}\}}}{\sum_{w=1}^{t}\delta_{\{h=a_i^{w*}\}}}, \quad \forall r, \forall i \in I_r, \forall h \in A_i \tag{11.17}$$

In case the denominator is zero, $p_i^t(y_i^t = h|a_i^{t*} = h) = 0$.

3. Computes the belief regarding the reliability of the information system (Eq. 11.14) based on Eqs. (11.15) to (11.17). The choice probability over any route h for traveller i on day t+1 is then defined as

$$\begin{cases} \text{If } h = y_i^t & p_i^{t+1}(h) = p_i^{t+1}(a_i^{t*} = h|y_i^t = h) \\ \text{If } h \neq y_i^t & p_i^{t+1}(h) = \dfrac{e^{\theta\bar{u}_{a_i=h}^t}}{\sum_{h' \in A_i \setminus y_i^t} e^{\theta\bar{u}_{a_i=h'}^t}}\left[1 - p_i^{t+1}(y_i^t)\right]. \end{cases} \tag{11.18}$$

where $\bar{u}_{a_i=h}^t$ is carried out according to Eq. (11.9), and θ is a sensitivity parameter. $p_i^{t+1}(a_i^{t*} = h|y_i^t = h)$ is calculated by (11.14)–(11.17). Note the behaviour principle of (11.19) is based on a stimulus-response mechanism and probabilistic choice behaviour under stochastic payoffs.

Note that the upper part of Eq. (11.18) performs the conditional probability of choosing the recommended route y_i^t on day t+1. The choice probability for any other route is determined by the logit model (lower part of Eq. 11.18) based on the average payoff obtained by player i up to day t.

The proposed model assumes travellers use Bayes' rule to update their perception of the reliability of recommended routes, instead of updating travel time distribution. From a bounded rationality point of view, this model is more plausible compared to the complicated updating of the distributions of travel time perception of each route in their choice set.

11.2.3.2 BL based on the descriptive information (Jha *et al.*, 1998)

The model of Jha *et al.* (1998) assumes each traveller has prior knowledge about the payoffs of their route choice in the network. Travellers update their travel time perception at the beginning of each day after receiving travel time information (pre-trip updating). At the end of each day, their experienced travel times are revealed to them and they update their travel time perception *a posteriori* (post-trip updating). There are two stages of travel time perception updating on each day. Let u'_{a_i}, t denote a normal-distributed random variable representing the payoff on route a_i provided by the ATIS to individual i on day t, $u'_{a_i}, t \sim N(\lambda'_{a_i}, \sigma'_{a_i})$ where λ'_{a_i} and σ'_{a_i} is the mean and variance of u'_{a_i}, t, respectively. Similarly, we denote u''_{a_i}, t, a normal-distributed random variable, as the experienced payoff on the route a_i of traveller i on day t, $u''_{a_i}, t \sim N(\lambda''_{a_i}, \sigma''_{a_i})$ where λ''_{a_i} and σ''_{a_i} is the mean and variance of u''_{a_i}, t, respectively. We assume the perceived payoff $\widehat{u}^{,t}_{a_i}$ of the route a_i of traveller i on day t is a random variable following a normal distribution with mean $\widehat{\lambda}_{a_i}$and variance $\widehat{\sigma}_{a_i}$, i.e., $\widehat{u}^{,t}_{a_i} \sim N(\widehat{\lambda}_{a_i}, \widehat{\sigma}_{a_i})$. Let $\widehat{u}^{,t,pre}_{a_i}$ and $\widehat{u}^{,t,post}_{a_i}$ denote travel time perception updates after the pre-trip and post-trip updating stages, respectively.

The two-stage updating of travel time perception based on Bayes' rule is defined as:

(i) Pre-trip stage updating: the mean (expectation) of the payoff perception is updated by travel time information provided by the ATIS as

$$E(\widehat{u}^{,t,\,pre}_{a_i}) = \frac{\widehat{\lambda}^{t}_{a_i}\mathrm{Var}(u'^{,t}_{a_i}) + \lambda'_{a_i}, t\mathrm{Var}(\widehat{u}^{,t}_{a_i})}{\mathrm{Var}(u'^{,t}_{a_i}) + \mathrm{Var}(\widehat{u}^{,t}_{a_i})} \tag{11.19}$$

where $\mathrm{Var}(u'^{,t}_{a_i})$ and $\mathrm{Var}(\widehat{u}^{,t}_{a_i})$ are the variance of payoffs provided by ATIS and the variance of perceived payoff of route a_i of traveller i, respectively.

The variance of the perceived payoff after pre-trip travel time provision is updated as

$$\mathrm{Var}(\widehat{u}^{,t,\,pre}_{a_i}) = \left(\frac{\mathrm{Var}(u'^{,t}_{a_i}) + \mathrm{Var}(\widehat{u}^{,t}_{a_i})}{\mathrm{Var}(u'^{,t}_{a_i}) + \mathrm{Var}(\widehat{u}^{,t}_{a_i})}\right) \tag{11.20}$$

(ii) Post-trip stage updating, the mean (E) and variance (Var) of the perceived payoff for each route are updated as

$$E(\widehat{u}^{,t,\,post}_{a_i}) = \frac{\widehat{\lambda}^{t,\,pre}_{a_i}\mathrm{Var}(u''^{,t}_{a_i}) + \lambda''_{a_i}, t\mathrm{Var}(\widehat{u}^{,t,\,pre}_{a_i})}{\mathrm{Var}(u''^{,t}_{a_i}) + \mathrm{Var}(\widehat{u}^{,t,\,pre}_{a_i})} \tag{11.21}$$

$$\mathrm{Var}(\widehat{u}^{,t,\,post}_{a_i}) = \left(\frac{\mathrm{Var}(u''^{,t}_{a_i}) + \mathrm{Var}(\widehat{u}^{,t,\,pre}_{a_i})}{\mathrm{Var}(u''^{,t}_{a_i}) + \mathrm{Var}(\widehat{u}^{,t,\,pre}_{a_i})}\right) \tag{11.22}$$

After the post-trip stage updating at the end of day t, the mean and variance of travel time perceptions of any route a_i of travel i at the beginning of day t+1 is assigned as $\lambda_{a_i} = E(\widehat{u}_{a_i}^{t,\,post})$ and $\widehat{\sigma}_{a_i} = \mathrm{Var}(\widehat{u}_{a_i}^{t,\,post})$, respectively. The variance of the perceived payoff reflects the uncertainty of travel time perception.

The choice probability of route a_i for traveller i on day t is computed by the logit model based on the travel time perception after pre-trip updating $\widehat{u}_i^{t,pre}$ as

$$p_i^t(a_i) = \frac{e^{\theta \widehat{u}_{a_i}^{t,pre}}}{\sum_{a_i \in A_i} e^{\theta \widehat{u}_{a_i}^{t,pre}}} \quad \forall r, \forall i \in I_r, \forall a_i \in A_i \tag{11.23}$$

where θ is the sensitivity parameter, assuming the same across all travellers. Note that this equation is based on a similar behaviour principle to Eq. (11.19).

11.3 Modelling pre-trip choice behaviour

In a totally preventive approach, choice behaviour is usually schematised as a choice between a discrete number of alternatives. Choice models may be behavioural, if based on theoretical paradigms that attempt to simulate, in a more or less simplified fashion, the user's choice process, or non-behavioural, if they make no interpretative hypothesis and use generic mathematical tools (e.g., regressions or artificial neural networks). In path choice, behavioural models are preferable both due to the nature of the phenomenon and because they allow interpretation of the phenomenon through calibration of their parameters.

As emphasised above, in the hypothesis of the behavioural approach, it is necessary to construct a formal paradigm that simulates the results of choices according to the decisional process hypothesised[#].

The phases that lead to the choice may be schematised into:

(i) information acquisition and elaboration,
(ii) choice making.

Representation of the various phases requires the implementation of a system of interrelated models, especially *relevant information perception models*, *information use models*, and *alternative choice models*.

11.3.1 Modelling the cognitive process to acquire and use the information

The process underlying path choice may be structured into the following phases:

(a) the user acquires information;
(b) the user decides to use it;

[#]The decisional process is not necessarily a simulation of the user's choice process. The decisional process could be a pure abstraction of the analyst. However, it is explicitly represented, and the model is wholly congruent with such an explicit representation. Behavioural, in other words, does not mean that one simulates the real behaviour of users, but only that there is 'an' explicit paradigm of behaviour that allows us to reproduce the results of user choices through an explicit decisional structure.

(c) the user chooses according to the process of information acquisition and use.

If it is assumed that:

- l is the generic source of information (e.g., variable message panels) and I_l is the set of all available sources of information;
- U_{al} is the perceived utility associated with information acquisition at the time t in which it is supplied;
- U_{rl} is the perceived utility (*information reference*)
- a is a binary variable of value 1 if the user obtains information from resource l; 0 otherwise.
- r is a binary variable of value 1 if the user makes reference to information received from source l in the process of choosing his/her path.

then the user's decisional process can be schematised with a hierarchical structure: first, he/she chooses whether to acquire information and then decides to use it. The process could be interpreted by using random utility theory. The choice at the higher level requires that a utility (U_{al}) be associated with individual information sources:

$$U_{al} = S_m(a_{ml}h_m) + lL_{al} + e_{al}$$
$$U_{rl} = S_n(b_{nl}trip_n) + S_j(g_{jl}source_{jl}) + e_{rl}$$

where,

- h_m are latent variables that measure the cognitive involvement and capacity to process information;
- a_{ml} and b_{nl} are parameters that homogenise the latent variables;
- *trip* represents all the variables that characterise the trip undertaken up to that moment, i.e., the user's experience regarding the variables of the trip about to be made. Such variables include trip purpose, path characteristics and level of en-route congestion encountered;
- *source* represents variables characterising the information source (e.g., purpose, accuracy etc...);
- e_{al} and e_{rl} are random residuals of perceived utilities.

In this context it is hypothesised that the traveller uses information l, i.e., receives information ($a = 1$) and uses it ($r = 1$) if utilities U_{al} and U_{rl} exceed two threshold values S_{al} and S_{rl}, that is:

$$acquisition = 1 \text{ if } U_{al} \geq S_{al}$$
$$reference = 1 \text{ if } U_{rl} \geq S_{rl}$$

where,

S_{al} is the threshold relative to the process of acquisition and may be expressed as a random variable consisting of a systematic quantity q_{al} and a random residual x_{al}.

S_r is the threshold relative to the process of information use and may be expressed as a random variable consisting of a systematic quantity q_{rl} and a random residual x_{rl}.

Modelling the phenomenon may yield the probability of acquiring information for the probability of using it, conditional upon having acquired it.

$$P[a,r] = Pr[U_{rl} \geq S_{rl}] \cap Pr[U_{al} \geq S_{al}] = P[r] \times P[a/r]$$

where,

$$P[r] = \Pr[U_{rl} \geq S_{rl}] = \Pr[V_{rl} + e_{rl} \geq q_{rl} + x_{rl}] = \Pr[V_{rl} - q_{rl} \geq x_{rl} - e_{rl}]$$

$$P[a/r] = \Pr[U_{al}S_{al}] = \Pr[V_{al} + e_{al} \geq q_{al} + x_{al}] = \Pr\left[V_{al} - q_{al}{}^{3}x_{al} - e_{al}\right]$$

Hypothesising that random residuals (x_{rl}, e_{rl}, x_{al}, e_{al}) are distributed like identically and independently distributed Gumbel random variables, the probabilities may be formulated as follows:

$$P[r] = \frac{\exp(V_{rl})}{\exp(V_{rl}) + \exp(\theta_{rl})}$$

$$P[a/r] = \frac{\exp[V_{al}]}{\exp[V_{al} + \lambda \ln(\exp V_{rl} + \exp\theta_{rl})] + \exp[\theta_{al}]}$$

One thereby obtains a wholly similar mathematical formulation to that of a hierarchised logit model. The above model allows us to estimate the propensity to act according to the information supplied. This must therefore be supplemented by a model that simulates choice following the information process.

In the event of prescriptive information, it is reasonable to think that P is precisely the degree of propensity to follow the advice or suggestions supplied. Having estimated the probability of diversion for alternative paths, the true probability may be obtained as:

$$P[\text{diversion}] = P[\text{alternative path}]\ P[a,r]$$

In the case in which information is not prescriptive, the user forms his/her own expectations, defines possible alternatives and chooses among them. In this scenario, the user does not receive advice but only more detailed information on the state of the network and his/her behaviour may be schematised as if the perception error of path costs were reduced. The information does not increase the propensity to diversion, brings the user closer to deterministic behaviour and, if there are minimum-cost alternative paths, the user will decide to modify his/her preventive choice. A straightforward, effective way to take account of the phenomenon is to define a variable that intervenes within the systematic utility of the available alternatives.

11.3.2 Modelling pre-trip choice behaviour

As clarified above, preventive path choice involves the user choosing among a discrete number of alternatives. In this sense, the specification of a choice model requires:

(a) definition of the choice alternatives,
(b) definition of the choice set perceived by each decision-maker (class of decision-makers),

(c) definition of the modelling solution:
  (i) as approach
  (ii) as mathematical formalisation of the problem
  (iii) as choice of the representative variables of the phenomenon

(a and b) definition of the choice alternatives and the choice set
As regards the definition of the choice alternatives, the hypothesis usually adopted is that the user, prior to making the trip, considers the various paths available to reach the set destination. However, to reduce the computational complexity of the problem, one prefers to adopt heuristic approaches which may be classified into four types.

(i) *Exhaustive approaches.* All existing paths on the network in question, whether elementary or without circuits, are to be considered admissible.
(ii) *Selective approaches.* Only some paths amongst those topologically admissible are hypothesised as admissible.
(iii) *Mixed approaches (exhaustive and selective).*
(iv) *Modelling approaches that estimate the probability that a choice set is the choice set perceived by the user.*

(c) definition of the modelling solution
The following problems can be distinguished:

(i) *which approach to simulate the phenomenon,*
(ii) *which theoretical paradigm to use,*
(iii) *which variables characterise the phenomenon.*

(i) *which approach*
As regards path choice and in a fairly preventive choice context, approaches may be behavioural or non-behavioural and aggregated or disaggregated. The approach usually adopted in simulating preventive choices is aggregate behavioural (in variables representing the phenomenon and in the experimental data available for model calibration) or disaggregate.

While disaggregate choice models on the one hand introduce greater complexity in the phase of model specification and calibration, they allow better and more realistic interpretation of the phenomenon. The better interpretation is due to the possibility of specifying more structured modelling formulations both in specifying the distribution function of the random residuals and in specifying systematic utility functions. The greater realism stems from direct observation of user choice behaviour. Contrasting with the above benefits, it cannot be overlooked that disaggregated information is difficult to obtain, costly, and often unreliable. Another aspect not to be underestimated is the need to simulate, both in the calibration phase and in the model application phase, the choice set of the individual users, that is, the need to explicitly enumerate potential paths.

(ii) *which theoretical paradigm*
Random utility theory which is the most consolidated theoretical paradigm in applications both due to the good predictive capacity of the models obtained from it, and the simplicity of calibration and application:

I = the choice set
$P[j/I]$ = the probability of choosing alternative j belonging to set I
$\boldsymbol{p}$ = the vector of choice probabilities, of dimension (m × 1), with elements $P[j/I]$;
$\boldsymbol{U}$ = the vector of perceived utility values of dimension (m × 1), with elements U_j;
$\boldsymbol{V}$ = the vector of systematic utility values of dimension (m × 1), with elements V_j;
ε = the vector of random residuals, of dimension (m × 1), with elements ε_j;
$f(\varepsilon)$ = the joint probability density function of the random residuals with the variance-covariance matrix Σ;
$F(\varepsilon)$ = the joint distribution function of random residuals.

The choice model may be posed as a choice map, which associates to each systematic utility vector $\boldsymbol{V}$ a choice probability vector $\boldsymbol{p}$:

$$p = p(V) \quad \forall V \in E^m$$

where,

$$P[j/I] = Pr[V_j - V_k > \varepsilon_k - \varepsilon_j] \quad \forall k \neq j; j, k \in I^i$$

If the variance–covariance matrix is not zero, $\Sigma \neq 0$, and is not singular, $\Sigma \neq 0$, the probabilistic models are obtained and the choice map $p = p(V)$ is a function. Moreover, if the joint probability density function of the random residuals, $f(\varepsilon)$, does not depend on the vector of systematic utilities V, the model will be termed additive.

(iii) *which mathematical model within the chosen theoretical paradigm*
The possible modelling solutions depend on the hypotheses about the distribution functions of the random residuals. It is a function of the characteristics of the phenomenon which one wants to simulate. In the context of the random utility models used in modelling path choices, the most widely used modelling formulations are:

(a) *Multinomial Logit (MNL),*
(b) *C-Logit,*
(c) *Multinomial Probit,*
(d) *Mixed-models,*

(a) The Multinomial Logit model (MNL)

For the MNL model, it is hypothesised that the random residuals ε_j are independently and identically distributed (i.i.d.) according to random variables of extreme value of type I with zero mean and parameter θ, also known as Gumbel variables.

Starting from the symbolic expression of the choice probability of a generic alternative j, explicit formulation of the same probability may be obtained.

$$P[j/I] = Pr[V_j - V_k > \varepsilon_k - \varepsilon_j] = Pr[U_j = \max_{k \neq j}[U_k]$$

$$P[j] = \frac{\exp(V_j/\theta)}{\sum^{m} \exp(V_k/\theta)} \quad \forall k \neq j; j,k \in I_i$$

The MNL is a homoscedastic model and is characterised by covariances among the perceived utilities of the null alternatives. The property of independently distributed residuals implies that the perceived utilities are not correlated. This hypothesis is reasonable when the available alternatives are sufficiently different, that is, the paths do not present many shared links (sub-paths).

(b) The C-Logit model

Through an appropriate change to the Logit model, it is possible to specify a closed-form path choice model which partly resolves the limits of the Probit model. We are referring to the *C-Logit* model (Cascetta, 2001; Cascetta *et al.*, 1996), which resolves the problems arising from the property of independence of the irrelevant alternatives of the Logit model by introducing an overlapping factor that reduces the systematic utility of a path according to its degree of overlap with other alternative paths.

The mathematical formulation of the model is as follows:

$$p[k/osdm] = \frac{\exp((-C_k - CF)/\theta)}{\sum_{h \in I_{odm}} \exp((-C_h - CF_h)/\theta)}$$

The term CF_k, is an inverse measure of the degree of independence of a path. It assumes the value zero if all the links of path k do not belong to any other path. The greater the number of paths that share the major links of path k, the greater is the term. The underlying concept is that the probability of choosing strongly overlapping paths is reduced.

Although it may be specified in various ways, the specification usually adopted is as follows:

$$CF_k = \beta_o \ln\left(1 + \sum_{h \neq k} \frac{X}{(X_h X_k)^{1/2}}\right)$$

where,

- X_h is a cost attribute of path h;
- X_k is a cost attribute of path k;
- X_{hk} is the cost attribute of links belonging to both paths

As for the Probit model, attribute X may differ from the attribute(s) used in specifying the vector of path costs. By the same token, it is recalled that these hypotheses ensure that the stochastic loading function of the network is monotone and non-increasing with regard to the congested costs of the link and has a symmetric Jacobian, ensuring the sufficient condition for the uniqueness of the stochastic equilibrium.

(c) The Multinomial Probit model

The Probit model is based on the hypothesis that utilities associated with the single alternatives are distributed according to a multivariate normal (*MVN*) random variable with a mean equal to V_j, any variances and covariances and, nevertheless, equal to those of the residuals ε_j, characterised by a dispersion matrix Σ. ($\boldsymbol{U} = \text{MVN}(\boldsymbol{V}, \boldsymbol{S})$)

$$E[U_j] = V_j$$
$$\text{Var}[U_j] = \sigma_j^2$$
$$\text{Cov}(U_j U_h) = \sigma_{jh}$$

The multivariate normal probability density function of the vector ε is supplied by:

$$f(\varepsilon) = \left[(2\pi)^m \det\left(\sum\right)\right]^{-1/2} \exp\left[-1/2\varepsilon^T \sum^{-1} \varepsilon\right]$$

The probability of decision-maker i choosing alternative j may be expressed formally as the solution of an integral with M dimensions

$$p^i[j] = \int_{U_1^i < U_j^i} \cdots \int_{U_j^i = -\infty}^{+\infty} \cdots \int_{U_m^i < U_j^i} \frac{\exp\left[-1/2(\mathrm{U}^i - \mathrm{V}^i)^T \sum^{-1} (\mathrm{U}^i - \mathrm{V}^i)\right]}{\left[(2\pi)^m \det(\sum)\right]^{1/2}} dU_1 ... dU_m$$

The Probit model does not lend itself to the analytical calculation of choice probabilities since a closed-form solution of the integral is not known. To calculate Probit choice probabilities one must therefore resort to approximate methods. It is possible to distinguish numerical approaches based on area methods and non-numerical approaches based on approximate hypotheses or on simulation (cit…)

(d) The Mixed models

Mixed-logit models allow us to simulate any type of correlation among the utilities of available alternatives, to simulate the difference in perception among users and do not constrain the model to only multivariate normal distribution. Indeed, a mixed-logit may approximate very different choice contexts, it may be simply heteroscedastic, it may be homoscedastic with non-zero covariances, or heteroschedastic with any covariance. As also demonstrated by McFadden and Train (2000), the appropriately specified mixed logit model can approximate any random utility model.

Let M be the available alternatives, $P[j]$ the choice probability of alternative j, Y($\boldsymbol{X}$, $\boldsymbol{b}$, W) the function that calculates the choice probability $P[j]$, and let $\boldsymbol{X}$, $\boldsymbol{b}$, W, be, respectively, the attributes of systematic utility, coefficients of the attributes of systematic utility and the covariance matrix of the probability density function of the random residuals.

If Y is a multinomial Logit model, the following may be expressed:

$$P[j] = \mathrm{Y}(\mathbf{X}, \mathbf{b}, \mathrm{W})_{\mathrm{Logit}} = \frac{\exp\left[V_j(\boldsymbol{\beta}, \boldsymbol{\theta})\right]}{\sum_{m=1}^{M} \exp\left[V_m(V_m(\boldsymbol{\beta}, \boldsymbol{\theta})\right]}$$

where *b* and *q* (their ratio) assume constant values, which may be estimated in the calibration phase. If the parameters cannot be hypothesised as constant and should they assume discrete values, with *f*(***b***/*q*) the relative probability function, then P[*j*] can be expressed as follows:

$$P[j] = Y(\boldsymbol{X}, \boldsymbol{b}, W) = \sum_{c=1}^{C} \left[\frac{\exp\left[V_j(\boldsymbol{\beta}/\boldsymbol{\theta})_c\right]}{\sum_{m=1}^{M} \exp\left[V_m(\boldsymbol{\beta}/\boldsymbol{\theta})_c\right]} \right] \cdot f(\boldsymbol{\beta}/\boldsymbol{\theta})_c$$

Therefore the choice probability is the average of the Logit choice probabilities with a certain vector of parameters, weighted on the probabilities of observing the same vector of parameters. Hypothesising that the vector of parameters is a continuous variable, the choice probability may be generalised in the following integral of dimension equal to the dimension of the vector (*b/q*):

$$P[j] = Y(\boldsymbol{X}, \boldsymbol{b}, W) = \int \left[\frac{\exp\left[V_j(\boldsymbol{\beta}/\boldsymbol{\theta})\right]}{\sum_{m=1}^{M} \exp\left[V_m(\boldsymbol{\beta}/\boldsymbol{\theta})\right]} \right] \cdot f(\boldsymbol{\beta}/\boldsymbol{\theta}) \cdot d(\boldsymbol{\beta}/\boldsymbol{\theta})$$

This may be further generalised:

$$\mathrm{P}[j] = \int L_j(\boldsymbol{\beta}/\boldsymbol{\theta}) \cdot f(\boldsymbol{\beta}/\boldsymbol{\theta}) \cdot d\boldsymbol{\beta}$$

The above mathematical formulation represents the mathematical formulation of the mixed logit model: 'mixed' because in statistics the weighted mean of functions is usually termed 'mixed'; 'logit' because the function whose weighted mean is calculated is the function that supplies the choice probabilities of a multinomial Logit model. In operative terms, the integral will always have to be solved with simulation methods, and the computational burdens are strictly linked to the hypotheses on L_j and/or *f*(*b/q*).

The widely used are random coefficient models and error components models.

Random coefficient models arise to simulate the heterogeneity of user choice behaviour following the different perceptions that each user (albeit being part of a homogeneous class of users) has of the utility associated with the individual alternatives.

They hypothesise that the coefficients of the attributes of systematic utility are distributed among the users (of a homogeneous class) according to a known probability function.

Error component models are based on the hypothesis of structuring the utility function into three distinct terms: a systematic utility constant for all the users of the class in question, a random utility function of a subset of attributes, and a random residual identically and independently distributed among alternatives. The model's rationale is to hypothesise an identical perception error for all utilities and for all users (logit type) and to introduce a non-homogeneous error distribution only in specific attributes.

(iv) *which variables characterise the phenomenon*
The effectiveness and efficiency of a random utility model are strongly influenced by the specification of systematic utilities. If variables with little significance can make a model poorly representative, on the other hand, variables which are complex to calculate can lead to models that are difficult to apply in real contexts. Specification of utility functions consists of specifying the functional form of utility in defining characteristic variables.

However, the problem of path choice is particularly complex due to many factors that influence the decision-maker (for an analysis of the problem, see Abdel-Aty, Kitamura and Jovanis, 1997).

The characteristic variables of the transport supply system, also known as the level of service, are all those variables that measure the disutility of a trip. What fall into this category are travel times and monetary travel costs. Alongside continuous variables, it is possible to imagine *Boolean* variables that simulate the existence, or otherwise, of specific characteristics of the supply system crossed (e.g., parking difficulty, restricted traffic zones).

The characteristic variables of the decision maker, also known as socio-economic variables, assume a very important role in the specification of systematic utility functions. Although demand models are calibrated and applied to homogeneous classes of users, homogeneity concerns macro-characteristics of decision makers such as origin (or destination) zones or trip purpose. Subdividing the population into homogeneous strata according to socio-economic characteristics such as gender, age or professional position would involve much more costly and complex sample surveys. Socio-economic variables in part allow such problems to be overcome, segmenting the population sample interviewed on the basis of variables that take the value 1 if one belongs to the socio-economic category in question. The choice of variables depends as much on information that can be obtained through a motivational survey as on information available to extend the calibrated models to the whole universe.

The characteristic variables of the alternative, also known conceptually as specifics of the alternative, measure all that influences choice behaviour but that the analyst is unable or cannot measure. In the case of path choice, given the number of alternatives, they may be used for some paths that present specific characteristics but that influence the user's choice process.

In general, path choice is influenced by various factors that may be classified into the following:

(a) level of service attributes (trip times and costs);
(b) attributes depending on infrastructure characteristics (or services), traffic conditions, and characteristics of the surrounding environment that can affect the working order of the same infrastructure;
(c) the user's socio-economic characteristics;
(d) the user's familiarity (Adler and McNally, 1994; Bonsall *et al.*, 1997; Lotan, 1997) with the existing infrastructure (services). We may distinguish topological familiarity and functional familiarity. The former envisages knowledge of physical and geometric characteristics of the transport network, and the latter knowledge of functional characteristics as possible operating conditions change. The concept of familiarity is important for the simulation of the choice process and, especially, for the choice of the interpretative and modelling paradigm of the phenomenon.
(e) characteristics of the trip undertaken (e.g., purpose, trip timetable, transportation mode, and destination);
(f) characteristics exogenous to the infrastructure and possible boundary conditions (e.g., meteorological conditions, day of the week, etc.);
(g) presence of information for users and user familiarity with the above information. In the presence of a system of user information, what must be taken into consideration are not only attributes linked to the road and the driver but also attributes of the same information (type, content, relevance in space and time), the capacity to process information, perceived reliability, and integration of information.

The universally adopted functional form is represented by a linear combination of variables (X_{kj}) through coefficients of homogenisation (β_j): $V_k = \sum_j \beta_j X_{kj}$. Linear formulation allows computational problems connected with model calibration to be appreciably simplified.

In studying long-distance trips, the hypothesis of systematic linear utility could prove an approximation that may lead to underestimating or overestimating the weight of certain attributes in calculating user choice probabilities. Simulation of non-linear phenomena (if existing) allows asymmetric choice probability functions and, enables the definition of more realistic substitution relationships among the attributes, should improve the reproductive capacity of the model, and reduce the effect of modal constants.

Simulation of non-linear phenomena may be obtained through explicit transformation of some Box-Cox variables (Box-Cox, 1964; Gaudry, 1981). Box-Cox transformations are monotone transformations of strictly positive variables characterised by the parameter λ. Application of the transformations may concern one or more attributes of the systematic utility function and may, or may not, involve all the alternatives. In turn, the values of the parameter λ of an attribute j may vary as the mode of the alternative changes. Typically, transformations are applied to the level of service attributes, but there is no reason why they cannot be extended to attributes related to the duration and type of activities to be carried out at the destination.

11.4 Modelling en-route choice behaviour

In a context in which the choice behaviour is preventive/adaptive, it is necessary to clarify what is preventive and what can be adaptive. As already seen above, two possible approaches may be envisaged:

'reference path',
'strategy'.

The first approach envisages specifying models of preventive choice and models that simulate the possibility of diversion at any specific node of the network. As regards preventive choice, the most widely adopted modelling solutions are based on random utility theory and eventually paired with generalised models of cost perception or with *information compliance* models.

As regards adaptive choices in the absence of user information, the models are analogous to solutions for preventive behaviour except that they use different generalised costs (instantaneous) or variables directly observable by the user and representative of transport system working conditions. In the presence of user information, it is possible to have recourse to the combined use of information compliance models and *switching* models (if the information is prescriptive) or a discrete choice model (if the information is descriptive).

Under the hypothesis of a '*strategy*' approach, the treatment of the problem, though it can be subdivided into a preventive choice and subsequent adaptive choices, is specific, and may be tackled by introducing the concepts of *hyperpaths*, of preventive choice among *hyperpaths* and adaptive choices at specific diversion nodes within the preventively chosen hyperpath.

11.4.1 Strategy approach

In the case of the strategy approach, the underlying hypothesis is that users do not know at the beginning of the trip all the information required to be able to decide their trip in its entirety. Thus users do not choose a predetermined path but rather a trip *strategy that* allows them to obtain the average trip cost deemed lowest. A strategy is usually defined by: (a) predefined choices and (b) behavioural rules to follow *en route* in the presence of random events or events not known to the user *a priori*. In this classification, two types of choice behaviour may be distinguished:

11.4.1.1 Preventive choice behaviour

This is what the user adopts before beginning the trip, comparing possible alternative strategies and choosing one on the basis of expected characteristics or attributes.

11.4.1.2 Adaptive choice behaviour

This is what the user follows during the trip, adapting to random or unpredictable events. The type of adaptive choice behaviour and the set of alternatives to which it applies define a trip strategy. Simulation of adaptive choices requires specification of the adaptive behavioural model, namely:

- the points in which the user can modify his/her own path (diversion nodes);
- what decisions he/she may take (e.g., no decision because the path is advised, alternative path, alternative strategy);
- what behaviour to adopt (e.g., reaction to the event, the path up to the destination, path to the next diversion node).

A consolidated approach to discontinuous transport supply systems consists of simulating the trip strategy (preventive choice alternative) using a sub-graph called the *hyperpath* of the transport service in question. All strategies that exclude adaptive choices are termed *simple hyperpaths*, while strategies that include diversion nodes, or adaptive choices, are termed *compound hyperpaths.*

To each diversion node, it is possible to associate feasible alternatives which may be used to continue the trip. According to the type of problem and the type of transport service concerned, various solutions are possible:

As with mixed preventive/adaptive path choice models, a complete specification of the choice model requires conceptually:

(a) definition of alternatives,
(b) definition of their set,
(c) definition of the model that simulates choice.

The considerations made for preventive choice models hold good for the definition of the alternatives and the choice set. In particular, starting from all the possible alternatives, we may distinguish:

1. an *exhaustive* approach, considering all strategies admissible (or the *hyperpaths* that they represent). This approach is typically associated with implicit enumeration of the *hyperpaths* and therefore requires specific algorithms for assigning demand flows to the transport network. It can be adopted with aggregate choice models and additive path costs.
2. a *selective* approach, considering admissible only *hyperpaths* that satisfy some conditions.

As regards modelling formalisation, the most consolidated approaches are based on random utility theory and such formalisation may thus be brought back to the definition of the functional form of the random utility model and its attributes. Also in this case it is assumed that for each hyperpath h, belonging to the set I_{odm} of hyperpaths that connect the pair (o,d), the perceived utility of the hyperpath U_h has negative mean (systematic utility) V_h equal to the mean cost G_h

$$U_h = V_h + \varepsilon_h = -G_h + \varepsilon_h \quad \forall h \in I_{odm}$$

In the same way, the mean cost of the *hyperpath* G_h may be expressed as the sum of an additive part G_h^{ADD} and of a non-additive as follows: $G_h = G_h^{ADD} + G_h^{NA}$.

Finally, as regards the model of choice among available alternatives (*hyperpaths*), it may be generally expressed as the probability $p_H[h]$ of *hyperpath h* being that of maximum perceived utility:

$$p_H[h] = Pr[-G_h + \varepsilon_h \geq -G_{h'} + \varepsilon_{h'}] \quad \forall h', h \in I_{odm}$$

For the *hyperpath* choice model, two approaches may be followed: that of deterministic utility, which assigns all the demand to the *hyperpath* of minimum generalised cost, or random utility models, typically Logit and Probit. However, in the former case, the independence of irrelevant alternatives (*IIA*) problem would represent itself even more significantly in the case of *hyperpaths* that may include many overlapping lines. Alternatively, one can use a Probit model with a structure of the variance-covariance matrix of the residuals similar to that described for the preventive path choice. Irrespective of the model to be used, once the *hyperpath* choice probabilities have been calculated, it is possible to obtain probabilities of using a certain path k by simulating diversion probabilities.

Strategy approaches are usually adopted to simulate the demand-supply interaction of public transport systems. For the latter, modelling diversion probabilities is linked to transport supply characteristics (frequencies) and to simplified but realistic behavioural hypotheses (the user boards the first vehicle that arrives). In actual fact, there are no applications of strategy approaches in the case of individual transport. The difficulties are connected to the following three problems:

- *Definition of alternative strategies.* If in a public transport system possible strategies could easily be determined from the topological and functional definition of paths of lines, in an individual transport system the network dimension makes such a task much more burdensome. This becomes a major problem at the moment in which the system of supply is very complex and a model needs to be specified on the basis of observed choice behaviour. In this case, the strategies from which the user can choose need to be defined, but the perception that the user may have of the strategies may differ substantially from the idea constructed by the analyst. In the study of public transport services, the two concepts almost coincide. A possible approach consists of identifying a limited number of diversion nodes and defining strategies with regard to them.
- *Specification of behaviour at diversion nodes in the absence of user information.* At each diversion node, the user chooses from the possible alternatives available within the chosen strategy. The problem may be boiled down to one of discrete choice and tackled with typical preventive choice models, if possible associated with perception models of generalised transport costs or to models of cost adjustment in relation to the available alternatives downstream of the diversion node (with regard to the remaining share of the trip). Also in this case, we may use either (i) the same attributes of level that the user employs in the process of choosing strategies (in the phase of preventive choices), or (ii) instantaneous level of service attributes or (iii) *actual* attributes. In the first case (i) adaptive choices within a strategy may be estimated preventively in the same way as discontinuous systems of transport supply; in the case of instantaneous or actual attributes (ii and iii), the

choice depends on system working conditions at the time of the adaptive choice. Using instantaneous times and costs is reasonable for systematic users who, as the type of congestion that they encounter along their path changes, are able to estimate travel times and costs downstream of the diversion node. Indeed, imagine that cumulative experience allows the systematic user to define a finite number of configurations of times and costs, that each configuration is repeated over the course of a year (periodically yet with different frequencies), and that the user manages to recognise the individual configuration during his/her trip and can construct an estimate of travel times and costs downstream of the diversion node. Alternative solutions could be adopted, specifying models based on vehicle flow characteristics that the user observes in the diversion node (link prior to the diversion node), that is, on the mean flow characteristics that the user encounters before reaching the diversion node.

- *Specification of behaviour at diversion nodes in the presence of user information.* The presence of user information may simplify the simulation of choice behaviour insofar as it can reduce user uncertainty regarding the travel times of each alternative. On the other hand, at each diversion node information has to be provided for all strategies that concern the same node and hence for all origin-destination pairs involved.

11.4.2 Reference path approach

Unlike the strategy approach, preventive behaviour can reasonably be imagined which leads users to choose a path (deterministic behaviour) or associate choice probabilities to more than one alternative path. With regard to such choices, and at specific nodes (even all of them), the user has the possibility to reconsider the path chosen (or rather, the share of the remaining path) and take a diversion towards an alternative path, i.e., associate choice probabilities to more than one alternative path.

The problems connected with the above approach are as follows:

(a) simulation of preventive choices;
(b) definition of diversion nodes;
(c) definition and modelling of behaviour at each diversion node.

Simulation of preventive choices within preventive/adaptive behaviour may be tackled by using the same modelling solutions proposed for totally preventive behaviour. Aggregate and disaggregate models may be envisaged, level of service variables may be instantaneous or actual in the case of simulation of within-period dynamics or obtained within equilibrium assignment procedures.

As regards diversion nodes, they may be defined as those nodes at which users may change the path they choose in advance. All network nodes from which alternative paths leave to reach the destination of the trip underway may be termed possible diversion nodes. Clearly, therefore, the set of diversion nodes varies with the variation in the o-d pair. Their definition may be:

- *exhaustive*, if all the possible diversion nodes for each o-d pair are considered;
- *selective*, if a subset of all possible nodes is considered according to empirical criteria. Indeed, one could think of excluding all those nodes close to the trip

origin or trip destination, all the nodes from which paths lead that cause distancing from the reference path over a certain threshold, all the nodes leading to possible alternative paths below a certain threshold or, finally, including all the nodes in which a user information system can be installed.

The definitive choice cannot but derive from a compromise between actual needs and computational practicality.

(**c**) At each diversion node, it is hypothesised that the generic user reconsiders his/her choice. Three possible modelling approaches can thus be hypothesised, as detailed in the subsections below:

(i) simulation of choice as a choice among a finite number of alternatives;
(ii) simulation as a 'switching' problem;
(iii) simulation of the cognitive and choice process.

11.4.2.1 Holding models

At the generic diversion node the user, on the basis of:

- consolidated experience (e.g., the day before),
- direct observation of network working conditions,
- possible information supplied exogenously,

updates the cost of the rest of the trip, defines a set of possible alternatives, associates a generalised cost to each of the alternatives, and chooses the least-cost alternative. The alternative chosen (it may also coincide with the preventive choice) becomes the new preventive choice which may be further modified at a subsequent diversion node.

The choice mechanism is entirely analogous to that of the preventive choice at the trip origin and thus may be implemented according to well-known modelling approaches. Clearly, however, such an approach may be very burdensome from a computational standpoint, requiring recalculation of minimum paths for each diversion node and, under a hypothesis of within-period non-stationarity, in each time interval in which the simulation horizon was discretised.

Special attention must be paid to the variables that affect choice and to the role of information to the user. As regards the variables, a separate consideration must be made for socio-economic and level of service variables. While the former, as emphasised above, appreciably affects adaptive choices (e.g., gender, age, and the degree of familiarity with the network), the latter represents the chief reason for a possible path diversion. In the absence of information, the user chooses according to his/her own experience and it is logical to suppose that the generalised trip costs differ from those perceived at the beginning of the trip (there would be no reason to change one's choice) and it would be thus preferable to use 'instantaneous' costs.

User information, in turn, may generate various modelling solutions:

- If the information is prescriptive, the problem can be considered one of binary *compliance* (following, or otherwise, information, to be tackled at a later stage) or as a problem of choice between proposed alternatives according to generalised perceived costs. Clearly, there are many problems to tackle. If the alternative is supplied exogenously (e.g., through a user information system), it must be a *better* alternative and it imposes a calculation of minimum paths starting from the diversion node towards the destination of the generic user. In turn, the calculation of minimum paths with the level of service attributes must be performed: *actual* costs or instantaneous costs. In the former case, the whole system needs to be simulated in advance; in the latter case, the information supplied may be unreliable and could reduce the user's degree of confidence towards the information system. If several alternatives are supplied exogenously, the problem is analogous, with the difference being that both the choice model and the information will be more complex to define.

- If the information is descriptive but qualitative, it may be supposed that the user will have a greater awareness of network working conditions and alternative paths. The derived choice behaviour may be simulated by using discrete choice models similar to preventive models possibly associated with models simulating the perception/availability of alternatives and perception of costs. In this case, the problem of 'actual cost' is not posed insofar as the descriptive information is defined starting from instantaneous costs.
- If the information is descriptive but quantitative, though filtering the information with his/her own experience, the user may be assumed to associate the information supplied with possible alternatives, exclude the alternatives where there is no information supplied and choose according to a preventive choice mechanism.

From a strictly modelling perspective, it may be hypothesised that the choice at each node is independent of choices already made at previous nodes, or that choices in a node are influenced by having already changed one's preventive choice at a previous node. It seems clear that there are various modelling solutions for simulating the two behaviours. A simple solution may be obtained by introducing a variable of inertia, representing the low propensity to change one's choice at a node if it has already been changed previously. More complex solutions may be obtained by modelling, explicitly simulating correlations between choices made in subsequent nodes. It is like observing choices between nodes, or 'longitudinal' ('panel-data') choices, and choices in the nodes, or 'transversal' ('cross-sectional'), and trying to interpret the phenomenon in its globality. In this perspective, Cross-Nested-Logit, Probit, Mixed-Logit or Ordered-Logit models appear the most appropriate modelling formulations. Lastly, it should not be overlooked that the specification of such models requires many test observations which are hard to acquire.

11.4.2.2 Switching models

A possible modelling alternative consists of implementing switching models, possibly combined with a limited rationality approach. Switching models allow us to simulate two distinct phenomena: the probability that a user changes his/her choice; and the probability of choosing a proposed alternative.

The behavioural mechanism envisages that user i will not change his/her choice (at time t and node j) as long as the travel time saved (TTS_{jt}) remains within an interval of indifference. The saving is calculated as the difference between the travel time on the preventive path (from node j to destination d) and the travel time of the best path (jd). The interval has a lower limit (zero) and an upper limit (IBR_{jt}). The upper limit is expressed as the maximum value assumed by the two thresholds:

- *threshold 1*: representing the threshold below which the user does not change his/her choice;
- *threshold 2*: representing the percentage of total travel time beyond which it is maintained that the user takes into consideration the idea of changing his/her preventive choice.

The problem consists of estimating the probability that, at time t, user i changes his/her choice before reaching node j**. This probability indicated below as the probability of 'switching', may be expressed as the probability that the time saved exceeds the maximum of the two random thresholds.

Having assigned a class of users i, the generic user preventively makes the path choice to travel from the origin o (diversion node j) to destination d. Assuming inter-period stationarity, the choice occurs on the basis of the level of service attributes associated with the available paths that the user perceives at time t (Following a day-to-day approach, the user can, instead, change his/her choice, updating the perceived costs of available paths and updating his/her choices.).

Having made the choice, knowing the path chosen or the choice probability, the following may be defined:

- $TTC_{k,jt}$ is the perceived travel cost associated with the chosen path k (or paths K, with non-zero choice probability) by user i, at time t, at node j, and destination d;
- TTB_{jt} is the minimum path travel cost, at time t and relative to node j and destination d.

The difference between the two costs supplies the savings which may be obtained by changing path k following the adoption of the minimum path:

$$TTS_{k,jt} = TTC_{k,jt} - TTB_{jt} \ \forall k \in I_K$$

User i changes path if, and only if, the saving in terms of travel costs exceeds an indifference threshold (IBR_{jt}). The problem may be interpreted according to a deterministic or probabilistic approach.

Following a deterministic approach, the switching phenomenon may be schematised by means of a binary variable, f_{jt}, which assumes the value 1 if the user changes his/her preventive choice otherwise -1:

$$f_{k,jt} = -1 \quad if\ 0 \leq TTS_{k,jt} \leq IBR_{jt}$$

**Clearly, node j may both coincide with the trip origin, and be any node in which the user is believed to be able to make a path diversion.

$f_{k,jt} = 1$ otherwise

In the probabilistic approach hypothesis, one speaks of diversion probability, hence:

$$P[f_{k,jt} = -1] = Pr[0 \leq TTS_{k,jt} \leq IBR_{jt}]$$

$$P[f_{k,jt} = 1] = 1 - P[f_{k,jt} = -1]$$

In both cases, it is necessary to calculate both $TTS_{k,jt}$ and IBR_{jt}.

As regards the estimate of $TTS_{k,jt}$ the problem is basically the estimate of the minimum path cost (TTB_{jt}) from node *j* to destination *d*. TTB_{jt}, in turn, may be the new minimum cost perceived by the user, or the minimum cost calculated and supplied exogenously to the user.

In the former case, the user observes and elaborates vehicle flow characteristics (speed on the link entering node *j*, queue length or variable message panels with travel time indications for some links) and recalculates his/her trip costs.

TTB_{jt} = g (vehicle density, speed, queue, info)

Specification of such an attribute requires the far from easy implementation of a model of travel cost construction. The main difficulties concern the need to have a significant number of test observations and reliable measurements of vehicle flow characteristics at the time of the interview.

In the hypothesis of costs supplied exogenously, the problem involves estimating the minimum path cost congruent with the cost that the user will encounter en route. This attribute may be expressed according to two approaches:

(a) *instantaneous*: according to the characteristics of the transport system at instant *t*;
(b) *predictive*: on forecasts of future network conditions.

As specified above, instantaneous information is based on the configuration of network flows at the time instant in which the user arrives at diversion node *j*. In this case, the user is provided with instantaneous information which does not predict the system evolution in the next instants. The advantage lies in the level of service attributes being easier to calculate, while the main drawback consists in the incongruence of information vis-à-vis the costs that the user will actually encounter en route. Non-congruent information may lead to user indifference to the information; it may make the inter-period evolution of the system very unstable and make the system further removed from its state of equilibrium.

Predictive information may be obtained by simulating: (a) current system evolution without path diversions; and (b) system evolution, taking account of possible diversions. The former hypothesis, undoubtedly less computationally burdensome, leads to costs that are not perfectly congruent with the costs that the user will encounter as far as the destination. Under the latter hypothesis, there is perfect congruence between the costs supplied at the diversion nodes and the costs that the user will encounter on the network. The computational burden of such a solution is far from negligible in the presence of many diversion possibilities.

As regards the estimate of the indifference interval (IBR_{jt}), it may be interpreted as a deterministic variable or as a random variable. In the former case, the upper limit of the above interval may be estimated by means of experimental observations and consolidated statistical techniques. In the latter case, it is assumed that IBR_{jt} is distributed with a known probability distribution function and parameters that may be estimated using statistical inference techniques. The latter approach, which is undoubtedly preferable, allows us to take account of the dishomogeneity among user perceptions (transversal heteroscedasticity), the dishomogeneity among perceptions of the same user due to factors that cannot be simulated explicitly (longitudinal heteroscedasticity), and errors made in the estimation phase of the level of service attributes which contribute to define travel costs.

Generally speaking, the band of indifference may be expressed as the maximum of the two variables (if more than one is needed), deterministic or random:

$$IBR_{jt} = \max\left[h_{jt}TTC_{jt}, p_{jt}\right]$$

where,

- h_{jt} is the percentage of preventive travel time $TTC_{od,t}$ that the user deems significant to change his/her choice;
- p_{jt} indicates the minimum time that the user expects to save following a change in his/her preventive path.

In practice the user, to be able to change his/her preventive choice, wishes to perceive a time saving which is greater than or equal to a share of the current trip time and no lower than a minimum threshold p_{jt}. Indeed, if the quantity h_{jt} is constant with the variation in trip time, clearly, for small trip times, the propensity for change could be overestimated (e.g., TTC = 10 minutes; TTB = 8.5 minutes; h_{jt} = 10% = 1 minute, TTS = 1.5 minutes, and the user changes path).

In the case of a band of *probabilistic* indifference it is reasonable to hypothesise h_{jt} and p_{jt} as two random variables, characterised by a systematic component and by a random component:

$$h_{jt} = g_r\left(\boldsymbol{X}, \boldsymbol{Z}_{jd,t}, \boldsymbol{q}_{jd,t}\right) + x_{jd,t}$$

$$p_{jt} = g_m\left(\boldsymbol{X}, \boldsymbol{Z}_{jd,t}, \mathbf{q}_{jd,t}\right) + z_{jd,t}$$

where,

- $\boldsymbol{X}$ is a vector representing socio-economic attributes for the purpose of segmenting the class of users i in question;
- $\mathbf{Z}_{\mathrm{jd,t}}$ is the vector of the level of service attributes relative to the preventive choice;
- $\boldsymbol{q}$ is the vector of parameters of functions g_r and g_m that have to be estimated;
- $x_{jd,t}$ and $z_{jd,t}$ are random components.

Also in this case it may be imagined that the diversion probabilities are independent or dependent on one another. In the second case, the diversion probability may expressed as the probability of diversion conditional upon having made a diversion at the previous node, in other words, having already made a diversion during the current trip. In practice:

- p diversion
- p diversion/diversion previous node
- p diversion/previous diversion

The problem may be tackled by using a modelling approach, specifying models that allow the correlation between choices made at different times to be taken into account (e.g., probit and/or mixed logit).

11.4.2.3 Compliance models

Simulation of the response rate to indications supplied by an ATIS is very important for realistic simulation of choices whether preventive or preventive/adaptive. The response rate may be interpreted as the result of a cognitive process with which users, combining their past experience with available information, choose whether or not to follow the supplied indication.

The problem may be framed within the theoretical paradigm of random utility, hypothesising that there are two alternatives available and that the non-response alternative has zero utility[††].

If:

- i is the generic user or class of users
- $d^i \in [0,1]$ is a binary variable of value 1 if the user follows the indication given
- U^i is the perceived utility associated with the alternative of following the indication supplied
- $V^i = V^i(\mathrm{SE}^i, \mathrm{ALS}^i)$: is the systematic utility where SE^i are the socio-economic variables and ALS^i the level of service variables that may influence the choice of user i
- e^i is the random residual for the hypotheses made:

$$d^i = 1 \quad \text{if } U^i > 0$$

There may be many possible modelling solutions, albeit within random utility theory. To include them, it is worth schematising the problem as a succession of user choices in time (with respect to one day and another) and space (with respect to possible diversion nodes). Every day, the user makes his/her own choice and, having begun the trip is called upon to make as many choices as there are possible diversion nodes. The choices made are strongly correlated (in space and time) and influenced by many aspects that are hard to simulate which causes a dispersion of behaviour between days and within the same days. There thus emerges a need for

[††]In the context of random utility theory and in the hypothesis of additive distribution functions of random residuals, it is always possible to reformulate the problem according to the differences of utilities perceived vs. the perceived utility of an alternative of reference. This allows us to hypothesise the utility of the reference alternative as zero, obtaining a further absolutely equivalent formulation.

modelling formulations that allow for behavioural inhomogeneity and correlations between the many choices made by the user. In addition, the phenomenon, due to practical requirements, is studied by observing the behaviour of more than one user which, albeit socio-economically similar, adds further sources of inhomogeneity. In this context, the problem is no longer a matter simply of binary choice. Rather, it is much more complex. Possible modelling formulations may be as follows:

if user choices are independent of each other in time and space, the most effective modelling solution consists of a binomial logit model. Such a solution is correct and coherent if the behaviour of a sample of users is observed at a generic node (origin or diversion) and on the same day. The choices observed are independent in space and time. Indeed, the choices made at previous nodes and on previous days are overlooked (hypothesis of within-period stationarity).

Alternative modelling formulations may be the binomial probit or a mixed logit model to simulate inhomogeneity among users. A mixed logit is preferable due to the possibility of interpreting heteroscedasticity by calibrating the variances of the coefficients of systematic utility attributes.

If the phenomenon has been studied by observing behaviour on several days and/or if there is more than one diversion node, it is more realistic to interpret the phenomenon according to a process of choice among a discrete number, no longer binary, of alternatives. Indeed, suppose that each origin or diversion node is a node of decision j (with $j = 1 \ldots J$) let each node be characterised by a time index representing the generic day t (with $t = 1 \ldots T$). At each node jt the user associates a perceived utility U^i_{jt} and makes a binary choice. The choice probability is a probability conditional upon other choices already made at other decision nodes and on previous days. The perceived utilities U^i_{jt} are thus inter-correlated and, hypothesising that the utilities are distributed like a multivariate normal random variable, the problem can be formalised rigorously. A simplification may be obtained by introducing some error components that allow us to simulate correlations between utilities of the same node but with respect to different days, or between utilities at different nodes but crossed on the same day.

Simulation of the '*compliance*' phenomenon may be attempted to simplify the choice simulation at a diversion node. Hypothesising a reference path approach, '*compliance*' may be interpreted as a variable within the choice process of the generic user. In this sense if:

- i is the generic user (or class of users)
- j is the possible diversion node (at which the user arrives at a generic time t)
- K_j are the possible alternatives to reach destination d starting from node j
- c_j is the alternative chosen in advance
- b_j is the best alternative (exogenously supplied by the information system)
- h_j is the generic alternative of the remaining choice alternatives

The user's choice process, following an approach based on random utility theory, occurs between alternatives K_j and may be specified as a choice between alternative b_j, alternative c_j, and the remaining alternatives. Supposing b_j and c_j can coincide, each of them may be associated with a perceived utility which, once the random residuals probability distribution has been defined, leads to a known modelling formulation. In reality, the information supplied exogenously ensures that alternative b_j is an alternative with different characteristics from the others, characteristics that can be schematised by introducing a utility component (U_{b2}), like a latent variable, into the formulation of the utility function perceived. In practice, knowing that one alternative is the best changes the utility that the user associates with the alternative. This better information is not necessarily either a positive or negative component, since it depends on the degree of the user's compliance with the information concerned. If there is no degree of adjustment, then there is no reason why the information supplied should increase the utility of the alternative indicated.

Thus the perceived utilities may be expressed as:

$$U_h = V_h + e_h \forall h \neq b, c$$

$$U_b = U_{b1} + U_{b2} = V_{b1} + e_b + V_{b2} + h_b$$

$$U_c = V_c + e_c$$

where U_{b2} is the latent variable hypothesised as the sum of systematic utility (V_{b2}) and the random residual (h_b)[‡‡]. The random residual is inserted to simulate the dishomogeneity among users typically assumed from the latent variables. While systematic utilities (V_{b1}, V_h, V_c) express typical characteristics that influence path choice (e.g., socio-economic variables, level of service, and details of the alternative), the systematic utility V_{b2}, in turn, may be expressed as a linear combination of variables representing the degree of compliance with the information. It is worth using four types of variables:

- variables representing the user's previous experience. Consider the delay observed the previous day following information or otherwise, the user's attitude to changing his/her departure time so as to arrive sooner at the destination (utility).
- variables representing the quality of information. Consider the travel time estimation error (overestimate or underestimate). This variable is obtained by subtracting the time reported by the indication and the time actually experienced. The values may be calculated with mean values or with values observed the day before, and generally have a negative impact, albeit with differing intensities. Experimental studies have verified that the relative intensities differ greatly if the best path coincides with the current one or if they are different.
- variables representing the congestion level on the network. Usually represented by aggregate measures for the links crossed or by means of symbolic variables that assume different values on a pre-defined and pre-calibrated scale.
- variables representing the benefits to be obtained by being compliant with information. This usually stands for travel time-saving.

[‡‡]Note that the formulation ensures that the mathematical formulation leads to a mixed logit model if it is hypothesised that the residuals ε are distributed like Gumbel random variables of zero mean and parameter θ.

Having specified the systematic utilities, it is necessary to define the probability distribution functions of the random residuals. Apart from the simulation of heteroscedasticity among the alternatives, it must be stressed that the utilities of alternatives h (not 'c' or 'b') may be assumed inter-correlated. At this point, it may be worth studying separately the two possible situations that could arise:

(a) if the best path coincides with the path currently chosen ($b=c$)
(b) if the two paths do not coincide ($b \neq c$)

In the first case (explained above) it might be necessary to simulate the correlation between the remaining alternatives (h) by introducing an error component which, besides simulating heteroscedasticity between the two groups of alternatives ($b=c$ and h), allows a correlation between the above alternatives to be introduced. In the second case (b), in the same way, it could be worth introducing an error component that simulates the correlation between the utilities of alternative c and alternative h.

Clearly, the random utility models obtainable are not closed-form and belong to the class of Mixed Logit or Probit models.

Alongside the phenomenon of compliance is the phenomenon of *inertia*. By inertia, we mean the mechanism that leads the user not to change his/her preventive choice. This phenomenon, often simulated with the simple introduction of shadow variables, may be tackled with a modelling approach quite similar to that seen for models of information compliance.

Since *inertia* influences choice, it is correct to introduce into the perceived utility of the alternative chosen in advance a latent variable that simulates the utility that the user associates with abandoning his/her choice. Analogously, the perceived utilities may be expressed as follows:

$$U_h = V_h + e_h \forall h \neq b, c$$

$$U_b = V_b + e_b$$

$$U_c = U_{c1} + U_{c2} = V_{c1} + e_c + V_{c2} + h_c$$

where U_{c2} is the latent variable hypothesised as the sum of systematic utility (V_{c2}) and a random residual (h_c). The systematic utility V_{c2} may be expressed as a linear combination of variables representing the degree of inertia to change. Also in this case one can hypothesise the same four types of variables seen above and make the same considerations on the residuals and the possible correlations existing (among alternatives h and b if $b \neq c$).

Although the two phenomena have been introduced separately, there is no reason why they cannot be studied at the same time. Distinguishing the case in which ($b \neq c$) and ($b=c$), we may write:

if ($b \neq c$)

$$U_h = V_h + e_h + x + w \quad \forall h \neq b, c$$

$$U_b = U_{b1} + U_{b2} = V_{b1} + e_b + V_{b2} + h_b + w$$

$$U_c = U_{c1} + U_{c2} = V_{c1} + e_c + V_{c2} + h_c + x$$

It is supposed that all the alternatives $h \neq (b,c)$ are inter-correlated, that the utilities of the alternatives $h \neq (b,c)$ and c through the error component x are correlated because they are not characterised by information and, lastly, that the utilities of the alternatives $h \neq (b,c)$ and b through the error component w, are correlated because they are not characterised by the phenomenon of inertia.

if $(b = c)$

$$U_h = V_h + e_h \quad \forall h \neq b, c$$

$$U_{b=c} = U_{b=c1} + U_{b=c2} = V_{b=c1} + e_{b=c} + V_{b=c2} + h_{b=c}$$

with all the alternatives $h \neq (b,c)$ correlated between them,

11.5 Summary

This chapter focuses on key aspects of Advanced Traveler Information Systems (ATIS). Specifically, it introduces fundamental concepts, approaches, and solutions related to: modelling the learning process, pre-trip route choice behaviour, and en-route decision-making.

Prospect theory

Prospect theory makes it possible to describe how people make choices in situations where they must decide between alternatives that involve risk (e.g., in financial decisions). Underlying the theory is an evaluation of the potential *losses* and *gains* as perceived by the individual. In the original formulation the term *prospect*, referring to a lottery, concerns the relationship between the probability of losses and gains and the phenomena of *risk aversion* and *risk seeking*. The theory consists of two stages, namely *editing* and *evaluation*. In the first, people decide which outcomes they see as basically identical, and they set a *reference point;* the domain of the gains and the losses is defined by establishing that lower outcomes can be considered as *losses* and larger outcomes can be considered as *gains*. In the second phase, people compute a value (utility), based on the potential outcomes and their respective probabilities, and then choose the alternative with a higher utility.

By referring to the expected utility theory, the utility is obtained by summation of the outcome multiplied by its respective probability: When using the formula that Kahneman and Tversky (1979) assume for the evaluation phase, the utility is somewhat different and is computed as where are the potential outcomes and their respective probabilities, and v is a so-called *value function* that assigns a value to an outcome.

An example of a value function is depicted in Figure 11.1: the function passes through the *reference point*, is s-shaped and, as its asymmetry implies in the depicted case, there is a bigger impact of losses than of gains (loss aversion). In

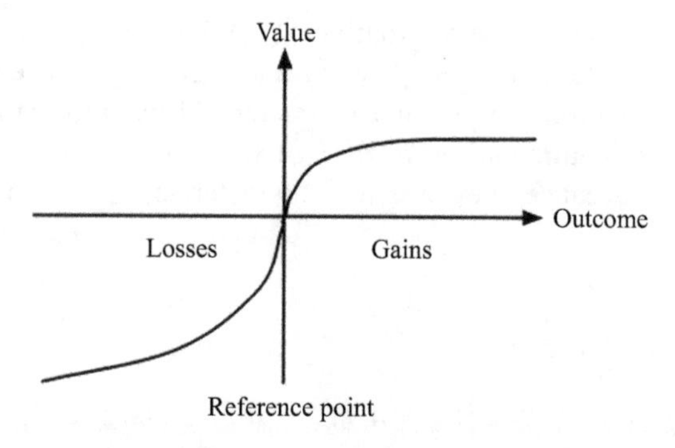

Figure 11.1 Example of value function based on Kahneman and Tversky

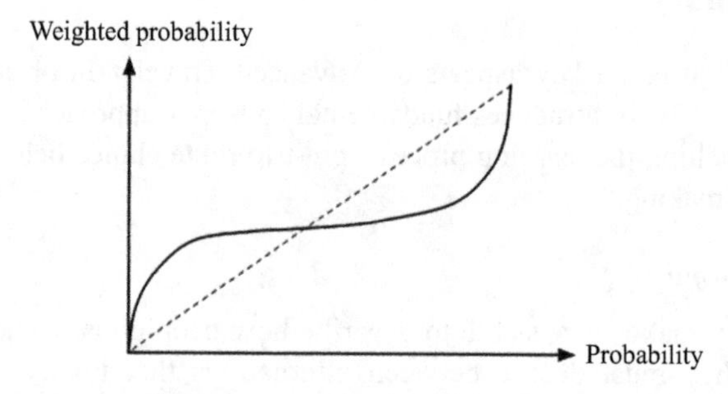

Figure 11.2 A typical weighting function in Prospect Theory

particular, the value function is assumed to be concave in gains and convex in losses, a pattern which is consistent with experimental evidence obtained by analysing the risk-sensitive preferences.

Function w is called the *weighting probability* function; as depicted in Figure 11.2, it expresses that people tend to overreact to small probability events and to under-react to medium and large probabilities.

Tversky and Kahneman later developed a new version of prospect theory that uses cumulative rather than separable decision weights; *Cumulative Prospect Theory* (*CPT*) is an improvement in Prospect Theory. This version applies the cumulative functional separately to gains and losses. In the case of CPT, the utility is computed as, considering separate gains and losses. The main contribution of Prospect Theory and Cumulative Prospect Theory is the introduction of the concepts of *risk-averse* and *risk-seeking*. In the context of travel choices, risk aversion concerns cases where, for instance, many routes have the same average expectations and the users more often choose the route with more reliable travel times (say, the route with a lower dispersion with respect to its average travel time). By contrast, risk-seeking can be applied to cases when the user more often chooses the route with smaller reliability (higher variability) thus aiming at great gains but also to lose more with respect to conservative behaviour. The unreliability effect on user behaviour has been analysed by several authors, especially by Avineri and Praskner (2006). They showed that, in the case of high payoff variability, user choices tend to move towards random choice; the unreliability of travel times also increases the traveller's inability to perceive actual differences among travel times.

An important study about user attitude to risk was carried out by Kastikopoulos *et al.* (2002). For every route (j) certain variables are defined, such as range (r_j), defined as the absolute value difference between maximum and minimum values (across days) of travel times and expected travel times (e_j), in turn defined as the average between minimum and maximum travel times. In a network, one of the routes is defined as the *reference route* (the main route – say the straighter route – that connects the origin and destination in question), and the others are defined as alternative routes. According to these preliminary considerations, estimated travel times are computed as: where is a random variable distributed like a normal function of distribution with average and variance. The diversion probability towards an alternative route is computed as $P(div) = P[(e+r) < c] = P[<(c-e)/r]$; it is also interpreted as the probability of exhibiting a *risk-seeking* attitude. The model proposes that drivers choose the route with the smallest value of *ETT*. Therefore, when $c>e$, $P(div)$ is decreasing in r (i.e., in the domains of *gains*) and $P(div)$ is increasing in r when $c<e$ (i.e., in the domain of *losses*).

In the above case, the travel time of the reference route was considered not distributed (for instance, the travel time is equal to 33 min; moreover, for the alternative route, on the basis of minimum and maximum travel times, parameters e and r can be calculated), but if for every route the values of e and r can be calculated, to be more precise minimum and maximum travel times are associated to both routes, then the probability of diversion can be computed as follows:

$$P(div) = P[< (e_R - e_A)/(r_R - r_A), r_R > r_A];$$

$$P(div) = P[> (e_R - e_A)/(r_R - r_A), r_R < r_A];$$

The computed probabilities refer to the defined conditions of risk as described in Figure 11.3.

In several cases, travel time unreliability on route choice is incorporated into modelling (Small, 1982; Bogers and van Zuylen, 2004).

Information accuracy

The investigation of the effects of information on travellers' behaviour at the individual level is a prerequisite for any analysis of the impacts of ATIS on traffic networks. It is widely expected that travellers' behaviour can be strongly influenced by the ability of the information system to make accurate estimations of the actual travel times they will experience on the network.

Accuracy can be defined as the ability of the information system to reduce the discrepancy between estimated travel times and the actual ones experienced by the

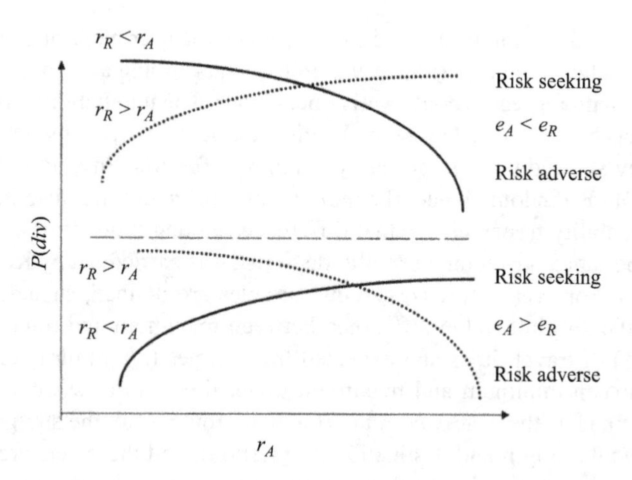

Figure 11.3 Computation of diversion probability

traveller. We refer to travel times estimated by the information system as descriptive information, to route suggestions made by the system as prescriptive information and to the actual travel times experienced by travellers as feedback information. The latter is assumed to be correct from the point of view of the traveller. In this context, we can define two types of travel time uncertainty in the choice environment. The first one depends on the network's performance and is related to actual variability in travel time, while the second relates to the ability of the information system to correctly estimate prevailing traffic conditions, a task that becomes more complex as congestion levels increase, particularly non-recurring congestion which is difficult to predict.

Travellers could well exhibit different behaviours in contending with this complex range of uncertainty depending on their risk attitudes. Risk-averse travellers are likely to prefer a more reliable route (i.e., a lower travel time variance) over an unreliable one with an average shorter travel time. Risk-seeking travellers are likely to prefer an unreliable route that provides on average a shorter travel time. An inaccurate ATIS may be perceived by travellers as corresponding to higher risk and also possibly affect the response rates to prescriptive information.

In general, lacking other sources of information, travellers will base their route choices on the travel times they had experienced on previous trips (Cascetta, 2001; Horowitz, 1984). In the presence of ATIS, route choice is influenced not only by experiential information but also by the information provided by the system (e.g., a visual description of average travel times). Depending on the type of information, different results have been documented, in both SP and Revealed Preference (RP) settings, when comparing behaviour between informed (i.e., users assisted by an information system) and non-informed users (i.e., users who only receive experiential or feedback information).

In the case of descriptive information, several SP experiments (e.g., Abdel-Aty *et al.*, 1997, Avineri and Prasker, 2006) assert that travellers will tend to exhibit risk aversion when faced with travel time information. In a series of repeated choices, Avineri and Prashker (2006) compared the effect of providing respondents *a priori* with static pre-trip information describing average expected travel times in addition to receiving feedback information about the chosen alternative. They demonstrated that informed respondents were more risk-averse and preferred a more reliable route compared to the control group who could only learn through past experience. Their result illustrates indirectly how the effect of information accuracy might be explained by the perceived difference between the provided expected 'average' and the experienced outcomes. Conversely, Ben-Elia *et al.* (2013), in a different repeated choice experiment, internalised inaccuracy by providing respondents with dynamic en-route information describing the ranges of travel times. They show that compared to non-informed respondents, informed ones learn faster and exhibit risk-seeking behaviour (i.e., prefer a shorter and riskier route) in the short run which dissipates in the long run as experiential information becomes more dominant. Since the actual travel times experienced were always drawn within the descriptive range, no apparent discrepancy should have been perceived here between the information provided by the system and the outcome of choice. The two aforementioned studies illustrate the possible associations that could exist between the accuracy of information on the one hand and changes in travellers' risk attitudes on the other hand. Regarding feedback information, Bogers *et al.* (2007) showed that respondents who were provided with foregone payoffs (i.e., feedback on chosen and non-chosen alternatives alike) performed better in terms of travel time savings, though these benefits decreased over time as more experience was accumulated.

Overall, the impact of information's accuracy is a very challenging task, and, in particular, if there does not exist a high degree of compliance. Since compliance is induced if the given information is consistent between the forecasted traffic conditions and actual travel time, the task is more than ever complex since the predicted state of traffic conditions should also consider travellers' reactions to the information itself (the anticipatory-route-guidance problem).

In this context, if ATIS could reduce the perception of travel time uncertainty, the accuracy of the provided information could affect this perception. As demonstrated by Ben-Elia *et al.* (2013) the (in)accuracy negatively affects compliance, and different levels of accuracy significantly change compliance behaviour. Thus it can be asserted that discrepancies between descriptive information and experience can lead to greater risk aversion, thus travellers prefer to anchor their decisions even to worthless information (as in low accuracy). Moreover, it has been demonstrated how a decrease in the level of accuracy increases the likelihood of choosing the reliable route, confirming risk aversion when uncertainty increases. When accuracy is low, even a useless alternative may appear attractive since it inhibits learning and causes dispersed behaviour.

Overall, inaccurate information increases the number of risk-averse travellers, thus limiting the role and the utility of ATIS.

In conclusion, the design of the information system is crucial, both in terms of how the information is presented and in terms of prescriptive/descriptive information. If on the one hand, it could be advisable to provide information in the form of ranges of travel times; on the other hand, it is more than ever crucial to develop a predictive modelling framework that should address several behavioural issues and forecast the evolution of the system over time. In particular, it will be necessary to deal with the compliance phenomenon, but also with concordance (Di Pace *et al.*, 2023).

A compliant-with-ATIS traveller chooses the suggested route. A concordant-with-ATIS traveller chooses the suggested route either because he/she trusts in the system (is compliant) or because he/she would have chosen the route in any case, independently of ATIS indications. Compliant-with-ATIS traveller is always concordant-with-ATIS, thus the set of concordant travellers contains the set of compliant travellers, and the probability of being concordant is greater than the probability of being compliant.

With respect to such a framework, travellers' behaviour may be modelled within the theoretical paradigm of random utility theory. Compliant behaviour has been already introduced in previous sections; concordant behaviour ('to be concordant' or 'to be discordant') may be interpreted as a *switching* or *holding* behaviour, and be modelled through random utility models.

The *holding* approach assumes that in each trial the traveller can choose if 'to be' or 'not to be' concordant, independent of the previous trial choice.

The *switching* approach assumes that the probability a traveller is concordant [*discordant*] at a given day t can be computed through the probability he/she is concordant [*discordant*] at previous day $t-1$ and he/she remains in his/her *concordant* [*discordant*] status or switches.

Within this behavioural framework, the choice probabilities are affected by ATIS accuracy and the actual travel times' reliability. To this aim, different indicators (potential attributes in the systematic utility functions) have been shown to be statistically significant and the reader may refer to Bifulco *et al.* (2011).

References

Some suggestions for further readings are reported below among the huge literature available on these topics, see also the reference list at the end of each of the following chapters.

Abdel-Aty, M. A., Vaugnh, K. M., Kitamura, R. K., Jovanis, P. P., and Mannering, F. (1994). Models of Commuters' Information Use and Route Choice: Initial Results Based on Southern California Commuter Route Choice Survey. *Transportation Research Record*, 1453, 46–55.

Abdel-Aty, M. A., Kitamura, R., and Jovanis, P. P. (1997). Using stated preference data for studying the effect of advanced traffic information on drivers' route choice. *Transportation Research Part C: Emerging Technologies*, 5(1), 39–50.

Adler, J. L., and McNally, M. G. (1994). In-laboratory experiments to investigate driver behavior under advanced traveler information systems. *Transportation Research Part C: Emerging Technologies*, 2(3), 149–164.

Arentze, T. A., and Timmermans, H. J. P. (2003). Modelling learning and adaptation processes in activity-travel choice. *Transportation*, 30(1), 37–62.
Arentze, T. A., and Timmermans, H. J. P. (2005). Information gain, novelty seeking and travel: a model of dynamic activity-travel behavior under conditions of uncertainty. *Transportation Research part A*, 39(2–3), 125–145.
Arthur, W. B. (1994). Inductive Reasoning and Bounded Rationality. *The American Economic Review*, 84(2), 406–411.
Avineri, E., and Prashker, J. N. (2005). Sensitivity to travel time variability: Travelers' learning perspective. *Transportation Research Part C*, 13, 157–183.
Avineri, E., and Prashker, J. N. (2006). The impact of travel time information on traveller's learning under uncertainty. *Transportation*, 33(4), 393–408.
Ben-Akiva, M. E., and Lerman, S. R. (1985). *Discrete Choice Analysis: Theory and Application to Travel Demand*. The MIT Press.
Ben-Elia, E., Di Pace, R., Bifulco, G. N., and Shiftan, Y. (2013). The impact of travel information's accuracy on route-choice. *Transportation Research Part C: Emerging Technologies*, 26, 146–159.
Ben-Elia, E., and Shiftan, Y. (2010). Which road do I take? A learning-based model of route-choice behavior with real-time information. *Transportation Research Part A*, 44, 249–264.
Bierlaire, M., and Crittin, F. (2001). New algorithmic approaches for the anticipatory route guidance generation problem. In: *1st Swiss Transportation Research Conference*.
Bifulco, G. N., Cantarella, G. E., de Luca, S., and Di Pace, R. (2011, October). Analysis and modelling the effects of information accuracy on travellers' behaviour. In *2011 14th International IEEE Conference on Intelligent Transportation Systems (ITSC)* (pp. 2098–2105). IEEE.
Bogers, E. A. I., and van Zuylen, H. J. (2004). The Importance of reliability in route choice in freight transport for various actors on various levels. In: *Proceedings of the European Transport Conference*. Strasbourg, France.
Bogers, E. A. I., Bierlaire, M., and Hoogendoorn, S. P. (2007). Modelling learning in route choice, *Transportation Research Record*, 2014, 1–8.
Bonsall, P., Firmin, P., Anderson, M., Palmer, I., and Balmforth, P. (1997). Validating the results of a route choice simulator. *Transportation Research Part C*, 5(6), 371–387.
Bottom, J., Ben-Akiva, M. E., Bierlaire, M., and Chabini, I. (1998). *Generation of consistent anticipatory route guidance Paper presented at the Tristan III Conference*, San Juan, Puerto Rico.
Box, G. E. P., and Cox, D. R. (1964). An analysis of transformations (with discussion). *Journal of Royal Statistical Society B*, 26, 211–252.
Brown, G. W. (1951). *"Iterative Solution of Games by Fictitious Play," in Activity Analysis of Production and Allocation*. New York: Wiley.
Cascetta, E. (2001). Transportation Systems Engineering: Theory and Methods, pp. 331–348.
Cascetta, E., Nuzzolo, A., Russo, F., and Vitetta, A. (1996). A modified logit route choice model overcoming path overlapping problems. Specification and some

calibration results for interurban networks. *In Transportation and Traffic Theory. Proceedings of The 13th International Symposium On Transportation And Traffic Theory*, Lyon, France, 24–26 July 1996.

Chen, R. B., and Mahmassani, H. S. (2009). Learning and Risk Attitudes in Route Choice Dynamics. In: Kitamura, *et al.* (eds.) *The Expanding Sphere of Travel Behaviour Research: The Proceedings of 11th Conference on Travel Behaviour Research*. Emerald Publishing.

Chorus, C. G., and Dellaert, B. G. (2012). Travel choice inertia: the joint role of risk aversion and learning. *Journal of transport, Economics and Policy*, 46(1), 139–155.

Chorus, C. G., Arentze, T. A., and Timmermans, H. J. P. (2009). Traveller compliance with advice: A Bayesian utilitarian perspective. *Transportation Research Part E*, 45, 486–500.

Cominetti, R., Melo, E., and Sorin, S. (2010). A payoff-based learning procedure and its application to traffic games. *Games and Economic Behavior*, 70(1), 71–83.

de Luca, S., and Di Pace, R. (2015). Evaluation of risk perception in route choice experiments: an application of the Cumulative Prospect Theory. In *2015 IEEE 18th International Conference on Intelligent Transportation Systems* (pp. 309–315). IEEE.

de Luca, S., Di Pace, R., Memoli, S., and Pariota, L. (2020). Sustainable traffic management in an urban area: An integrated framework for real-time traffic control and route guidance design. *Sustainability*, 12(2), 726.

Dhivyabharathi, B., Hima, E. S., and Vanajakshi, L. (2018). Stream travel time prediction using paper filtering approach. *Transportation Letters*, 10(2), 75–82.

Di Pace, R., Bruno, F., Bifulco, G. N., and De Luca, S. (2023). Modelling travellers' behaviour in a route choice experiment with information under uncertainty: Calibration, validation, and further refinements. *Transportation Letters*, 15(3), 211–226.

Dong, J., Mahmassani, H. S., and Lu, C.-C-. (2006). How reliable is this route? Predictive travel time and reliability for anticipatory traveler information systems. *Transportation Research Record*, 1980, 117–125.

Erev, I., and Barron, G. (2005). On Adaptation, Maximization, and Reinforcement Learning Among Cognitive Strategies. *Psychological Review*, 112(4), 912–931.

Erev, I., and Roth, A. E. (1998). Predicting how people play games: reinforcement learning in experimental games with unique, mixed strategy equilibria. *The American Economic Review*, 88(4), 848–881.

Feltovich, N. (2000). Reinforcement-based vs. belief-based learning models in experimental asymmetric-information games. *Econometrica*, 68(3), 605–641.

Fudenberg, D., and Levine, D. K. (1995). Consistency and cautious fictitious play. *Journal of Economic Dynamics and Control*, 19, 1065–1089.

Fudenberg, D., and Levine, D. K. (1998a). Learning and Evolution: Where Do We Stand? Learning in games. *European Economic Review*, 42, 631–639.

Fudenberg, D., and Levine, D. K. (1998b). *The theory of learning in games*. Cambridge, MA: MIT Press.

Gaudry, M. (1981). The inverse power transformation Logit and Dogit mode choice models. *Transportation Research B*, 15, 97–103.

Hart, S., and Mas-Colell, A. (2000). A simple adaptive procedure leading to correlated equilibrium. *Econometrica*, 68, 1127–1150.

Horowitz, J. (1984). The stability of stochastic equilibrium in a two-link transportation network, *Transportation Research B*, 18(1), 13–28.

Hurkens, S. (1995). Learning by forgetful players. *Games and Economic Behavior*, 11(2), 304–329.

Jha, M., Madanat, S., and Peeta, S. (1998). Perception updating and day-to-day travel choice dynamics in traffic networks with information provision, *Transportation Research Part C*, 6(3), 189–212.

Jordan, J. S. (1995). Bayesian learning in repeated game. *Games and Economic Behavior*, 9, 8–20.

Kalai, E., and Lehrer, E. (1993). Rational learning leads to Nash equilibrium. *Econometrica*, 61(5), 1019–1045.

Katsikopoulos, K. V., Fisher, D. L., Duse-Anthony, Y., and Duffy, S. A. (2002). Risk attitude reversals in drivers' route choice when range of travel time is provided. *Human Factors*, 44(3), 466–473.

Kahneman, D., and Tversky, A. (1979). Prospect Theory: An analysis of decision under risk. *Econometrica*, 47, 263–291.

Lambert III, T. J., Epelman, M. A., and Smith, R. L. (2005). A Fictitious Play Approach to Large-Scale Optimization. *Operations Research*, 53(3), 477–489.

Laslier, J. F., Topol, R., and Walliser, B. (2001). A Behavioral Learning Process in Games. *Games and Economic Behavior*, 37, 340–366.

Lotan, T. (1997). "The effects of familiarity on route choice behaviour in the presence of information". *Transportation Research C*, 5, 225–243.

Lu, R., Chorus, C., and van Wee, B. (2014). Travelers' use of ICT under conditions of risk and constraints: an empirical study based on stated and induced preference. *Environment and Planning B*, 41(5), 928–944.

Ma, T. Y., and Di Pace, R. (2017). Comparing paradigms for strategy learning of route choice with traffic information under uncertainty. *Expert Systems with Applications*, 88, 352–367.

Mahmassani, H., and Chang, G. (1986). Experiments with departure time choice dynamics of urban commuters. *Transportation Research B*, 20(2), 297–320.

Mahmassani, H. S., and Liu, Y. H. (1999). Dynamics of commuting decision behaviour under advanced traveller information systems. *Transportation Research Part C*, 7, 91–107.

Marden, J. R., Arslan, G., and Shamma, J. S. (2009). Joint Strategy Fictitious Play with Inertia for Potential Games. *IEEE Transactions on Automatic Control*, 54(2), 208–220.

McFadden, D. (1974). Conditional Logit Analysis of Qualitative Choice Behavior. In: Zarembka, P. (Ed.) *Frontiers in Econometrics*, Academic Press, pp. 105–142.

McFadden, D., and Train, K. (2000). Mixed MNL models for discrete response. *Journal of Applied Econometrics*, 15, 447–470.

Monderer, D., and Shapley, L. (1996). Fictitious play property for games with identical interests. *Journal of Economic Theory*, 68, 258–265.

Parry, K., and Hazelton, M. L. (2013). Bayesian inference for day-to-day dynamic traffic models. *Transportation Research Part B*, 50, 104–115.

Raiffa, H., and Schlaifer, R. (1961). *Applied Statistical Decision Theory*. Harvard University Press, Boston, MA.

Roth, A. E., and Erev, I. (1995). Learning in extensive-form games: Experimental data and simple dynamic models in the intermediate term. *Games and Economic Behavior*, 8(1), 164–212.

Selten, R., Chmura, T., Pitz, T., Kube, S., and Schreckenberg, M. (2007). Commuters route choice behavior. *Games and Economic Behavior*, 58, 394–406.

Shanks, D. R., Tunney, R. J., and McCarthy, J. D. (2002). A re-examination of probability matching and rational choice. *Journal of Behavior Decision Making*, 15, 233–250.

Simon, H. A. (1955). A behavioural model of rational choice. *Quarterly Journal of Economics*, 69 SU, 99–118.

Small, K. A. (1982). The scheduling of consumer activities: work trips. *The American Economic Review*, 72, 467–479.

Srinivasan, K. K., and Mahmassani, H. S. (2002). Dynamic Decision and Adjustment Processes in Commuter Behavior under real-time Information, Research Report SWUTC/02/167204-1 prepared by the Centre for Transportation Research Bureau of Engineering Research University at Austin, pp. 97–112.

Van der Mede, H. J., and Van Berkum, E. C. (1996). Modelling Car Drivers' Route Choice Information Environments, Delft University of Technology, Transportation Planning and Traffic Engineering Section Rudimental Contributions.

Young, H. P. (2004). *Strategic learning and its limits*. OUP Oxford.

Chapter 12
Advanced driving assistance systems

Roberta Di Pace[1] *and Luigi Pariota*[2]

Young men's love then lies not truly in their hearts, but in their eyes.

William Shakespeare

Outline. *This chapter provides an overview of the advanced driving assistance systems. In particular, starting from the preliminary discussion of the intelligent speed adaptation, a further discussion and classification of advanced driving assistance systems is provided.*

Over the past two decades, the concept of personalisation – tailoring something to meet the specific needs and preferences of an individual – has garnered significant interest across various disciplines. Personalisation can be implemented in two primary ways: *explicitly* or *implicitly* (Fan and Poole; 2006).

Explicit personalisation involves allowing drivers to select their preferred options from a range of predefined system settings. This approach gives drivers direct control but restricts choices to a limited number of standard settings that are easy to understand. However, this method requires drivers to focus and exert effort, which can be highly distracting during normal driving.

Implicit personalisation, on the other hand, involves estimating drivers' preferences by observing their behaviour. This method allows for a more nuanced and complex system adaptation, but there is a risk that drivers may not fully understand the system's behaviour and configuration. Personalisation can apply to various categories, such as infotainment, which involves developing adaptive user interfaces. For example, a driving route recommendation system, Rogers *et al.* (1997) and Rogers and Langley, P. (1998) can generate routes with the driver's input, create a model of the driver's preferences, and refine this model through interaction. Additionally, the operation and driving range of electric vehicles depends heavily on actual driving behaviour, such as speed and acceleration, and the road profile.

[1]Department of Civil Engineering, University of Salerno, Italy
[2]Department of Civil, Architectural and Environmental Engineering, University of Naples 'Federico II', Italy

12.1 Intelligent speed adaptation

Intelligent Speed Adaptation (ISA) is a road safety device that alerts the driver to speeding, discourages the driver from speeding, or prevents the driver from exceeding the speed limit (Brookhuis and De Waard, 1999) and, therefore, can be considered an ADAS (Advanced Driving Assistance Systems) application.

An Intelligent Speed Adapter (ISA) takes local restrictions into account and adjusts the maximum driving speed to the posted maximum speed. Concerning the speed limitation of private vehicles, the use of intelligent speed limiters is preferred due to the further differentiation of speed limits for private cars compared to heavy goods transport vehicles. A non-intelligent speed limiter is set to the maximum allowed speed for highways (120 km/h), whereas most of the safety benefits of a speed limiting system can be achieved on rural roads or A-type roads (limit 80 km/h) and in urban areas (50 km/h).

Intelligent speed limiters require communication between the vehicle and a suitable information source. For the exchange of information on road class or local restrictions, one option is to equip road signs with transmitters, signallers, or tags.

In general, a standard speed device is a system that limits speed control, meaning it sets the maximum possible driving speed, whereas an intelligent speed limiter can set this maximum speed in accordance with locally posted legal limits. A less invasive device is a system that provides the driver with feedback on local limits, for example, on a small display.

The acceptance of feedback-type systems is expected to be higher than that of a standard speed limiter, as the behaviour is less constrained. Another advantage of feedback systems, compared to actual speed restriction, is that speeding violations can sometimes be beneficial for road safety; for instance, there are cases where a critical overtaking manoeuvre (perhaps poorly judged) is executed faster and more safely if the limit is exceeded for a short period.

Observing driver reactions to these systems is of primary importance. Besides individual reactions, the interaction with other vehicles not equipped with speed limiters is also crucial. In a 'mixed traffic' situation, cars with speed limiters might easily annoy drivers of non-limited cars and vice versa. Providing feedback by displaying the speed limit inside the car ensures that the information remains continuously visible rather than being visible only when passing a sign. This could reduce speeding due to a lack of awareness of the limit.

Intelligent Speed Adaptation (ISA) is any system that constantly monitors the vehicle's speed and the local speed limit on a road and acts when it detects that the vehicle exceeds the speed limit. This can be done through a warning system, where the driver is alerted, or through an intervention system, where the vehicle's driving systems are automatically controlled to reduce the vehicle's speed.

ISA uses information about the road the vehicle is travelling on to make decisions about what the safe speed should be. This information can be obtained through the use of digital maps that incorporate road coordinates and speed limit data for that road at that location, through general information on speed limits for a defined geographic area (e.g., an urban area that has a single defined speed limit), or through feature recognition technologies that detect and interpret speed limit signage. The purpose of

ISA is to assist the driver in always complying with the speed limit, especially when traversing different speed zones; this is particularly useful when drivers are in unfamiliar areas or when passing through zones with variable speed limits.

There are many types of ISA systems, but GPS-based ISA systems are the most recent and effective.

The GPS-based ISA system uses a GPS receiver to locate the vehicle's geographical position and a wireless transmitter to receive data, which consists of a table of coordinates and speed limits.

Overall, the system contains the essential components of the ISA system with a wireless transmitter: highway toll booths have a transmitter capable of communicating with the vehicle's transmitter. When the vehicle, equipped with the proposed system, approaches the toll booth, the data transmitted from the booth is received by the vehicle module and stored in the ISA system's memory. The transmitted data constitutes a table of information on coordinates and speed limits. They contain the latitude and longitude values of a location and the corresponding speed limit or highway traffic symbols.

As soon as the data is stored in memory, the GPS receiver tracks its position and compares its coordinates with the received coordinate table. As soon as the first pair of coordinates and the current coordinates from the GPS match, the corresponding speed limit listed in the table is chosen as the speed limit from that point. The ISA system then warns the driver through an acoustic signal, displaying the speed limit and indicating the speed limit of the location.

A GPS receiver is used with the ISA system to detect the current position of the vehicle on the earth as coordinates, and these coordinates are used to find its location via a preloaded map containing information about the roads and their corresponding highway signs and speed limits.

In more detail, the algorithm steps are described below:

Step 1: The vehicle's wireless device is turned on as soon as the vehicle is started and waits to receive data.
Step 2: When the vehicle approaches the highway toll booth and enters the booth's wireless coverage area, a communication link is established.
Step 3: Based on the communication link and direction, data is sent to the vehicle's wireless node. The data is stored in the memory stack.
Step 4: The stored data is a table of information based on the vehicle's geographical coordinates (longitude and latitude) and the corresponding speed limits and highway signs, arranged in the correct sequence.
Step 5: The data in the stack is compared with the received data; when the received data matches the stack data, the corresponding data is used.
Step 6: The signal is used as the current signal for the position, and the corresponding action is performed, either to display, warn, or control the vehicle.
Step 7: Now, the data in the second row of the table is selected and compared again with the current position.

Concerning the Dynamic Data Update System for the ISA, the system includes transceivers in the vehicle and transceivers at highway toll booths. The transceivers at the highway toll booth should be able to communicate with any vehicle equipped with a transceiver. When a car approaches the highway toll booth, as soon as both transceivers enter each other's coverage area, the data transaction is initiated, and then the data from the toll booth is sent to the vehicle and further saved in the ISA system's memory. The primary goal of the Dynamic Data Update System used is to dynamically update the vehicle with information about the road situation. Usually, in GPS-based ISA systems, the road map and speed limit data are very outdated, and many times the details of new road conditions are not available in the vehicle's ISA database. Therefore, updating the information from the toll booth will provide more recent details and emergency instructions such as 'Take Detour', 'Road Work in Progress', 'Accident Zone, Drive Slowly', etc.

12.2 Driver Assistance Systems

12.2.1 Overview

The acronym ADAS stands for Advanced Driver Assistance System. ADAS systems are additional devices installed in vehicles that are considered as a subset of DAS: Driver Assistance Systems, systems whose main function is to ensure the safety of the vehicle, the driver, pedestrians, or cyclists. ADAS systems are capable of informing the driver, issuing warnings that can indicate potentially dangerous situations, improving driving comfort, ensuring greater road safety, performing operations to manoeuvre the vehicle autonomously to bring it back to safe conditions, reducing fuel usage, and so on. ADAS systems are closed-loop control systems that add a greater use of complex processing algorithms to simple DAS by evaluating the environment surrounding the vehicle based on data collected through a variety of signals from a multitude of sensors.

In Figure 12.1, the general architecture describing a vehicle equipped with ADAS systems can be observed. The first block encompasses all sensory

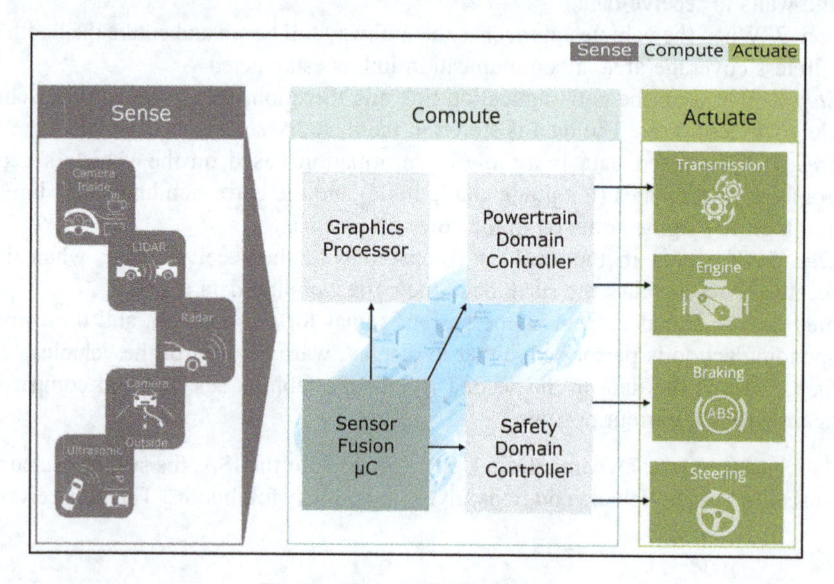

Figure 12.1 ADAS overview

technologies that can be utilised, such as radar, cameras, etc.; the central part, the computing section, includes an Electronic Control Unit (ECU) on which a sensor fusion algorithm is implemented to condense all received data into a single reliable datum, along with a control system that uses the output data from the sensor fusion as a reference and appropriately commands all actuating organs shown in the third block.

This architecture is crucial in several aspects. Throughout this work, simulation environments have been used to test the logic of the designed devices. In this case, the actual programming of the various algorithms does not pose major constraints; one is free to choose all the technologies offered by the market. However, the effort will be in programming these systems with the intention of being tested using Hardware In-the Loop (HIL) techniques, meaning that the code has been written to be installed on a development controller connected to simulators that emulate the vehicle and its environment.

The sensors most commonly used for the development of ADAS systems are radar (Radio Detection And Ranging), lidar (Light Detection and Ranging or Laser Imaging Detection and Ranging), ultrasonic sensors, and night vision cameras, all devices that allow a vehicle to monitor nearby and distant areas in every direction and to evolve and improve sensor fusion algorithms that ensure the safety of vehicles, drivers, passengers, and pedestrians based on factors such as traffic, weather conditions, hazardous conditions, etc.

Sensor data fusion is one of the fundamental concepts behind the major advancements in ADAS design. This process allows for the creation of a map of potential obstacles around the vehicle and keeps track of all possible hazardous situations. Sensor fusion is essential to 'fuse' measurements from instruments that would already be capable of detecting an obstacle on their own; for example, both radar and cameras can detect a vehicle, but depending on environmental conditions, one of the measurements may be effective. Therefore, by combining all data from various sensors, a single, more reasonable measurement can be obtained. Below are listed some of the most well-known ADAS applications:

- ACC – Adaptive Cruise Control: This system allows the driver to choose a cruise speed at which the vehicle is automatically capable of maintaining, while also adjusting its speed to keep a safe distance from the vehicle ahead, by acting on the brakes and accelerator.
- AEB – Automatic Emergency Braking: The vehicle is capable of automatically braking when a sudden obstacle (e.g., a pedestrian) appears in front of the vehicle.
- RCW – Rear Collision Warning: Warns the driver of a possible rear collision due to the presence of an obstacle detected with rear cameras.
- FCW – Forward Collision Warning: The system uses cameras and radar sensors to monitor the front area and detect obstacles and/or vehicles on the path.
- LDW – Lane Departure Warning: An LDW system alerts a driver on the highway if their vehicle is departing from the current lane without activating the turn signal.
- DSM – Driver System Monitoring: A DSM system tracks the direction of the driver's face and alerts if the driver's gaze is not forward. Systems based on the same principle capable of detecting driver fatigue are already being developed.

- LKS – Lane Keeping System: Similar to LDW, but in this case, in addition to alerting the driver, if the vehicle is not brought back into the correct lane, the vehicle is capable of autonomously correcting the trajectory.
- BSD – Blind Spot Detection: Monitors blind spots, i.e., the rear side areas not easily visible to the driver; the system alerts the driver if a vehicle approaches the rear blind spots to prevent collisions in case of lane changes.
- CTA – Cross Traffic Alert: Monitors traffic at intersections using cameras and radar. It activates when approaching an intersection, warning the driver if a vehicle is approaching dangerously from around the corner or from another direction that the driver cannot see.
- TJA – Traffic Jam Assistant: Traffic jam assistant is an advanced function of ACC that requires automatic transmission, as it is capable of stopping and restarting the car in stop-and-go traffic situations, on highways or expressways, when stop-and-go traffic situations occur.

12.2.2 Approaches to personalisation

Current personalisation approaches in the automotive field primarily focus on the technical implementation of personalised functionalities. These approaches are typically *data-driven*, meaning a model of the driver is developed from driving data. This model then serves as a proxy for the driver. Main steps in the personalisation process are:

1. *Observe the driving behaviour:* The fundamental, though implicit, assumption in personalisation is that drivers are most comfortable with a driving style similar to their own. Consequently, driving data are collected from a group of drivers in a field study using an instrumented vehicle.
2. *Build a model of human driving behaviour:* A driver model is developed from the data of an individual driver and used directly as part of the controller. The controller is often divided into two parts: a high-level controller, which models the driving behaviour and whose parameters are adapted to the specific driver during personalisation, and a low-level controller, which is responsible for actuating the vehicle according to the input from the high-level controller.
3. *Validate the model:* The personalised system is validated and compared to a standard system to demonstrate that it effectively adapts to different driving styles. Depending on the maturity of the approach, this is done in three steps:
   (a) *Off-line playback*: Recorded driving data are fed into the personalised controller to verify that it accurately reproduces the observed driving behaviour.
   (b) *Simulation in a traffic simulator*: The personalised controller is evaluated by drivers in controlled traffic situations and often compared with a standard controller.
   (c) *Field test*: Finally, the personalised controller is implemented in a vehicle and tested in real traffic.

We refer to approaches that follow this sequence of steps as personalisation with a pre-trained driver model.

Driver models play a central role in personalised ADAS. These models represent the driver and convert information about the driving situation into actions for the vehicle's actuators. In ADAS, they are used to mimic or predict the driver's intent and behaviour to provide relevant assistance. Currently, most driver models represent the average driver, have fixed parameters, and cannot adapt to different drivers. Given that human behaviour is inherently non-deterministic and exhibits significant inter- and intra-driver variability, accurately modelling driver behaviour is a challenging task studied across various disciplines.

12.2.3 Equipments

The main sensors used in the automotive field, especially in the development of ADAS (Advanced Driver Assistance Systems), are radar, lidar, and cameras. Below is a more detailed analysis of all the mentioned technologies.

12.2.3.1 RADAR

As discussed previously, in the not-so-distant future, autonomous vehicles are destined to become a commercial reality. When that day arrives, it will also be attributable to the versatility of RADAR (Radio Detection And Ranging), which car manufacturers are increasingly using in BSD (Blind Spot Detection) and side impact systems, as well as in ACC design apparatus.

One of the main features that differentiate RADARs used for vehicle safety from those used in more traditional applications is the much higher operating frequencies (between 76 GHz and 80 GHz). These frequencies were chosen due to their signal propagation characteristics and are termed millimeter waves due to their extremely short wavelengths (from 10 mm to 1 mm). The millimeter wave region starts at frequencies above 30 GHz (up to 300 GHz).

There are various reasons for this choice. Millimeter wave signal propagation is characterised by a limited range that decreases with increasing frequency. The laws of physics teach us that the shorter the wavelength, the shorter the transmission range at the same power level. Millimeter wave signals are also sensitive to attenuation caused by any element in front of them – from rain to snow, fog, foliage, to any solid structure. Even under good mutual visibility conditions, the range is much lower than that of lower frequencies used in applications such as wireless communications and broadcasting of radio and television signals. However, automotive applications do not require very long ranges; we are talking about hundreds of meters.

Traditionally, millimeter wave systems are expensive to implement because mechanical components such as antennas are very small and require extremely precise adjustment. There are not many semiconductor devices capable of delivering acceptable performance at such high frequencies. All these disadvantages are only overcome when a market is identified that, thanks to enormous volumes, allows costs to be reduced and innovation to be financed.

Despite the above-mentioned characteristics making millimeter waves unsuitable for many applications, their use in automotive safety systems brings numerous advantages. For example, limitations in terms of range are not valid for every frequency in the millimeter wave spectrum because atmospheric absorption at some frequencies is lower than others. For radar systems used in the automotive sector, the most suitable frequencies are between 71 GHz and 81 GHz. Another advantage of millimeter waves is that radar systems require very low output power, a factor particularly important for the automotive industry where costs are critical, not to mention the inherent difficulty in generating high power levels.

Radars are used to detect obstacles near the vehicle, particularly exploiting the phenomenon of reflection of electromagnetic waves by measuring the distance to the obstacle by evaluating the time of flight of the wave, τ, which is the time elapsed from the emission of the wave to the instant of reflection (resulting from the wave impacting the obstacle) plus the time taken by the reflected wave to return to the source according to the following equation:

$$D = (c \cdot \tau)/2$$

where c is the speed of wave propagation in the medium (air). It is usually a fixed value that in a vacuum is equal to the speed of light but varies depending on the propagation medium and environmental conditions. Of course, the device will need to be equipped with a time measurement system, and a clock. Radar systems are classified based on two parameters:

Beamwidth of the emitted wave;
Maximum reachable distance (range);

12.2.3.2 LIDAR

Lidar (an acronym for Light Detection and Ranging or Laser Imaging Detection and Ranging) is a remote sensing technique that allows determining the distance of an object or surface using a laser pulse. It can also determine the concentration of chemical species in the atmosphere and bodies of water. Similar to radar, which uses radio waves instead of light, the distance to the object is determined by measuring the time elapsed between the emission of the pulse and the reception of the reflected signal.

The source of a lidar system is a laser, which emits a coherent beam of light at a specific wavelength towards the system to be observed. Lidar technology has applications in geology, seismology, archaeology, and remote sensing, and is now also becoming prevalent in the automotive industry for the development of new ADAS (Advanced Driver Assistance Systems). The main difference between lidar and radar is that lidar uses ultraviolet, visible, or near-infrared wavelengths; this makes it possible to locate and obtain images and information about very small objects with sizes comparable to the wavelength used.

For an object to reflect an electromagnetic wave, it must produce a dielectric discontinuity; at radar frequencies (radio or microwave), a metallic object produces a good echo, but non-metallic objects such as rain and rocks produce much weaker reflections, and some materials do not produce reflections at all, resulting in being invisible to radar. This is especially true for very small objects like dust, molecules, and aerosols.

Regarding the automotive field, with the advent of ADAS systems falling into levels of automation 3, 4, and possibly even 5, lidar should play an increasingly important role. The integration of lidar offers a suite of sensors with improved performance and the redundancy necessary to perform autonomous functions at a sophisticated level using appropriate sensor fusion algorithms.

The main components that make up a lidar system are:

Laser scanner
High-precision clock
GPS (Global Positioning System) – measures the position of the scanner
IMU (Inertial Measurement Unit) – measures orientation relative to its own reference system
Data management system

It is noted that the distance D of an object and its resolution ΔD are measured as follows

$$D = ct/2 \quad \Delta D = c\Delta t/2$$

where c is the speed of light (approximately 300,000 km/s but variable depending on environmental conditions), t is the time elapsed between the transmission of the wave and the reception of the reflected wave measured through the clock, Δt is time measurement resolution, the minimum measurable time interval, or the clock period.

Lidar emits pulses with frequencies ranging from 50 kHz to 200 kHz with wavelengths falling within the following ranges: infrared (1,500–2,000 nm), near-infrared (1,040–1,060 nm), blue-green (500–600 nm), and ultraviolet (250 nm).

They are safe for human vision and dissipate little power (we are talking about powers below 1 W). Depending on the frequency of the waves used, we will have a different Maximum Unambiguous Range (MUR), which is the range within which various objects in the scene can be recognised without ambiguity.

Another fundamental property of such a tool is the Field of View (FOV). It depends on the divergence angle of the light beam (the angle with respect to the propagation axis). Small Field of View is used to accurately map small areas. Whereas lidars with wide Fields of View are used for more comprehensive sampling and to detect as many objects as possible, even though they may yield a lower signal-to-noise ratio (SNR).

12.2.3.3 CAMERA

The main functions for which cameras are used are obviously to capture an image and recognise objects of interest within this image (pedestrians, other vehicles, cyclists, etc.) and assess their distance from the vehicle on which they are installed. This information can be used to design systems such as AEB (Automatic Emergency Braking), ACC (Adaptive Cruise Control), CCC (Cooperative Cruise Control), etc. In Figure 12.2, the possible positions of a camera on board a vehicle and the corresponding systems that use it are shown.

Cameras play an important role in safety systems applied to automotive vehicles. However, cameras with generic technologies are not always suitable for

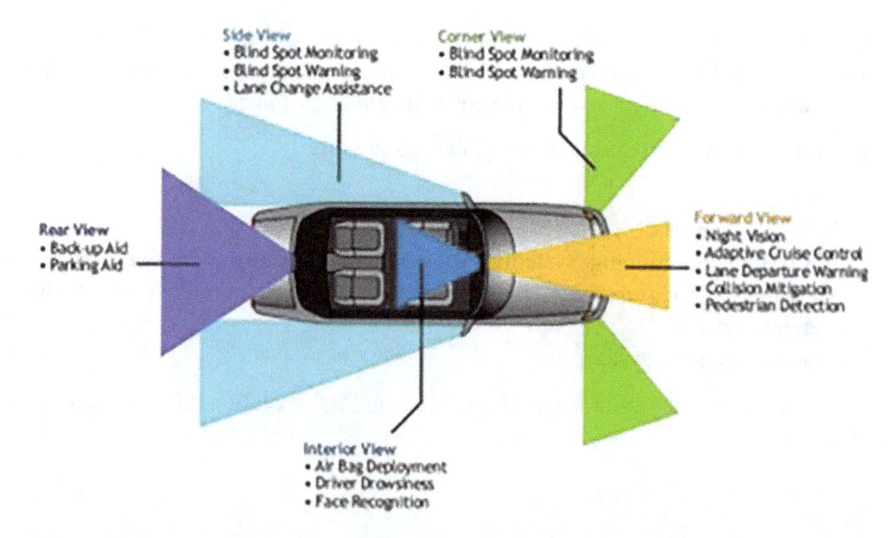

Figure 12.2 Cameras on board distribution

such systems; in fact, to work correctly in these applications, they must meet stringent requirements. The camera must be able to perform the same functions as a generic camera under all lighting conditions, for any wavelength of light in the scene, and for any speed of the object to be detected.

The dynamic range of the camera is the ratio between the highest and lowest light intensities that the sensor can capture simultaneously. For standard cameras, it is about 60 dB, while for automotive cameras, it must reach values around 120 dB. Thus, we speak of cameras with wide dynamic range (WDR).

The main technologies used for the development of automotive cameras are CCD and CMOS. Regarding cameras based on CCD (Charge Coupled Devices) technology, most do not meet the requirements regarding dynamic range. Standard CCD image sensors are designed for 'serial reads', meaning all previous pixels must be read before the next one can be read. This architecture limits the frame frequency, which is a significant disadvantage for automotive applications where fast image capture is crucial. Two other disadvantages of CCD technology are:

Limited subsampling: This phenomenon, which is not fully exploited by CCDs, allows returning an image similar to the original, with slightly less detail. This implies less data to transfer and a higher frame rate.

Limited sub-windowing: Similarly, this process allows selecting only a portion of interest of an entire image to decrease the volume of data to transfer and increase the frame rate. This technique, like the previous one, is not fully utilised by CCD technology.

Regarding cameras based on CMOS (Complementary Metal Oxide Semiconductor) technology, they are capable of meeting the requirements of wide dynamic range, high image sensitivity, and wide spectral range. CMOS image sensors use a random readout technique, meaning pixels can be addressed randomly, thus increasing the frame rate. This significantly improves the subsampling and sub-windowing processes compared to CCD technology. Given these observations, the most widely used technology in the automotive field is certainly based on CMOS.

12.2.4 Modelling approaches

In the personalisation of ACC, we can distinguish between group-based and individual-based approaches. In the group-based approach, drivers are assigned to one of a small number of representative driving styles for which an ACC control strategy is implemented. In the individual-based approach, the ACC control strategy aims to best reproduce the driving style of an individual driver.

Rosenfeld *et al.* (2012, 2015) present a group-based approach to predicting the driver's preferred ACC gap setting and their tendencies to engage and disengage ACC. They cluster drivers from a field test of driving behaviour with ACC to create three general driver profiles and use these profiles along with demographic information to predict gap settings. Their focus is on data analysis using regression models and decision trees, rather than the practical application of the derived models. These models are not validated.

A more comprehensive group-based approach to personalising adaptive cruise control with stop-and-go functionality is proposed by Canale *et al.* In this approach, drivers are assigned to one of three predefined clusters based on observations of their driving style. The cluster membership determines the parameters of a reference acceleration profile, which serves as input to the low-level controller of the ACC. This approach uses data from field experiments and is validated by offline playback.

Bifulco *et al.* (2013) propose an ACC that adapts in real-time to individual drivers based on their observed driving style. Their ACC controller framework is based on a linear car-following model solved by a recursive least squares filter, which reproduces the time gaps observed in a short manual driving session. The vehicle trajectory is calculated from this personalised car-following model using a linear, time-invariant dynamic system with acceleration and jerk as state variables. Vehicle actuation is then managed by a low-level controller. The personalised ACC has been validated through offline playback with satisfactory results. This approach features two modes for achieving personalisation: a 'learning mode', activated on-demand by the driver, in which the current driving style is observed, and the corresponding parameters of the car-following model are learned, and a 'running mode', in which the newly learned car-following model is deployed to the controller.

12.2.4.1 An example of individual approach

The model architecture and the supporting data

In this section, the four-layer framework is briefly presented. The approach was explicitly conceived for ACC applications. We argue that it can be employed for other kinds of applications but here we do not discuss this opportunity.

The inception idea is that the driver's behaviour in car-following conditions can be identified by means of the time series of spacing (intervehicle separation distance, in metres), sampled at a given frequency (e.g., at 1 Hz). Computation of this sequence is the role of the sampler. Within the sampler, the vehicle dynamics are neglected, as is consistency among different kinematic variables (position, spacing, speed, acceleration, etc.). The second modelling layer (the profiler) recovers the full (and consistent) representation of vehicle trajectory by adopting a time-continuous state-space approach. Of course, the profiler must ensure that the resulting trajectory fits the points identified by the sampler; this is obtained by assuming these points as the requests supplied to the profiler. The profiler also checks whether these requests are admissible. Checking is based on general kinematic considerations and rough hypotheses about vehicle performance. For instance, if the sampler requests too high a variation of position to be satisfied with an admissible acceleration, the profiler limits the reached position consistently with a predefined maximum acceleration.

The trajectory produced by the profiler is continuous, consistent, and likely to be admissible; it represents a reference trajectory. However, the actual trajectory can be different from the reference one given that it results from the actuation (throttle, brake, engine, etc.) performed by the vehicle mechanics and electronics, as well as from interaction with the road (slope, grip, etc.). The simulation of how the vehicle and its actuators can match the profiler-supplied reference trajectory pertains to the performer. The development of a tool like the performer is a typical service for the automotive sector*. Hence the performer is here totally neglected, and the assumption is made that such a tool is available and able to actuate the reference trajectory.

Between the profiler and the performer, the tutor is inserted; it ensures that safety conditions are satisfied. It computes the maximum allowed speed (or spacing, or position increment) compatible with the safety, revealed by applying a safety model to real-time (and high-frequency – much higher than the sampler) sensor measurements. The performer then tries to apply the lowest position increment suggested by the profiler or the tutor.

The modelling layers (Figure 12.3) described here must be specified, calibrated, and validated. In particular, a great effort in terms of calibration must be devoted to the sampler which is responsible for reproducing a human-like cruise control logic. Thus, a considerable amount of data is required. In this case, required car-following data were collected by using an instrumented vehicle with radar-based sensors.

Data were employed to estimate the parameters of the sampler (first seconds of each trajectory), as well as to compare the results of the modelling framework. It is worth noting that the nature of the fully adaptive approach excludes the calibration of a set of parameters common to all the trajectories and the calibration of the dispersion of these parameters. Rather, the parameters are intended to be calibrated for every single driver and for each single driving session (in real-time and on-

*Some commercial tools already exist to this aim (e.g., CarSim – https://www.speedgoat.com/).

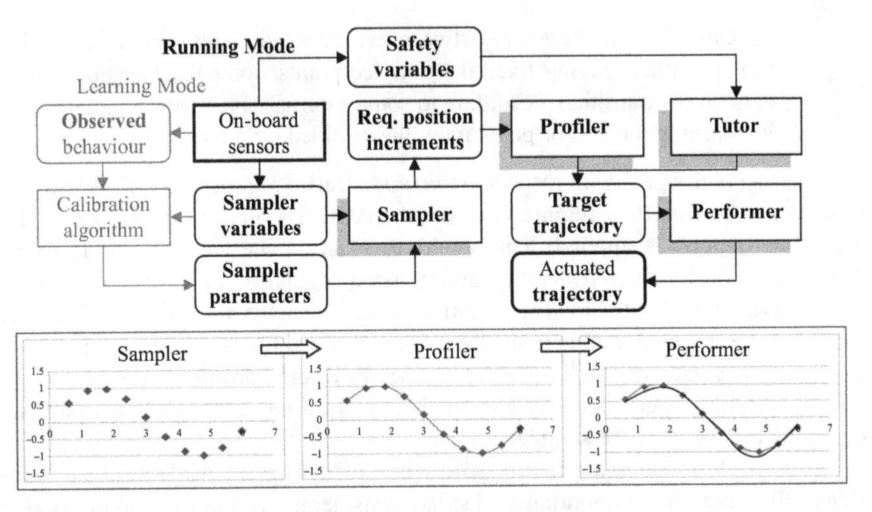

Figure 12.3 Modelling architecture

demand), according to the learning-mode phase of the ACC system. From this point of view, each calibration session made available by the data we collected should be treated separately.

The sampler: a car-following model for the human-like ACC problem

The sampler is responsible for estimating a time series for spacing. From this, a time series of driven distances can be derived. It reproduces as closely as possible the sampling of the car-following trajectory that a driver would have applied in manual driving conditions. The adopted sampling frequency is in the order of magnitude of 1 Hz; this was chosen based on some major considerations:

- the typical reaction time of car-following models proposed in the literature is around 1 second; from that, a sort of human-like refreshing frequency of 1 Hz can be argued; in other words, we assume that drivers habitually distinguish with a granularity no more detailed than 1 second (of course, if things go smoothly);
- even if the human likeness of the trajectory is sampled at a 1 Hz frequency, the ACC system checks the safety at a much higher frequency (say, 10 Hz); this is the task assigned to the tutor; as a result, the system is able to react to stimuli, if dangerous, much more promptly than the driver (say, human likeness is excluded in the case of danger);
- the time step between sampled points is a trade-off between opposite interests as expressed by the following points:
  - the car-following model implemented by the sampler is based on some approximate assumptions on how the leading vehicle moves; a shorter time step between two successive sampled points bounds the errors introduced by this approximation;

– the car-following trajectory between two sampled points evolves according to the profiler; having fixed the sampled points, some optimisation can be done in the transition, according to some external objectives (e.g., reduction in consumption and/or pollutants); this is aided by a longer time step.

The sampler works with respect to two main traffic regimes: free-flow and car-following. The free-flow regime is where the ACC actually acts as a CCC and a pre-defined speed is applied. This could vary along the route, possibly being associated with location-aware (dynamic) speed regulation policies and onboard speed navigators. That said, the free-flow speed is not a modelling task in the context of the research and is treated as a known, fixed (constant within each time step), and exogenously given value. Of course, the desired position increment in the case of free-flow speed is easily computable. Thus, a model for the car-following regime is the focus of the sampler.

It is worth noting again that no safety considerations are made to avoid collisions. Given the capital importance of safety considerations, these are applied to the vehicle at a higher frequency and are superimposed upon any other consideration. For such a reason, safety and emergency considerations are not included as sampler tasks. Rather, they are postponed between the profiler and the performer and constitute the main task of the tutor.

The ACC-oriented car-following approach proposed here is based on a simple linear model, able to relate the instantaneous speeds of the leader and the follower and their spacing with the target spacing the follower desires for the next time-step. The output of the sampler aims to be a reproduction of driver behaviour, but the analytical structure of the model is not intended to mimic behavioural processes. From this point of view, it might be said that any behavioural interpretation has been distanced from our approach. Drivers' behaviours are reproduced by a purely descriptive (regressive) model. Hence the first modelling layer is here called the sampler (and not, for instance, the modeller).

The choice of reproducing spacing is in some respects original. In ACC applications, it is more common to control the vehicle's acceleration directly, while speeds and distances are influenced indirectly. However, assume that the speed is almost perfectly reproduced (due to an even more accurate reproduction of acceleration), but for 1 or 2 seconds the actuated speed is only slightly wrong. After that, the speed is again reproduced with no errors. The few seconds of the wrong speed have induced erroneous spacing; this error does not further increase when the speed is recovered, but it will never be corrected if the goal is to reproduce speed (or acceleration). This does not happen if the spacing is directly controlled, errors in terms of spacing can be recovered. Moreover, if the time series is correct in terms of spacing (and so is vehicle advancement, given the leader's trajectory), it is even more correct in terms of spacing derivatives (speed and acceleration).

The advantage of using a more-than-linear formulation for the regressive model was estimated at 9% in terms of better RMSE of the simulated vs. observed spacing. Similarly, the advantage obtained by an even more flexible regression model based on artificial neural networks was measured at 15%. These advantages are not enough, in the authors' opinion, to forego the great efficiency of the linear approach, especially in our case where real-time and on-demand estimation of the parameters is required.

Let

$\Delta\xi(k, k+1)$ be the target spacing (intervehicle separation) estimated at time step k for time step $k+1$;
$\Delta x(k)$ be the spacing at time step k;
$\Delta v(k)$ be the relative speed (difference of speeds) at time k between the leader and the follower;
$v_L(k)$ be the speed of the leader at time step k.
$s_L(k)$ be the position of the leader at time step k;
$s_F(k)$ be the position of the follower at time step k;
$v_F(k)$ be the speed of the follower at time step k.

The main equation of the sampler is:

$$\Delta\xi(k,\ k+1) = \beta_0 + \beta_1\Delta x(k) + \beta_2\Delta v(k) + \beta_3 v_L(k) \tag{12.1}$$

By definition:

$$\begin{aligned} s_L(k) &= \Delta x(k) + s_F(k) \\ v_L(k) &= \Delta v(k) + v_F(k) \end{aligned} \tag{12.2}$$

To reach the target spacing the vehicle computes at time step k a target position increment for time step $k+1$:

$$\begin{aligned} \Delta s_F(k,k+1) &= \sigma_F(k+1) - s_F(k) = (\sigma_L(k,k+1) - \Delta\xi(k,k+1)) - s_F(k) \\ &= \sigma_L(k,k+1) - \Delta\xi(k,k+1) - (s_L(k) - \Delta x(k)) \\ &= (s_L(k) + \Delta\sigma_L(k,k+1)) - \Delta\xi(k,k+1) - s_L(k) + \Delta x(k) \\ &= \Delta\sigma_L(k,k+1) - \Delta\xi(k,k+1) + \Delta x(k) \\ &= \Delta\sigma_L(k,k+1) - (b_0 + b_1\Delta x(k) + b_2\Delta v(k) + b_3 v_L(k)) + \Delta x(k) \end{aligned} \tag{12.3}$$

where $\sigma_F(k+1)$ is an estimate of the position the follower will reach at time $k+1$, $\sigma_L(k, k+1)$ is an estimate of the position the leader will reach at time $k+1$ and, consistently, $\Delta\sigma_L(k, k+1)$ is an estimate of the space driven by the leader from time k to time $k+1$.

In particular, the sampler estimates at time step k the distance driven by the leader in the next future by considering a uniformly accelerated motion equation:

$$\Delta\sigma_L(k,\ k+1) = 0.5a_L(k)\,\Delta T2 + v_L(k)\,\Delta T$$

where $a_L(k)$ is a direct measure or an estimate of the acceleration of the leader at time step k. Since the controlled vehicle will apply a driven distance increment computed according to Equation (12.3) and in this equation the driven distance increment by the leader has been estimated, the actual driven distance increment at time step $k+1$ will differ from the target one by a quantity Err.

This is the error in estimating the leader progression, deliberately introduced in the model because of the uniformly accelerated motion. Formally:

$$\begin{aligned} \mathrm{Err}(k+1) &= \Delta\xi(k,\ k+1) - \Delta x(k+1) = \sigma_L(k,\ k+1) - s_L(k+1) \\ &= s_L(k) + \Delta\sigma_L(k,\ k+1) - s_L(k+1) \end{aligned} \tag{12.4}$$

Equation (12.3) is the output of the sampler, as well as the input for the profiler. It can be rearranged to make the dynamic process explicit in terms of the position progression of the follower:

$$\begin{aligned} \sigma_F(k+1) = {} & \beta_1 s_F(k) + s_L(k)(1 - \beta_1) + v_L(k)(\Delta T - \beta_3) + \tfrac{1}{2} a_L(k)\Delta T^2 \\ & - b_2\Delta v(k) - \beta_0 \end{aligned} \tag{12.5}$$

We can also rearrange our formulation in a way like the well-known *stimulus–response* paradigm (see Chapters 2 and 3 about Traffic Flow Theory), where we define the response as the target variation of the spacing. In fact, Equation (12.6) below can be easily verified to be equivalent to Equation (12.1).

$$\Delta\xi(k,k+1) - \Delta x(k) = \beta_0 + (\beta_3 + \beta_2)\Delta v(k) + \beta_3 v_F(k) + (\beta_1 - 1)\Delta x(k) \tag{12.6}$$

Equation (12.6) states the stimulus–response approach in a non-classic way. The response is broadly defined in the literature as the acceleration applied by the follower; our definition refers to the target spacing which also depends on the distance driven by the leading vehicle in the meantime. Equation (12.6) can be specified for the case where the stimulus from the relative speed is null:

$$\xi^0(k,k+1) - \Delta x(k) = \beta_0 + \beta_3 v_F(k) + (\beta_1 - 1)\Delta x(k) \tag{12.6.a}$$

By also imposing that the response is null, we obtain pairs (speed, spacing) that represent equilibrium points:

$$\Delta x^* = \beta_0/(1-\beta_1) + \beta_3/(1-\beta_1)v^* \tag{12.7}$$

In other words, when the stimulus related to the relative speed is null, a car-following dynamics still can be observed (the response is not null), as in Equation (12.6a). This happens if the actual spacing is not suitable for the speed at which the vehicles cruise. However, if the system satisfies Equation (12.7), then it is at equilibrium, and the spacing no longer changes.

This phenomenon fits the expectations and confirms that the variables employed in the sampler have been correctly identified. It is worth noting that Equation (12.7) implicitly states the admissibility of parameters β of the model. It has to be ensured that for any feasible (non-negative) speed v^* a feasible (non-negative) value of Δx^* can be obtained. Moreover, it is expected that the regime spacing increases according to the regime speed. In practice, the (linear) curve $\Delta x^* = \Delta x(v^*)$ defined by Equation (12.7) must be a non-decreasing and non-negative function. Sufficient conditions for that are:

$$\begin{aligned} &\beta_0/(1-\beta_1) \geq 0 \\ &\beta_3/(1-\beta_1) \geq 0 \end{aligned} \tag{12.8}$$

Equation (12.8) above could be assumed as a constraint during the calibration of the sampler; however, they were heuristically employed within an unconstrained calibration algorithm.

The sampler should be calibrated in real-time, while the vehicle is driven by the real driver. It represents the learning mode of the ACC system. Without dealing with more complicated considerations, a simple explicit formulation can be applied that minimises the distance between the model outputs and the observed outputs by solving an ordinary least square (OLS) problem:

$$\widehat{\beta} = (\Gamma^T \Gamma)^{-1} \Gamma \Upsilon \tag{12.9}$$

where:

$\widehat{\beta}$ is the vector containing the desired estimates of the model parameters;
Γ is the matrix enlisting in each column the values observed at each time step k for the independent variables (Δx, Δv, v_L), plus a first column of unitary values, accounting for the estimation of the intercept of our model (β_0);
γ is the vector of all the observations of the dependent variables (observed target spacing).

Values in Γ and γ are the observed.

Several algorithms can be employed to solve the OLS problem. One of the most efficient for real-time applications is the recursive least squares (RLS) algorithm; it is widely applied in many areas, such as real-time signal processing. In the general formulation, it minimises a weighted least squares cost function related to the input signal under the hypothesis that a new sample of signals is received at each iteration. Given the incremental nature of the algorithm, computation takes a very reasonable time even if a considerable amount of observed data is processed. Compared with most of its competitors, the RLS exhibits extremely fast convergence. The algorithm was used here in a heuristic way and the stop criterion for terminating the estimation of the parameters was based on two conditions (occurring jointly): Equation (12.8) holds and the objective function of the algorithm improves negligibly. Of course, it is not guaranteed that the algorithm has reached stable solutions, but some empirical evidence can be claimed by considering all the successfully performed calibrations. Moreover, the effectiveness of the algorithm can be empirically accepted if good fitting of the calibrated model against the observed data is evidenced *a posteriori*, as happens in all our cases.

The profiler

The *sampler* estimates at each time step k the distance that the controlled vehicle should drive to reach at time step $k+1$ a human-like target spacing. The responsibility of the *profiler* is to produce a time-continuous trajectory consistent with the requests of the *sampler*. This is done by representing the trajectory of the controlled vehicle (and not the controlled vehicle itself) as a state-space dynamic system evolving from time step k to time step $k+1$. This evolution is controlled (forced) to impose the distance requested to be covered according to the estimates of the *sampler*. Note that the *profiler* is not in our case a car-following model, unlike other time-continuous differential-equation-based approaches. Within the *profiler*, some of the variables introduced in the previous section are redefined:

t_0 is the time instant corresponding to the *sampler* discrete time k;
ΔT is the duration of the time step defined in the discrete-time approach adopted for the sampler; as a consequence, the profiler is in charge of producing a continued trajectory in the time interval $[t_0, t_0+\Delta T]$ and the instant $t_0+\Delta T$ coincides with time $k+1$ of the sampler;
$u = \Delta s_F(k, k+1)$ is the driven distance the profiler should impose from t_0 to $t_0+\Delta T$, as supplied by the *sampler*; it is the instantaneous step-solicitation;
Δs is the distance actually covered by the vehicle in the time interval $[t_0, t_0+\Delta T]$; it could prove different from the requested one (u);
ΔV is the variation in speed actually attained in the time interval $[t_0, t_0+\Delta T]$;
$x_1(t)$ is the instantaneous acceleration at a generic time instant $t\in[t_0, t_0+\Delta T]$; it represents our first state variable;
$x_2(t)$ is the instantaneous jerk at a generic time instant $t\in[t_0, t_0+\Delta T]$; it represents our second state variable.

We assume that the trajectory of the vehicle can be described as a (linear, time-invariant in ΔT) dynamic system according to the following model:

$$\dot{x}_1(t) = \tilde{e}_1\, x_1(t) + \tilde{e}_2\, x_2(t) + \tilde{f}\, u$$
$$\dot{x}_2(t) = e_1\, x_1(t) + e_2\, x_2(t) + f\, u$$

Of course, the state variables have to respect the physical consistency between jerk and acceleration:

$$\dot{x}_1(t) = x_2(t)$$

One of the possible solutions that ensures consistency is:

$$\tilde{e}_1 \ = 0\, \tilde{e}_2 \ = 1\, \tilde{f} \ = 0$$

This is equivalent to rewriting the dynamic system in the form:

$$\dot{x}_1(t) = \ x_2(t)$$
$$\dot{x}_2(t) = e_1\, x_1(t) + e_2\, x_2(t) + f\, u$$

Or, by using a matrix notation:

$$\dot{x}(t) = A\,x(t) + c\,u$$

where

$$A = \begin{bmatrix} 0 & 1 \\ e_1 & e_2 \end{bmatrix} \; c = \begin{bmatrix} 0 \\ f \end{bmatrix} \; x(t) = \begin{bmatrix} x_1(t) \\ x_2(t) \end{bmatrix}$$

As usual for dynamic systems, it is important that the model parameters ensure the stability of the system. This depends on the eigenvalues of matrix A:

$$\lambda_1 = 1/2\left(e_2 - \sqrt{}(4\ e_1 + e_2)^2\right)$$

$$\lambda_1 = \frac{1}{2}\left(e_2 - \sqrt{4\ e_1 + e_2{}^2}\right)$$

$$\lambda_2 = \frac{1}{2}\left(e_2 + \sqrt{4\ e_1 + e_2{}^2}\right)$$

In particular, stability is ensured if the eigenvalues are real, distinct, and negative; this enables the fixed point (regime) to be viewed as a so-called *sink-node* (in the gradient field). This is ensured if:

$$e_1 < 0\, e_2 < 0\, e_2{}^2 > \ -\ 4e_1 \tag{12.10}$$

Eigenvalues can be related to the so-called *time constants* (τ_1, τ_2) that can be used to compute (with excellent approximation) the so-called *settling time* (ts_1, ts_2), defined as the time elapsing from the application of an instantaneous step input to the time at which the state variable has entered a δ bound around the final (fixed-point) value. In formal terms:

$$\tau_i = \ -1/\lambda_i\, \forall\ i \in \{1,2\}$$
$$ts_i \cong 2\ \tau_i\ \ln\left(1/\varepsilon\right) \forall\ i\hat{\mathrm{I}}\{1,2\}$$

This means that if, for instance, δ = 10% then $ts_i \cong 4.6\ \tau_i$ and if δ = 5% then $ts_i \cong 6\ \tau_i$

In practice:

$ts_i \cong \alpha\ \tau_i\ \forall\ i \in \{1,2\}$ where $\alpha\in[4.5,\ 6]$ for an attained system response varying in the range [90%, 95%] of the final response.

On the other hand, the settling times can be imposed to be equal to a predefined part of the whole transition period ΔT. This can be set as a function of two parameters (ω and ψ):

$$ts_1 \ = \ \omega\Delta T\ \omega \in]0,1[ts_2 \ = \ \psi\, \Delta T\ \psi \in]0,1[$$

Finally, it results that:

$$\lambda_1 = -\frac{1}{\tau_1} = -\frac{1}{ts_1/\alpha} = -\frac{\alpha}{\omega\,\Delta T} = \frac{1}{2}\left(e_2 - \sqrt{4\ e_1 + e_2{}^2}\right)$$

$$\lambda_2 = -\frac{1}{\tau_2} = -\frac{1}{ts_2/\alpha} = -\frac{\alpha}{\psi\,\Delta T} = \frac{1}{2}\left(e_2 + \sqrt{4\,e_1 + e_2{}^2}\right)$$

It can be easily verified that previous equations are satisfied by:

$$e_1 = -\frac{\alpha^2}{\Delta T^2\,\psi\,\omega} \tag{12.11}$$

$$e_2 = -\frac{\alpha(\psi + \omega)}{\Delta T\,\psi\,\omega} \tag{12.12}$$

By using the previous values of e_1 and e_2 in Equation (12.10), it can be noted for stability conditions that:

$e_1 < 0$ and $e_2 < 0$, because $\psi\,\omega > 0$ and $\psi + \omega > 0$, being $\psi > 0$ and $\omega > 0$;
$e_2{}^2 > -4\,e_1$ is ensured if $(\psi - \omega)^2 > 0$ that is if $\psi \neq \omega$; this can be verified by substituting Equations (12.11) and (12.12) in $e_2{}^2 + 4\,e_1$, thus obtaining $e_2{}^2 + 4\,e_1 = \varepsilon^2\,(\psi - \omega)^2$, where $\varepsilon^2 = \alpha^2 / (\Delta T^2\,\psi^2\,\omega^2)$.

It results that the stability of the trajectory is ensured by very mild assumptions on the parameters (ψ and ω) governing the settling time of the system. In practice, they just have to be admissible (values from 0 to 1) and distinct.

Stability is evaluated around the fixed point (the regime). Regime values for the state variables can be evaluated as:

$$0 = A\,x^* + c\,u \quad \rightarrow \quad x^* = -A^{-1}\,c\,u$$

where:

$$-A^{-1}\,c\,u = \begin{bmatrix} -f/e_1 \\ 0 \end{bmatrix}$$

It results that the jerk assumes at the fixed point a null value, while the acceleration is finite and assumes a value that depends on the step (u) requested by the *sampler*:

$$x_1^* = -f/e_1\,u \quad x_2^* = 0$$

Given the imposed stability of the system and assuming that the settling times have been properly set by means of parameters ψ and ω, the status of the system at time $t_0 + \Delta T$ can be reasonably considered as being attained:

$$x_1(t_0 + \Delta T) = x_1^* = -f/e_1\,u \tag{12.13}$$

$$x_2(t_0 + \Delta T) = x_2^* = 0 \tag{12.14}$$

Now, consider the variation of speed (Δv) in the time interval $[t_0, t_0{+}\Delta T]$:

$$\Delta v = \int_{t_0}^{t_0+\Delta T} a(t)\, dt = \int_{t_0}^{t_0+\Delta T} x_1(t)\, dt = \int_{t_0}^{t_0+\Delta T} \left[\frac{1}{2}(\dot{x}_2(t) - e_2\, x_2(t) - f\, u)\right] dt =$$

$$= \int_{t_0}^{t_0+\Delta T} \left[\frac{1}{2}(\dot{x}_2(t) - e_2\, \dot{x}_1(t) - f\, u)\right] dt =$$

$$= \frac{1}{e_1}[x_2(t_0 + \Delta T) - \; x_2(t_0)] - \frac{e_2}{e_1}[x_1(t_0 + \Delta T) - \; x_1(t_0)] - \frac{f}{e_1} u\, \Delta T$$

From Equation (12.14) it is also evident that $x_2(t_0) = 0$, because it represents the regime status of a previous time step: $t_0 = (t_0 - \Delta T) + \Delta T$

Then:

$$\Delta v = -\frac{e_2}{e_1}\left[x_1^* - \; x_1(t_0)\right] - \frac{f}{e_1} u\, \Delta T = -\frac{e_2}{e_1}\left[x_1^* - \; x_1(t_0)\right] + x_1^*\, \Delta T \qquad (12.15)$$

Equation (12.15) above, once x_1^* is known, allows the computation of the variation in speed imposed by the *profiler* when the step requested by the *sampler* is imposed. Now consider the actual variation of space (Δs) in the time interval $[t_0, t_0{+}\Delta T]$:

$$\Delta s = \int_{t_0}^{t_0+\Delta T} v(t)\, dt$$

To compute it, the expression of the speed must be derived:

$$v(t) = v(t_0) + \int_{t_0}^{t} a(z)\, dz = v(t_0) + \int_{t_0}^{t} x_1(z)\, dz$$

$$= v(t_0) + \int_{t_0}^{t} \left[\frac{1}{e_1}(\dot{x}_2(z) - e_2\, x_2(z) - f\, u)\right] dz =$$

$$= v(t_0) + \int_{t_0}^{t} \left[\frac{1}{e_1}(\dot{x}_2(z) - e_2\, \dot{x}_1(z) - f\, u)\right] dz =$$

$$= v(t_0) + \frac{1}{e_1}(x_2(t) - x_2(t_0)) - \frac{e_2}{e_1}(x_1(t) - x_1(t_0)) - \frac{f}{e_1} u\, (t - t_0)$$

Therefore:

$$\Delta s = \int_{t_0}^{t_0+\Delta T} v(t)\, dt$$

$$= \int_{t_0}^{t_0+\Delta T} \left[v(t_0) + \frac{1}{e_1} x_2(t) - \frac{e_2}{e_1}(x_1(t) - x_1(t_0)) - \frac{f}{e_1} u\,(t - t_0) \right] dt =$$

$$= \int_{t_0}^{t_0+\Delta T} \left[v(t_0) + \frac{1}{e_1} \dot{x}_1(t) - \frac{e_2}{e_1} x_1(t) + \frac{e_2}{e_1} x_1(t_0) - \frac{f}{e_1} u\,(t - t_0) \right] dt =$$

$$= v(t_0)\,\Delta T + \frac{1}{e_1} \int_{t_0}^{t_0+\Delta T} \dot{x}_1(t)\, dt - \frac{e_2}{e_1} \int_{t_0}^{t_0+\Delta T} x_1(t)\, dt + \frac{e_2}{e_1} x_1(t_0)\,\Delta T$$

$$- \frac{f}{e_1} u \int_{t_0}^{t_0+\Delta T} (t - t_0)\, dt =$$

$$= v(t_0)\,\Delta T + \frac{1}{e_1} x_1^* - \frac{1}{e_1} x_1(t_0) - \frac{e_2}{e_1} \Delta v + \frac{e_2}{e_1} x_1(t_0)\,\Delta T + x_1^* \frac{\Delta T^2}{2}$$

Substituting the expression of Δv (Equation 12.15), it is possible to obtain:

$$\Delta s = v(t_0)\,\Delta T + \frac{e_2}{e_1} x_1(t_0)\,\Delta T - \frac{({e_2}^2 + e_1)}{{e_1}^2} x_1(t_0)$$
$$+ x_1^* \left(\frac{\Delta T^2}{2} - \frac{e_2}{e_1} \Delta T + \frac{({e_2}^2 + e_1)}{{e_1}^2} \right) \tag{12.16}$$

Solving with respect to x_1^* and considering that the aim of the profiler is to force the driven distance to the step requested by the sampler ($\Delta s = u$):

$$x_1^* = \frac{2[{e_1}^2(u - v(t_0)\,\Delta T) + {e_2}^2\, x_1(t_0) + e_1\, x_1(t_0)\,(1 - e_2\,\Delta T)]}{2\, {e_1}^2 + \, e_2\,(2 - 2\, e_1\,\Delta T + e_2\,\Delta T^2)} \tag{12.17}$$

All the calculations stated above are applied by the profiler in the following way.

Step 1.
Fix values for settling time parameters ψ and ω, use admissible values ($\in]0,1[$), ensure stability ($\psi \neq \omega$) and, if appropriate, try to have mild transitions (e.g., $\psi, \omega > 0.70$).

Step 2.
Compute parameters e_1 and e_2 from Equations (12.11) and (12.12) and compute the regime acceleration (x_1^*) by using Equation (12.17) that ensures the driven distance equals that requested by the sampler ($\Delta s = u$).
Step 3.
Check the resulting regime acceleration (x_1^*). Only if the resulting value is inadmissible and/or judged to be inappropriate (e.g., $\|x_1{}^*\| \geq 2$ m/s^2) fix an appropriate value for x_1^* and compute by Equation (12.16) the driven distance (Δs) the profiler will actually set (different from that – u – requested by the sampler).
Step 4.
Compute parameter f by solving from Equation (12.13): $f = -x_1^* e_1 / \Delta s$
Step 5.
Run the profiler as a continuous state-space dynamic system from time t_0 to time t_0 $+\Delta T$; the resulting trajectory is the output of the profiler; in the overall multilayer architecture, this is the input for the tutor.

The output is formally expressed in terms of acceleration and jerk (state variables); if required, other variables describing the trajectory (e.g., speed, position, etc.) can be easily derived from the state variables by using standard integration techniques. Note that the main purpose of step 3 above is to ensure the development of a robust algorithm. In theory, the step requested by the *sampler* imitates a human-like behaviour, that is it should be intrinsically consistent (for example) with admissible accelerations. If this is not the case, it is due to local errors in estimating the human-like spacing to be imposed, for instance, due to an instantaneous malfunctioning of the onboard sensors. The role of step 3 is to avoid the propagation of such an error over the controlled vehicle trajectory.

12.3 Summary

This chapter presents an overview of advanced driver assistance systems (ADAS). It begins with an initial discussion on Intelligent Speed Adaptation (ISA) and proceeds to further discuss and classify various types of advanced driving assistance systems.

Concerning the modelling, some models are also discussed in Chapter 3 about Dynamic Traffic Flow models and in particular in the section about microscopic traffic flow models.

Another topic to be discussed in further detail is the assessment of innovative driving solutions; this is a key problem and it is carried out at different stages of the development process in the automotive field, from design to pre-sale prototypes and Field Operational Tests (FOTs). Distance-based methods were initially used, but they proved inefficient due to the large number of testing kilometers required to assess the safety and reliability performance of Autonomous Driving Systems (ADSs) (Kalra and Paddock, 2016). Consequently, scenario-based testing has become the primary adopted methodology. The scenario-based approach focuses on intentionally varying and analysing the operating scenarios of ADSs, emphasising the more challenging conditions a vehicle may encounter. The SOTIF (Safety of The Intended Functionality) paradigm, based on the ISO standard (ISO 21448:2022, 2022), guides the verification/validation process for ADSs. This standard extends the functional safety model used for conventional vehicles, with a focus on hazardous events resulting from insufficiencies of the intended functionality. ISO 21448 demands the identification of unknown unsafe regions of operation in as much detail as possible. The ADSs' Operational Design Domain (ODD) establishes the boundaries of the entire operation region, identifying the conditions under which a given ADS is designed to operate. In any given ODD, a comprehensive and systematic safety assessment plays a vital role in the verification/validation process to ensure ADSs meet the requirements of ISO 21448.

ADAS provide innovation not only from a safety standpoint but also represent steps towards one of the main market trends of the coming years: autonomous driving.

The vision of autonomous driving has been around for almost as long as automobiles themselves, primarily as a scenario in science fiction novels and movies. However, as early as 1939, the vision was very close to reality. At the New York World's Fair, the idea was presented in the exhibit 'Futurama'. By 1960, it was said, 'this should be accomplished'.

Nevertheless, in the 1950s, engineers were already developing the first concepts for fully automated long-distance units on American highways. Their idea was to combine infrastructure measures and vehicle technology: before setting off, the driver simply had to inform the traffic control center of their destination, and the automated journey could begin.

Developments have progressed. Today's systems use signals from a wide variety of sensors installed on board the vehicle to support the driver. Modern cars are equipped with numerous advanced driver assistance systems (ADAS) that assist the driver in some driving and parking situations and can even take over some tasks entirely. For example, today's lane departure warning (LDW) system alerts the driver to an unintentional lane change, or more advanced systems automatically keep the vehicle in the lane (LKS).

The future seen yesterday is destined to become a reality today and/or tomorrow. Furthermore, automation is more than just the realisation of a long-appreciated vision. It lays the foundation for successfully overcoming the numerous and varied global mobility challenges.

Autonomous driving has the potential to help address the challenges accompanying global development trends. The primary goal is to make road traffic even safer. ADAS technology potentially offers further significant reductions in the number of accidents and traffic congestion. For example, adaptive cruise control (ACC) systems improve traffic flow and make a significant contribution to accident prevention.

Further details about this topic may be found in Chapter 13.

References

Some suggestions for further readings are reported below among the huge literature available on these topics.

Bifulco, G. N., Pariota, L., Simonelli, F., and Di Pace, R. (2013). Development and testing of a fully adaptive cruise control system. *Transportation Research Part C: Emerging Technologies*, 29, 156–170.

Brookhuis, K., and de Waard, D. (1999). Limiting speed, towards an intelligent speed adapter (ISA). *Transportation Research Part F: Traffic Psychology and Behaviour*, 2(2), 81–90.

Canale, M., Malan, S., and Murdocco, V. (2002). Personalization of ACC Stop and Go task based on human driver behaviour analysis. *IFAC Proceedings Volumes*, 35(1), 357–362.

Fan, H., and Poole, M. S. (2006). What is personalization? Perspectives on the design and implementation of personalization in information systems. *Journal of Organizational Computing and Electronic Commerce*, 16(3–4), 179–202.

Kalra, N., and Paddock, S. M. (2016). Driving to safety: How many miles of driving would it take to demonstrate autonomous vehicle reliability? *Transportation Research Part A: Policy and Practice*, 94, 182–193.

Rogers, S., and Langley, P. (1998). Personalized driving route recommendations. *In Proceedings of the American Association of Artificial Intelligence Workshop on Recommender Systems* (pp. 96–100).

Rogers, S., Langley, P., Johnson, B., and Liu, A. (1997). Personalization of the automotive information environment. *In Proceedings of the Workshop on Machine Learning in the real world* (pp. 28–33).

Rosenfeld, A., Bareket, Z., Goldman, C. V., Kraus, S., LeBlanc, D. J., and Tsimhoni, O. (2012). Towards adapting cars to their drivers. *AI Magazine*, 33(4), 46–46.

Rosenfeld, A., Bareket, Z., Goldman, C. V., LeBlanc, D. J., and Tsimhoni, O. (2015). Learning drivers' behavior to improve adaptive cruise control. *Journal of Intelligent Transportation Systems*, 19(1), 18–31.

Chapter 13

Connected, cooperative, and automated mobility ecosystem

Roberta Di Pace[1] *and Facundo Storani*[1]

It's the sides of the mountain which sustain life, not the top.

Robert M. Pirsig, Zen and the Art of Motorcycle Maintenance

Outline. *This chapter provides an overview of the main components of the Cooperative Connected and Automated Mobility. The standardisation, the types of communication, the technological architecture, the protocols, the automation, and the communication levels are presented. The details of the main use cases are summarised. Finally, also the glossary is briefly discussed.*

The CCAM (Connected, Cooperative, and Automated Mobility) ecosystem represents a transformative approach to modern transportation. This mobility revolution brings together various vehicles, infrastructures, and citizens that are *interconnected* to share real-time information. They use this information, process it, and respond *cooperatively,* aided by advanced systems that partially or completely *automate* some or all dynamic driving tasks. With the full integration of CCAM into the transport system, the anticipated positive impacts for society relate to four key areas:

Safety: reducing the number of road fatalities and accidents caused by human error;
Environment and efficiency: decreasing transport emissions and congestion by optimising capacity, smoothing traffic flow, and avoiding unnecessary trips;
Inclusiveness: ensuring inclusive mobility and access to goods for all;
Competitiveness: strengthening the competitiveness of industries through technological leadership, ensuring long-term growth and job creation.

13.1 Intelligent Transportation Systems Services

An Intelligent Transportation System Service (ITS-S) is a service designed to achieve a positive transport-related impact for a transport user (such as drivers or other non-ITS automated systems). This service is delivered to the user of ITS by

[1]Department of Civil Engineering, University of Salerno, Italy

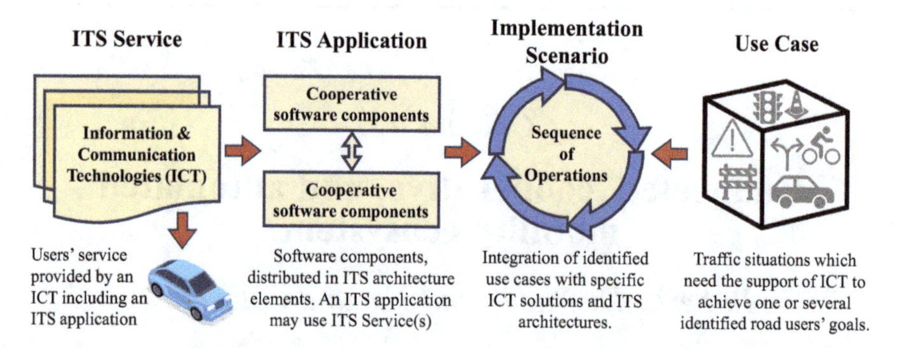

Figure 13.1 Relationship between ITS service, ITS application, use cases, and implementation scenarios

an application, which is a software component installed on Information and Communication Technology (ICT) available at an ITS Station, which is a device or system functioning as a specified entity within its architecture (e.g., vehicle ITS-S, roadside ITS-S, central ITS-S, personal ITS-S).

To provide a service, an application can use data obtained through its own sensors or data from other sources that provide their own services on the same ITS Station. In the case of Cooperative ITS (C-ITS), the information is generated on one ITS-S and used by another ITS-S. For example, one ITS-S may use an application to send a warning to other stations, which eventually reaches a receiving ITS-S that uses another application to provide the information to a user.

Use cases in ITS refer to specific traffic situations where ICT support is required to help road users achieve particular goals. These include both dynamic objects, such as vehicles and pedestrians, and static objects used for traffic management or road hazard identification.

The implementation scenarios integrate use cases with specific ITS architecture and ICT solutions. They outline various environmental conditions and describe in detail the time, position, and relationships between all participants involved. Since ITS systems are typically distributed, effective communication standards are essential for ensuring interoperability between various elements, and these standards must be defined within the relevant implementation scenarios. Figure 13.1 presents the relationship between ITS Service, ITS Application, use cases, and implementation scenarios.

ITS services are expected to function across various situations and cover different use cases. An ITS service could support a single or multiple use cases: for example, the Road Work Warning (RWW) service could support a use case focused on safely guiding drivers through a construction zone (safety-related use case). Alternatively, it could support a use case that simply informs drivers about the road work, allowing them to choose an alternate route (non-safety-related use case). Each use case has its own system requirements.

13.2 Standardisation

In the CCAM ecosystem, numerous standards have been developed by various organisations, companies, associations, industry alliances, research institutes, and universities. These standards focus on specific aspects, aiming to establish common interfaces and languages to advance the deployment of these technologies. They cover a wide range of technology domains, including:

- AD/ADAS functions
- Artificial Intelligence
- Big Data
- Connectivity
- Cybersecurity
- Ethics
- Fleet operations
- Human Machine Interface
- Infrastructure
- In-Vehicle systems, networks, data and interface definition
- Management/Engineering standards
- Map and positioning
- Privacy and security
- Safety
- Terms and definitions
- Testing, verification, and validation

These standards are constantly evolving as technology and research progress, improving the current state and developing new applications. Some of the organisations that develop these standards are international entities, such as:

- ASAM Association for Standardisation of Automation and Measuring systems
- ETSI European Telecommunications Standards Institute
- IEC International Electrotechnical Commission
- IEEE Institute of Electrical and Electronics Engineers
- ISO International Organization for Standardisation
- SAE Society of Automotive Engineers

Other standards have been developed by industry alliances, such as:

- CCC Car Connectivity Consortium
- Car 2 Car Communication Consortium
- 3GPP 3rd Generation Partnership Project
- 5GAA 5G Automotive Association

Specifically, SAE has published several important standards in the Terms and Definitions domain. Some of the most notable ones include:

- SAE J3016 Taxonomy and Definitions for Terms Related to Driving Automation Systems for On-Road Motor Vehicles
- SAE J3208 Taxonomy and Definitions of ADS V&V

- SAE J3206 Taxonomy and Definition of Safety Principles for Automated Driving System (ADS)
- SAE J3216 Taxonomy and Definitions for Terms Related to Cooperative Driving Automation for On-Road Motor Vehicles
- SAE J3259 Taxonomy and Definitions for Operational Design Domain (ODD) for Driving Automation Systems

Other organisations, such as the 5G Automotive Association (5GAA) and the European Telecommunications Standards Institute (ETSI), have developed standards that describe and set the requirements for different applications of V2X (Vehicle-to-Everything) communications. These 'use cases' are high-level procedures of particular traffic situations that need the support of information and communication technologies (ICTs) to achieve specific purposes, described by one (or several) identified road users' goals. Some of these standards include:

- ETSI TR 102 638 Intelligent Transport Systems (ITS); Vehicular Communications; Basic Set of Applications; Definitions
- 5GAA TR T-200111 C-V2X Use Cases and Service Level Requirements Volume I
- 5GAA TR T-210021 C-V2X Use Cases and Service Level Requirements Volume II
- 5GAA TR T-210022 C-V2X Use Cases and Service Level Requirements Volume III
- C-ROADS C-ITS SaUCD C-ITS Service and Use Case Definitions

13.3 Types of communication

The basis of this mobility is the communication and exchange of information, commonly referred to as V2X *(Vehicle-to-Everything)*. V2X encompasses scenarios where vehicles transmit information to any other entity that may affect or be affected by the vehicle. This communication can be classified into different categories according to the entities involved:

- **Vehicle-to-Device (V2D):** when the vehicle communicates with any electronic device connected to it, such as a mobile phone to open and start a car, or connect to the media console via Bluetooth or WiFi-Direct to use specific applications.
- **Vehicle-to-Network (V2N):** when the vehicle communicates with a traffic network management system, for example, a traffic control centre, to receive alerts about traffic congestion, accidents, etc. or to send information about the local state of traffic. Sometimes, 'V2N' also refers to communication with an existing mobile phone network to enable cloud services or other uses. In this case, the mobile phone network acts as the medium to connect to other services, referred to as 'Vehicle-to-Network-to-Everything' V2N2X approach.
- **Vehicle-to-Infrastructure (V2I):** when the vehicle exchanges information with specific road infrastructure around it, such as traffic lights, parking meters, lane markings, road signs, etc.
- **Vehicle-to-Vehicle (V2V):** when vehicles transmit information among themselves;

- **Vehicle-to-Pedestrian (V2P):** when vehicles transmit or receive information from pedestrians, or other vulnerable road users (VRUs) such as cyclists, people using wheelchairs or other mobility devices, commuters, children in strollers, etc.
- **Vehicle-to-Cloud (V2C):** when vehicles exchange information and store data with a dedicated system in the cloud, such as over-the-air updates to vehicle software, remote vehicle diagnostics, etc.
- **Vehicle-to-Grid (V2G):** in the case of electric vehicles, when they communicate with a smart electrical grid to balance electrical loads more efficiently. If the shared (or cooperated) electricity is constrained within a building, it is called Vehicle-to-Building (V2B) or Vehicle-to-Home (V2H). When it is connected to a specific device (or load), it is called Vehicle-to-Load (V2L).

13.4 Technological architecture

In the CCAM framework, several specific devices are necessary to collect data from sensors, process and interpret it, share it, receive other information, decide and choose the following course of action cooperatively or independently, and act accordingly. Numerous on-board systems in vehicles have been developed primarily to increase safety, reduce fuel consumption, and improve transport efficiency. These systems are generally referred to as ADAS (Advanced Driver Assistance Systems) and include features such as Lane-Keeping Assistance (LKA), Forward Collision Warning (FCS), Blind-Spot Warnings (BSW), Adaptive Cruise Control (ACC), Pedestrian Detection, Road Sign Recognition, Automatic Emergency Braking (AEB), Automatic Emergency Steering (AES), etc. However, the systems operate 'isolated' within the vehicle, without V2X communication with external entities, and therefore, without cooperation. In a CCAM context, additional specific devices are needed to facilitate communication:

- **Vehicle side**: On-board units (OBU) reside in the vehicle and manage V2X communications. They receive information from Road Side Units (RSUs), other vehicles, or other on-board units, and share information collected from various subsystems and sensors within the vehicle.
- **Infrastructure side:** Roadside units (RSUs) are installed in permanent locations or temporarily (e.g., at a construction zone or near an accident location). They may collect data from the surrounding environment using specific sensors and cameras, process the information, and receive and transmit information to equipped vehicles and infrastructure devices. The information provided by RSUs is related to traffic, safety messages, adverse weather conditions, accidents, etc. Generally, RSUs are interconnected and linked to a traffic control centre using wired or wireless networks.

13.5 Communication types and protocols

V2X communications utilise physical systems that can wirelessly transmit and receive data between different devices manufactured by various companies. For data to be 'understood' and accessible to connecting devices, a set of known and shared standard protocols is necessary to define the structure and format of the exchanged data. These protocols ensure that data is transmitted reliably and securely between devices, enabling them to make informed decisions based on the received information.

Below, the two most common modes to transmit data are analysed: *Dedicated Short-Range Communications* (DSRC) and *Cellular Vehicle-to-Everything* (C-V2X).

- *Dedicated Short-Range Communications (DSRC):* is a short to medium-range wireless communication channel specifically designed for automotive use. It uses WLAN technology to create a *vehicular ad-hoc network* (VANET) as two V2X senders come within each other's range, facilitating direct communication between vehicles (V2V) as well as vehicles and traffic infrastructure (V2I), without requiring any additional communication infrastructure. This type of communication allows for the transmission of critical information at high speeds. DSRC is a standard proposed by the U.S. *Federal Communication Commission* (FCC), which reserves a specific bandwidth in the 5.9 GHz frequency range. The reference communication standard for DSRC is IEEE 802.11p, which includes improvements to IEEE 802.11 (the standard for Wi-Fi) to support Intelligent Transportation System (ITS) applications.
- *Cellular Vehicle-to-Everything (C-V2X):* it uses cellular networks based on the 3GPP standardised 4G LTE or 5G technology to transmit data over short or long ranges between vehicles, roadside units, cloud services, or generally other V2X devices. It has two modes of operation:
  Device-to-device: direct communication (called 'PC5 or sidelink') without the use of network scheduling for vehicle-to-vehicle (V2V), vehicle-to-infrastructure (V2I), and vehicle-to-pedestrian (V2P) applications.
  Device-to-network: this mode uses a network indirect interface (called 'uU') that allows connected vehicles to access cellular networks to communicate among themselves. It also enables long-range connections to access remote servers and cloud connectivity.

The reference communication standard for C-V2X is the 3GPP communication protocols of releases 14 and above, which define communication standards for cellular mobile networks, including data transmission protocols, network access modes, and security requirements.

13.6 Automation levels

The standard J3016, published by the international association SAE (Society of Automotive Engineers), defines six levels of driving automation. These levels range from no driving automation (Level 0) to full driving automation (Level 5) in the context of motor vehicles and their operation on roadways. The levels depend on the respective roles of the human user and the driving automation system in relation to each other (see Table 13.1).

- **Level 0 (No Automation)**: The human driver is responsible for all aspects of driving, including steering, braking, accelerating, and monitoring the vehicle and roadway.
- **Level 1 (Driver Assistance)**: The vehicle can assist with either steering or acceleration/deceleration using information about the driving environment. The human driver is responsible for the remaining aspects of driving.
- **Level 2 (Partial Automation)**: The vehicle can control both steering and acceleration/deceleration. The human driver must continue to monitor the driving environment and be ready to take over at any time.
- **Level 3 (Conditional Automation)**: The vehicle can perform all driving tasks within certain conditions. The human driver must be available to take over when the system requests.
- **Level 4 (High Automation)**: The vehicle can perform all driving tasks and monitor the driving environment in certain conditions or geofenced areas. The system can bring the vehicle to a safe state if the driver does not take over when requested.
- **Level 5 (Full Automation)**: The vehicle is capable of performing all driving tasks, under all conditions that a human driver could manage. No human intervention is required.

Table 13.1 Levels are determined by the respective roles of the human user and the driving automation system in relation to one another

		Level	Name	**DDT** Dynamic Driving Task		**DDT-F** Dynamic Driving Task Fallback	**ODD** Operational Design Domain
				Vehicle Motion Control	**OEDR** Object and Event Detection and Response		
Driver Support Driver performs part or all of the DDT		0	**No Driving Automation**	*Driver*	*Driver*	*Driver*	Not available
		1	Driver **Assistance**	*Driver* and System	*Driver*	*Driver*	Limited
		2	**Partial** Driving Automation	System	*Driver*	*Driver*	Limited
ADS Automated Driving System ADS performs the entire DDT (while engaged)		3	**Conditional** Driving Automation	System	System	Fallback-ready user	Limited
	ADS-DV Dedicated Vehicle	4	**High** Driving Automation	System	System	System	Limited
		5	**Full** Driving Automation	System	System	System	Unlimited

13.7 Cooperation levels

The standard J3216 of the SAE International describes machine-to-machine (M2M) communication that enables cooperation between two or more participating entities or communication devices possessed or controlled by those entities. This cooperation supports or enables the performance of the dynamic driving task (DDT) for a subject vehicle with driving automation features engaged. Other participants may include vehicles with driving automation features engaged, shared road users (e.g., drivers of manually operated vehicles or pedestrians or cyclists carrying personal devices), or road operators (e.g., those who maintain or operate traffic signals or work zones).

As a result, a vehicle can cooperate with other entities if it has some automated system (level 1 to level 5), and V2X communication to exchange information. This cooperation can be classified into four different classes, from A to D, based on the increasing amount of cooperation (see Table 13.2).

For driver support features (SAE driving automation Levels 1 and 2), only limited cooperation may be achieved. This limitation arises because the automation does not perform complete object and event detection and response (OEDR), relying on the human driver to perform at least some of these functions and supervise feature performance in real-time.

For Automated Driving Systems (Levels 3 through 5), more substantial cooperation may be achieved, as these systems are capable of performing the complete dynamic driving task (DDT) under defined conditions. The Cooperative-Automated Driving Systems (C-ADS) in these levels can leverage full cooperation to enhance driving performance and safety.

Table 13.2 Cooperation classes and their definitions

Class	Cooperation	Phrase	Definition
A	Status-sharing	"Here I am, and here is what I see."	Perception information about the **traffic environment** and information about the **sending entity.** Provided by the sending entity for **potential** utilization by receiving entities.
B	Intent-sharing	"This is what I plan to do."	Information about **planned future actions** of the sending entity Provided by that entity for **potential** utilization by receiving entities.
C	Agreement-seeking	"Let's do this together."	A **sequence** of collaborative messages among **specific CDA devices.** Intended to influence **local planning** of specific DDT-related actions.
D	Prescriptive	"I will do as directed."	The direction of **specific action(s)** to **specific traffic participants** for **imminent** performance of the **DDT** or performance of a **particular task** by a road operator (e.g., changing traffic signal phase), provided by a **prescribing CDA device agent(s)** and **adhered** to by a **receiving CDA device agent(s).**

Cooperative Driving Automation (CDA) enhances the functionalities of devices at two levels:

- Supporting level: when a capability is enhanced (e.g., improving the level of accuracy and reliability of situational awareness) via an A-status sharing or B-intent sharing cooperation classes.
- Enabling level: when the cooperative driving automation overcomes operating limitations and adds new functions (e.g., line of sight, field of view, the ability to directly coordinate specific decision-making, and control processes with other actors), for all the cooperation classes.

13.8 C-V2X use cases

The use cases can be found in various technical reports and standards. These documents describe and define the conditions, requirements, objectives, and other pertinent details for each use case. Particularly, the 5GAA groups some use cases based on the expected benefits into the following categories:

1. Safety,
2. Vehicle operations management,
3. Convenience,
4. Autonomous driving,
5. Platooning,
6. Traffic efficiency and environmental friendliness,
7. Society and community.

Other technical reports or standards may use different classification schemes, such as those based on the entities involved, the goal of the use case, the area of application, or the type of communication (V2V, V2I, etc.).

Below is a non-exhaustive list of some use cases in cooperative V2X interactions. This list may evolve with the development of new technologies, standards, and protocols.

13.8.1 Safety

These use cases provide enhanced safety for vehicles and drivers. Examples of use cases include emergency braking, intersection management, assisted collision warning, and lane change.

- **Cross-traffic left-turn assist:** Assist a host vehicle attempting to turn left across traffic approaching from the opposite, left, or right direction.
- **Intersection movement assist:** Stationary host vehicle proceeds straight from a stop at an intersection. Host vehicle is alerted if it is unsafe to proceed through the intersection.
- **Emergency brake warning:** Alert a host vehicle that a lead remote vehicle is undergoing an emergency braking event.

- **Traffic jam warning and route information:** Alert a host vehicle of an approaching traffic jam, which may be occurring 'on road' (i.e., on the same road as the vehicle) or 'on route' (i.e., on the planned navigation route).
- **Real-time situational awareness and high-definition maps: Hazardous location warning:** An autonomous or semi-autonomous vehicle is driving on a road (route), heading towards a road segment, which presents unsafe and unknown conditions ahead. A host vehicle is made aware of situations detected and shared by remote vehicles. Situations may include such things as accidents, weather, traffic, construction, etc.
- **Cooperative Lane Change (CLC) of automated vehicles: Lane change warning:** A host vehicle signals an intention to change lanes.
- **Vulnerable Road User:** Alert the host vehicle of approaching a vulnerable road user (VRU) on the road or crossing an intersection and warn of any risk of collision.
- **Cooperative traffic gap:** A vehicle tries to pull into a certain lane of a multi-lane road. To do so, it needs to cross multiple lanes. It asks vehicles in the traffic flow to cooperate in forming a gap to support the host vehicle's manoeuvre. If enough vehicles are (opportunistically) found to support this, the host vehicle is informed by the supporting group.
- **Interactive VRU crossing:** A VRU is preparing to cross the street. After signalling this intent, nearby vehicles acknowledge to reassure the VRU that it is safe to cross. As the VRU is crossing, communicating continues with stopped vehicles: the VRU tells vehicles when it has cleared the zone in front of them so that they may continue driving. The VRU double-checks with vehicles just before moving in front of them that they are clear to move forward.
- **Cooperative Adaptive Cruise Control (CACC):** A host vehicle wants to reduce the gap to the leading vehicle (remote vehicle – RV) while keeping an adapted and safe distance to the RV. The goal is to achieve more fluid and controlled speed adaptation to the RV, which creates smoother manoeuvring of the vehicle but also enables a safe reaction to currently not detectable behaviours such as acceleration and breaking from vehicles in the lane ahead.
- **Dynamic Speed Limits (DSL):** A system adjusting speed limits based on real-time traffic and weather conditions to improve safety and reduce congestion. DSL technology is installed on infrastructure devices like electronic signage and can communicate with compatible vehicles.
- **Curve-Speed Warning (CSW):** CSW applications are developed to inform drivers of upcoming curves when their current speed is considerably high and not safe for navigating the curve. The application incorporates data provided by the infrastructure along with vehicle sensor data to provide a recommended safe speed. The recommended speed depends on the geometric shape of the curve, real-time road and weather conditions, vehicle dynamics, and telematics stability information.
    - **Road Works Warning (RWW):** Road Works Warning provides in-vehicle information and alerts about road construction, changes to road layouts, and applicable driving regulations. Road construction usually affects road layout and driving rules. Despite dedicated signage ahead of

roadwork zones, such modified conditions are often a surprise to drivers. This can lead to increased risk and sometimes even accidents, both for road users and workers. The goal of the services is to encourage more careful driving when approaching and passing through a work zone by providing information and warnings onboard the vehicle about road construction, changes to road layout, and applicable driving rules.

- **Road Hazard Warning (RHW):** The road hazard warning service aims to promptly inform drivers of imminent and potentially hazardous events and locations. This allows drivers to be better prepared for impending hazards and make necessary changes and manoeuvres in advance. Participating actors in road hazard warning are vehicle drivers who receive information about the hazardous location on the vehicle's dashboard display, road operators who can report the existence of a hazardous area, service providers who disseminate information about the hazardous location to vehicle drivers, and other organisations responsible for repair, maintenance, and/or cleaning may act on the hazardous location information.
- **Signal Violation Warning (SVW):** SVW aims to reduce the number and severity of reported intersection collisions by warning drivers who, due to high speed, may run a red light or when another vehicle is likely to violate a red light. This service aims to increase drivers' vigilance at signalled intersections to reduce the number of accidents or reduce the impact of accidents when a collision is imminent.
- **Traffic Sign Recognition (TSR):** A technology that enables a vehicle to recognise road signs, such as speed limits, no-turn signs, or tunnel presence.
- **In-Vehicle Signage (IVS):** IVS displays both static and dynamic road sign information inside the vehicle. The service aims to inform drivers through in-vehicle information systems about signages.

13.8.2 Vehicle operations management

These use cases provide operational and management value to the vehicle manufacturer. Use cases in this group would include sensor monitoring, ECU software updates, remote support, etc.

- **Software update:** Vehicle manufacturer updates electronic control module software for targeted vehicles.
- **Vehicle health monitoring:** Owners, fleet operators, and authorised vehicle service providers monitor the health of the host vehicle and are alerted when maintenance or service is required.

13.8.3 Convenience

These use cases provide value and convenience to the driver, not mandated from a safety point of view. Examples for this group can include infotainment, assisted and cooperative navigation, and autonomous smart parking.

- **Automated valet parking – Joint authentication and proof of localisation:** A manual driver drops off the vehicle in a designated transition zone for autonomous parking in a parking facility. The transition zone is separated from the parking area, for example by a barrier that should open for validated vehicles only. After successful authentication and proof of the vehicle's position, the vehicle is accepted for autonomous parking (i.e., the barrier opens for this vehicle).

- **Automated valet parking (Wake up):** A parked sleeping vehicle in a parking facility should be autonomously moved for (re)parking (e.g., charging) or pick-up. For this purpose, the vehicle receives a wake-up call upon which the autonomous drive is prepared.
- **Awareness confirmation:** A host vehicle sends out messages, which can include basic safety messages and in the future, more evolved message types. It indicates which of these messages it would like to receive confirmation for and sets the corresponding properties. It receives confirmations on the messages and processes these to perceive awareness of remote vehicles.
- **Cooperative curbside management:** A vehicle and pedestrian are attempting to meet at a crowded curbside area. A certified infrastructure node designates a pick-up area based on other active pick-ups. It shares this information with vehicles and pedestrians to facilitate the interaction. The vehicle and pedestrian meet in the area, and they confirm the pick-up at the infrastructure node.
- **Cooperative lateral parking:** A vehicle would like to perform longitudinal parking and needs more space. It asks surrounding vehicles to 'squeeze together' and thus make more space temporarily; this is a recursive scheme, if one vehicle cannot create sufficient space, it asks the next vehicle to move as well. Once completed, all remote vehicles return to their original space, except for leaving space for the newly parked vehicle.
- **Obstructed view assist:** When faced with an obstructed view, a host vehicle is provided with an alternate view via a streaming video from the onboard camera of another vehicle or from a fixed camera.
- **Vehicle decision assist**: A host vehicle detects a stationary or a slow vehicle in front. It enquires whether this remote vehicle is only stopping for a short period of time or if it has broken down or similar, making it necessary to overtake the vehicle. The host vehicle decides to overtake it or not, and shares the decision.
- **High-definition sensor sharing:** Vehicle uses its own sensors (e.g., HD camera, lidar), and sensor information from other vehicles, to perceive its environment (e.g., come up with a 3D model of the world around it) and safely performs an automated driving manoeuvre.
- **See-through for passing:** Driver of the host vehicle that signals an intention to pass a remote vehicle using the oncoming traffic lane is provided a video stream showing the view in front of the remote vehicle.
- **Traffic Sign Recognition (TSR):** A technology that enables a vehicle to recognise road signs, such as speed limits, no-turn signs, or tunnel presence.

13.8.4 Autonomous driving

These use cases are relevant for level 4 and 5 autonomous vehicles. Examples in this group are Control if autonomous driving is allowed or not, Tele-operation (potentially with Augmented Reality support), handling of dynamic maps (update/download), and some of the Safety UCs that require cooperative interaction between vehicles to be efficient and safe.

- **Automated intersection crossing:** An autonomous vehicle goes through an intersection with a set of traffic lights. AV goes through or stops taking signal timing into account. When stopping at the intersection, the AV can readjust its position.
- **Autonomous vehicle disengagement report**: When an autonomous HV virtual driver system disengages, it submits a disengagement report containing a time-windowed recording of vehicle systems data, rich sensory information, and dynamic environmental conditions to OEM and government data centres.
- **Cooperative lane merge:** A host vehicle accommodates a remote vehicle travelling ahead in an adjacent lane that is merging with the HV's traffic lane.
- **Cooperative manoeuvres of autonomous vehicles for emergency situations:** An obstacle is detected by an autonomous vehicle in its lane and a manoeuvre is needed to avoid a crash with the obstacle, e.g., a sudden lane change. However, this could result in an accident with a neighbouring or approaching AVs in the adjacent lane. The emergency manoeuvre, together with the actions (e.g., emergency braking, manoeuvre) of neighbouring vehicles are agreed and planned in a cooperative manner.
- **Coordinated, cooperative driving manoeuvre:** A main traffic participant wants to perform a certain action (e.g., lane change, exit highway, U-turn, etc.). Participant shares this intention with other traffic participants potentially involved in the manoeuvre. The traffic participants indicate to the main traffic participant whether they support or plan to decline the planned manoeuvre. The main traffic participant informs a superset of the traffic participants whether it plans to perform the manoeuvre.
- **Data collection and sharing for HD maps:** Vehicles equipped with LIDAR or other sensors can collect environment data around themselves, and share the information classified as objects (with precise position, 3D geometry, but also temporary and variable road signs, road conditions, ...) with an HD map provider. The HD map provider analyses the information collected and merges or combines it to build a regional HD map. This helps to build HD maps that are dynamically updated and more accurate to reflect the actual environment conditions with added accurately positioned objects. Dynamic information on moving vehicles and VRUs not related to the map is not gathered and shared.
- **Infrastructure-assisted environment perception:** When an automated vehicle enters a section of the road covered by infrastructure sensors, it enrols to receive information from the infrastructure containing environment data provided by dynamic and static objects on the road. This data is used to increase the trust level of the car's own sensor observations and extend its viewing range.
- **Infrastructure-based tele-operated driving:** When an automated vehicle detects a failure in a critical subsystem it prepares a status report and using its

geo-position performs the necessary safety function (e.g., slow down or stop) and transmits all information to a tele-operator. Assuming the incident location is covered by infrastructure sensors, the teleoperator retrieves a real-time picture of the road environment centred around the host vehicle. Based on the perceived situation and the capabilities of the car, the remote driver can provide the appropriate trajectory and manoeuvre instructions to help the autonomous vehicle move to a safer location.

- **Tele-operated driving:** A temporary health issue (e.g., illness, headache) of a driver impairs his/her concentration, reactions, and judgement, and consequently affects his/her ability to drive safely. The driver of the vehicle (with some autonomous capabilities) asks a remote driver to take control of the vehicle and drive the vehicle in an efficient and safe manner, from the current location to the destination.
- **Tele-operated driving for automated parking:** When a vehicle arrives at its destination parking area, the driver leaves the vehicle and it is parked by a remote driver located in a tele-operation centre.
- **Hazard information and road event collection for AVs:** Whenever a vehicle detects a hazard or road event based on its own sensor data, the corresponding information (including hazard or event location) is collected for the purpose of sharing with other vehicles, especially AVs and V2X application server.
- **Data sharing of dynamic objects:** Vehicles collect information on dynamic objects on the road and other traffic participants based on vehicle sensor data. They share the relevant information as processed data, so that the environmental perception of the other vehicles is extended.
- **Automated Valet Parking (AVP):** When a vehicle arrives at a designated hand-over zone, the driver leaves the vehicle, and the vehicle is parked being operated by an Automated Valet Parking System (AVPS) after being authorised by the driver.

13.8.5 Platooning

These use cases are those relevant for platooning of vehicles, such as platoon management (e.g., collect and establish a platoon, determine a position in the platoon, dissolve a platoon, manage distance within a platoon, leave a platoon, control of platoon in steady state, request passing through a platoon, etc.).

- **Vehicles platooning in steady state:** Platooning enables a group of vehicles of the same vehicle class (e.g., cars, trucks, buses, etc.) to drive in close proximity in a coordinated manner (e.g., high-density platooning). The head of the platoon (host vehicle) is responsible for coordinating other vehicles in the group (member vehicles) and, potentially, for coordinating with cloud assistance and overall support of the platoon. By sharing status information (such as speed, heading, and intentions such as braking, acceleration, etc.) between the members, and with the support of the platoon head, the distances between vehicles can be reduced, the overall fuel consumption and emissions are also reduced, together with the overall cost. Moreover, platooning enhances safety and efficiency by reducing the influence of unanticipated driving behaviour, small speed variations, and road capacity issues. Having low latency information exchange, the platoon can achieve string stability.

13.8.6 Traffic efficiency and environmental friendliness

These use cases provide enhanced value to infrastructure or city providers, where the vehicles will be operating. Examples of this Use Case group include green light optimal speed advisory (GLOSA), traffic jam information, and routing advice, e.g., smart routing.

- **Speed harmonisation:** Notify the host vehicle of recommended speed to optimise traffic flow, minimise emissions, and to ensure a smooth ride, based on traffic, road conditions, and weather information.
- **Bus lane sharing request:** In order to improve road usage and traffic efficiency a (temporary) access to bus lanes can be granted by the road authority/city. This access could be granted to certain vehicles, e.g., to create an incentive for electric and/or autonomous vehicles.
- **Bus lane sharing revocation:** A bus lane usage application detects an upcoming interruption of public transport by authorised vehicles on the bus lane. The bus usage application revokes the usage granted and instructs the relevant vehicle(s) to free up the bus lane.
- **Continuous traffic flow via green lights coordination:** A series of traffic lights are dynamically coordinated to allow continuous traffic flow over several intersections in one main direction. The benefits of traffic flow optimisation via dynamic traffic signal and phase changes are: reduction of CO_2 emissions, reduction of fuel consumption, reduction of cars' waiting time at side roads, pedestrians receive more time to cross (and help to cross streets with vehicles travelling in platoons), and increased traffic efficiency in urban areas.
- **Group start:** A traffic control centre or host vehicle identifies several vehicles which intend to cross an intersection on a similar path at a similar time. The candidate vehicles are placed into groups following the same paths, guided through the intersection by their corresponding 'group lead' vehicle. After the manoeuvre is executed, the groups are dissolved. The lead vehicle reports the manoeuvre to the traffic control centre.
- **Green Light Optimised Speed Advisory (GLOSA):** Calculates the ideal speed to catch a 'green wave' through the city, suggesting a gradual reduction in speed, up to 250 m before the traffic light, allowing the vehicle to reach the intersection just as the light turns green.
- **Time-to-Green:** Calculates the remaining time until the traffic light turns green.

13.8.7 Society and community

These use cases are of value and interest to society and the public in general, e.g., public services such as road authorities, the police force, fire brigade, and other emergency or government services. Examples in this group are emergency vehicle approaching, traffic light priority, patient monitoring, and crash reporting.

- **Accident report:** When host vehicles are involved in an incident, an accident report containing a time-windowed recording of vehicle systems data, rich sensory information, environmental conditions, and any available camera views is sent to government and private data centres.
- **Patient transport monitoring:** Paramedics, patient monitoring equipment, trauma centres, and doctors share vital patient telemetry data, images, voice, and video during patient transport.
- **Green Priority (GP):** GP modifies the state of traffic signals along the route of a priority vehicle (e.g., public transport or emergency vehicles), stopping conflicting traffic and granting the right-of-way to the vehicle to help reduce service and response times and improve traffic safety. For safety, environmental, traffic flow, or other reasons, it may be advantageous to prioritise specific classes of vehicles. This service enables drivers of priority vehicles to have the right of way at signalled intersections.

13.9 Glossary (based on SAE J3016-202104 and SAE J3216-202107)

In the following, a glossary of some terms is provided, based on the SAE publications J3016-202104 and SAE J3216-202107. A reference to a more comprehensive and updated glossary and taxonomy for CCAM is provided in the further readings.

Active safety system: (SAE J3063) *vehicle* systems that sense and *monitor* conditions inside and outside the *vehicle* for the purpose of identifying perceived present and potential dangers to the *vehicle*, occupants, and/or other *road users*, and automatically intervene to help avoid or mitigate potential collisions via various methods, including alerts to the *driver*, *vehicle* system adjustments, and/or active control of the *vehicle* subsystems (brakes, throttle, suspension, etc.).

Automated Driving System (ADS): The hardware and software that are collectively capable of performing the entire *DDT* on a *sustained* basis, regardless of whether it is limited to a specific *operational design domain (ODD)*; this term is used specifically to describe a Level 3, 4, or 5 *driving automation system*.

Cooperative Driving Automation (CDA): Automation that uses M2M communication to enable cooperation among two or more entities with capable communications technology and is intended to facilitate the safer, more efficient movement of *road users*, including enhancing the performance of the *DDT* for a *vehicle* with *driving automation feature*(s) engaged.

Cooperative Automated Driving System (C-ADS): An *ADS* capable of utilising *CDA*.

Cooperative automated driving system (*C-ADS*)-equipped vehicle: A *vehicle* equipped with Level 3, 4, or 5 *driving automation* and capable of utilising *CDA*. Note that Level 3 systems require human *driver* intervention upon *ADS* request.

CDA driver support feature: A Level 1 or 2 *driving automation system* capable of utilising *CDA* (Refer to SAE J3016 for additional information about *driving automation systems*.).

CDA cooperation classes: Classes of cooperation facilitated by M2M communications among *CDA* devices that may influence *DDT* performance and traffic operations, defined as Classes A through D based on the increasing amount of cooperation entailed in each successive class.

Status-sharing cooperation (Class A): Perception information about the traffic environment and information about the sending entity provided by the sending entity for potential utilisation by receiving entities ('Here I am, and here is what I see.').

Intent-sharing cooperation (Class B): Information about planned future actions of the sending entity provided by that entity for potential utilisation by receiving entities. ('This is what I plan to do.').

Agreement-seeking cooperation [among CDA device agents] (Class C): A sequence of collaborative messages among specific *CDA* devices intended to influence local planning of specific *DDT*-related actions ('Let's do this together.').

Prescriptive cooperation (Class D): The direction of specific action(s) to specific *traffic participants* for imminent performance of the *DDT* or performance of a particular task by a *road operator* (e.g., changing traffic signal phase), provided by a prescribing *CDA* device agent(s) and adhered to by a receiving *CDA* device agent(s) ('I will do as directed.').

CDA device: A device equipped with requisite M2M communication technology that is used by *traffic participants* to perform *CDA* features.

CDA device agent: A *traffic participant* that authorises its *CDA* device to send and receive communications enabling *traffic participants* to engage in *CDA*, and authorises *CDA*-related actions.

CDA feature: The design-specific functionality supported or enabled by M2M cooperation among CDA devices communicating with a *C-ADS* engaging in *CDA*.

Supporting CDA feature: A *CDA* feature capable of promoting cooperation among *traffic participants* intended to augment the performance of actions by *road users* and *road operators*.

Enabling CDA feature: A *CDA* feature capable of promoting cooperation among *traffic participants* intended to facilitate the performance of actions by *road users* and *road operators* that they would otherwise not be able to perform.

[Driverless operation] Dispatching entity: An entity that dispatches an *ADS-equipped vehicle*(s) in driverless operation.

Dispatch [in driverless operation]: To place an *ADS-equipped vehicle* into service in *driverless operation* by engaging the *ADS*.

Driving automation: The performance by hardware/software systems of part or all of the *DDT* on a *sustained* basis.

Driving automation system or technology: The hardware and software that are collectively capable of performing part or all of the *DDT* on a *sustained* basis; this term is used generically to describe any system capable of Level 1 to 5 *driving automation*.

[Driving automation system] Feature: A Level 1–5 *driving automation* system's design-specific functionality at a given level of *driving automation* within a particular *ODD*, if applicable.

Manoeuver-based feature: A *driving automation system feature* equipped on a *conventional vehicle* that either:

1. Supports the *driver* by executing a limited set of lateral and/or *longitudinal vehicle motion control* actions sufficient to fulfil a specific, narrowly defined use case (e.g., parking manoeuvre), while the *driver* performs the rest of the *DDT* and *supervises* the Level 1 or Level 2 feature's performance (i.e., Level 1 or Level 2 *driver support features*);
2. Executes a limited set of lateral and *longitudinal vehicle motion control* actions, as well as associated *object and event detection and response (OEDR)* and all other elements of the complete *DDT* in order to fulfil a specific, narrowly defined use case without human supervision (Level 3 or 4 *ADS* features).

Sub-trip feature: A *driving automation system feature* equipped on a *conventional vehicle* that requires a human *driver* to perform the complete *DDT* for at least part of every *trip*.

Full-trip feature: ADS features that *operate* a *vehicle* throughout complete *trips*.

Driver support [Driving automation system] feature: A general term for Level 1 and Level 2 *driving automation system features*.

Driverless operation [of an ADS-equipped vehicle]: On-road operation of an *ADS-equipped vehicle* that is unoccupied, or in which *on-board users* are not *drivers* or *in-vehicle fallback-ready users*.

Dynamic Driving Task (DDT): All of the real-time operational and tactical functions required to *operate* a *vehicle* in on-road traffic, excluding the strategic functions such as *trip* scheduling and selection of destinations and waypoints, and including, without limitation, the following subtasks:

1. *Lateral vehicle motion control* via steering (operational).
2. *Longitudinal vehicle motion control* via acceleration and deceleration (operational).

3. Monitoring the driving environment via object and event detection, recognition, classification, and response preparation (operational and tactical).
4. Object and event response execution (operational and tactical).
5. Manoeuvre planning (tactical).
6. Enhancing conspicuity via lighting, sounding the horn, signalling, gesturing, etc. (tactical).

Failure mitigation strategy: A *vehicle* function (not an *ADS* function) designed to automatically bring an *ADS-equipped vehicle* to a controlled stop in path following either: (1) prolonged failure of the *fallback-ready user* of a Level 3 *ADS* feature to perform the fallback after the *ADS* has issued a *request to intervene*, or (2) occurrence of a *system failure* or external event so catastrophic that it incapacitates the *ADS*, which can no longer perform *vehicle* motion control in order to perform the fallback and achieve a *minimal risk condition*.

[Dynamic Driving Task (DDT)] Fallback: The response by the *user* to either perform the *DDT* or achieve a *minimal risk condition* (1) after the occurrence of a *DDT performance-relevant system failure*(s), or (2) upon *operational design domain (ODD)* exit, or the response by an *ADS* to achieve *minimal risk condition*, given the same circumstances.

Fleet operations [Functions]: The activities that support the management of a fleet of *ADS-equipped vehicles* in *driverless operation*, which may include, without limitation:

- Ensuring operational readiness.
- Dispatching *ADS-equipped vehicles* in *driverless operation* (i.e., engaging the *ADSs* prior to placing the *vehicles* in service on public roads).
- Authorising each *trip* (e.g., payment, *trip* route selection).
- Providing fleet asset management services to *vehicles* while in use (e.g., managing emergencies, summoning or providing *remote assistance* as needed, responding to customer requests and break-downs).
- Serving as the responsible agent vis-a-vis law enforcement, emergency responders, and other authorities for *vehicles* while in use.
- Disengaging the *ADS* at the end of service.
- Performing *vehicle* repair and maintenance as needed.

Lateral vehicle motion control: The *DDT* subtask comprises the activities necessary for the real-time, *sustained* regulation of the y-axis component of *vehicle* motion.

Longitudinal vehicle motion control: The *DDT* subtask comprises the activities necessary for the real-time, *sustained* regulation of the x-axis component of *vehicle* motion.

Minimal risk condition: A stable, stopped condition to which a *user* or an *ADS* may bring a *vehicle* after performing the *DDT fallback* in order to reduce the risk of a crash when a given *trip* cannot or should not be continued.

[DDT performance-relevant] system failure: A malfunction in a *driving automation system* and/or other *vehicle* system that prevents the *driving automation system* from reliably performing its portion of the *DDT* on a *sustained* basis, including the complete *DDT*, that it would otherwise perform.

Monitor: A general term describing a range of functions involving real-time human or machine sensing and processing of data used to *operate* a *vehicle*, or to support its operation.

Monitor the user: The activities and/or automated routines are designed to assess whether and to what degree the *user* is performing the role specified for him/her.

Monitor the driving environment: The activities and/or automated routines that accomplish real-time roadway environmental object and event detection, recognition, classification, and response preparation (excluding actual response), as needed to *operate* a *vehicle*.

Monitor vehicle performance [For DDT performance-relevant system failures]: The activities and/or automated routines that accomplish real-time evaluation of the *vehicle* performance, and response preparation, as needed to *operate* a *vehicle*.

Monitor driving automation system performance: The activities and/or automated routines for evaluating whether the *driving automation system* is performing part or all of the *DDT* appropriately.

Object and Event Detection and Response (OEDR): The subtasks of the *DDT* include monitoring the driving environment (detecting, recognising, and classifying objects and events and preparing to respond as needed) and executing an appropriate response to such objects and events (i.e., as needed to complete the *DDT* and/or *DDT fallback*).

Operate [a motor vehicle]: Collectively, the activities performed by a (human) *driver* (with or without support from one or more Level 1 or 2 *driving automation features*) or by an *ADS* (Level 3 to 5) to perform the entire *DDT* for a given *vehicle*.

Operational Design Domain (ODD): Operating conditions under which a given *driving automation system or feature* thereof is specifically designed to function, including, but not limited to, environmental, geographical, and time-of-day restrictions, and/or the requisite presence or absence of certain traffic or roadway characteristics.

Plan: A sequence of tasks defined to achieve or maintain a *DDT*-relevant goal during a *trip*.

Receptivity [of the user]: An aspect of consciousness characterised by a person's ability to reliably and appropriately focus his/her attention in response to a stimulus.

Remote assistance: Event-driven provision, by a remotely located human, of information or advice to an *ADS-equipped vehicle* in *driverless operation* in order to facilitate *trip* continuation when the *ADS* encounters a situation it cannot manage.

Remote driving: Real-time performance of a part or all of the *DDT* and/or *DDT fallback* (including, real-time braking, steering, acceleration, and transmission shifting), by a *remote driver*.

Request to intervene: An alert provided by a Level 3 *ADS* to a *fallback-ready user* indicating that s/he should promptly perform the *DDT fallback*, which may entail resuming manual operation of the *vehicle* (i.e., becoming a driver again), or achieving a *minimal risk condition* if the *vehicle* is not operable.

Routine/normal [*ADS*] operation: Operation of a *vehicle* by an *ADS* within its prescribed *ODD*, if any, while no *DDT performance-relevant system failure* is occurring.

Supervise [driving automation system performance]: The driver activities, performed while operating a *vehicle* with an engaged Level 1 or 2 *driver support feature*, to *monitor* that feature's performance, respond to inappropriate actions taken by the feature, and to otherwise complete the *DDT*.

Sustained [operation of a vehicle]: Performance of part or all of the *DDT* both between and across external events, including responding to external events and continuing performance of part or all of the *DDT* in the absence of external events.

Traffic participant: Entities whose actions influence travel in the transportation environment, which may include *road users* engaged in travel upon or across publicly accessible roadways and *road operators*.

> *Road users:* A *traffic participant* on or adjacent to an active roadway for the purpose of travelling from one location to another.
>
> *Road operator:* A *traffic participant* who provides, operates, and maintains the roadways and supporting infrastructure that enable and support the mobility needs of *road users*.
>
> *Certificate authority:* An entity that issues digital certificates that confirm the authenticity of the certificate owner.
>
> *Communications service provider (CSP):* A *traffic participant* who provides and maintains the hardware and software necessary to support secure, low-latency communication M2M between and among *traffic participants*.

Trip: The traversal of an entire travel pathway by a *vehicle* from the point of origin to a destination.

Usage specification: A particular level of *driving automation* within a particular *ODD*.

[Human] User: A general term referencing the human role in *driving automation*.

> *[Human] Driver:* A *user* who performs in real-time part or all of the *DDT* and/or *DDT fallback* for a particular *vehicle*.
>
> > In-vehicle driver: A driver who manually exercises in-vehicle braking, accelerating, steering, and transmission gear selection input devices in order to *operate* a vehicle.
> >
> > Remote driver: A driver who is not seated in a position to manually exercise in-vehicle braking, accelerating, steering, and transmission gear selection input devices (if any), but is able to *operate* the vehicle.

Passenger: A *user* in a *vehicle* who has no role in the operation of that *vehicle.*

[DDT] Fallback-ready user: The user of a *vehicle* equipped with an engaged Level 3 *ADS* feature who is properly qualified and able to *operate* the *vehicle* and is receptive to *ADS*-issued requests to intervene and evident to *DDT performance-relevant system failures* in the *vehicle* compelling him or her to perform the *DDT fallback.*

In-vehicle fallback-ready user: A fallback-ready user of a *conventional vehicle* with an engaged Level 3 *ADS* feature who is seated in the driver's seat.

Remote fallback-ready user: A fallback-ready user of a Level 3 *ADS-equipped vehicle* in *driverless operation* who is not in the driver's seat.

Driverless operation dispatcher: A user(s) who dispatches an *ADS-equipped vehicle*(s) in driverless operation.

Remote assistant: A human(s) who provides *remote assistance* to an *ADS-equipped vehicle* in *driverless operation.*

[Motor] Vehicle: A machine designed to provide conveyance on public streets, roads, and highways.

Conventional vehicle: A *vehicle* designed to be operated by an *in-vehicle driver* during part or all of every *trip*.

[ADS-equipped] Dual-mode vehicle: An *ADS-equipped vehicle* designed to enable either *driverless operation* under routine/normal operating conditions within its given *ODD* (if any), or operation by an *in-vehicle driver*, for complete *trips*.

ADS-dedicated vehicle (ADS-DV): An *ADS-equipped vehicle* designed for *driverless operation* under routine/normal operating conditions during all *trips* within its given *ODD* (if any).

13.10 Summary

This chapter presents an overview of the key elements of Cooperative Connected and Automated Mobility. It covers standardisation, communication types, technological architecture, protocols, levels of automation, and communication, and provides summaries of major use cases. Additionally, it briefly discusses the glossary. The enhancements in terms of modelling approaches are not discussed in this chapter.

The European Union has funded several projects to accelerate the implementation of innovative CCAM technologies and services. One significant initiative under these projects is the creation of a platform designed to provide a common and searchable baseline to ensure the transferability of knowledge for future research, development, and testing of CCAM. This platform, known as the **Knowledge Base on Connected and Automated Driving** (CAD), stands as the ultimate repository consolidating knowledge and experiences related to Cooperative, Connected and Automated Mobility (CCAM) in Europe and worldwide. The CAD platform is continuously updated to reflect the evolving CCAM Research and Innovation (R&I) landscape, incorporating feedback and input from stakeholders. It centralises targeted and structured information on research and innovation projects, demonstration activities, common terminologies, and methodologies for testing and assessment of CCAM, as well as related regulations and procedures, standards, strategies, and action plans, both on European and national levels. It also contains evaluation methodology guidelines, CCAM data-sharing catalogue, glossary, and taxonomies for CCAM, news and events, and offers opinion leaders in the field the opportunity to share their perspective on CCAM.

Reference

Some suggestions for further readings are reported below among the huge literature available on these topics. In particular, the reference to the website reference of the CAD platform discussed in more detail in the remarks, is indicated.

Website of the CAD platform: https://www.connectedautomateddriving.eu/

Part IV

Impacts

Chapter 14
Introduction

Roberta Di Pace[1] *and Stefano de Luca*[1]

> If the present growth trends in world population, industrialization, pollution, food production, and resource depletion continue unchanged, the limits to growth on this planet will be reached sometime within the next one hundred years. The most probable result will be a rather sudden and uncontrollable decline in both population and industrial capacity
>
> The Limits to Growth Abstract established by Eduard Pestel.
>
> *A Report to The Club of Rome (1972)*

Outline. *The following chapter offers an overview of the motivation for examining externalities in relation to the design of traffic management strategies. Specifically, it identifies the main impact indicators and provides a summary of each chapter of Part IV, that discusses these indicators in detail.*

In general, the study of the impact of strategies and their comparison is associated with the analysis of common performance indicators (e.g., travel times, queues, number of stops, etc.).

More recently, due to the shift in context, different programmes reflect the EU's commitment to enhancing road safety through collaboration, research, and innovative practices (e.g., the Road Safety 2021–2030 Strategy, European Road Safety Charter, EU Road Safety Action Program, Road Safety Performance Index, etc.) while also aiming to achieve targets for combating climate change (e.g., the European Green Deal, Fit for 55, Climate Adaptation Strategy, European Climate Pact, etc.).

Safety programmes aim to significantly reduce the number of fatalities and serious injuries, fostering a safe, sustainable, and efficient transport system that meets the needs of all road users. This involves investing in safer road infrastructure and ensuring that existing roads adhere to safety standards. Additionally, the integration of new technologies, including connected and automated vehicles, is actively promoted. Efforts to enhance data collection and analysis are also emphasised, enabling a better understanding of road safety trends to inform policy decisions. One of the key objectives of these programmes is to raise public

[1]Department of Civil Engineering, University of Salerno, Italy

awareness about road safety issues through campaigns and educational initiatives, which aim to change user behaviour and improve compliance with traffic regulations.

In terms of key initiatives aimed at achieving climate neutrality and promoting sustainable development, the focus is on reaching net-zero greenhouse gas emissions by 2050. This includes setting a target to reduce emissions by at least 55% by 2030 compared to 1990 levels by incentivising more efficient and sustainable transport options, as well as supporting the transition to renewable energy sources and increased energy efficiency. New regulations to reduce vehicle emissions and promote electric mobility are also anticipated.

Reducing exposure to noise pollution is a crucial objective for city authorities aiming to improve public health and quality of life. In 2018, the World Health Organization (WHO Europe, 2018) stated that traffic noise harms the health of almost one-third of Europe's population. Indeed, living in an area affected by transport noise is associated with worse health, well-being, and quality of life. In 2017, approximately 18 million people in the European Union (EU) were affected by long-term high annoyance due to transport noise from road, rail, and aircraft sources (EEA, 2022). One of the key targets of the *zero-pollution* action plan is to reduce the number of individuals chronically disturbed by transport noise by 30% by 2030. The number of those subjected to high noise disturbance would therefore need to be reduced by roughly 5.3 million compared to 2017. The above objective may well prove particularly challenging, as the number of people exposed to harmful noise levels has remained stable over the last decade (EEA, 2022). Some major strategic measures concerning road traffic have been identified to meet the zero-pollution target on noise by 2030, such as fleet electrification and the use of low-noise asphalt and noise barriers on main roads. Other tactical measures based on developing enhanced mobility management policies that directly target noise exposure should also be considered. Designing specific traffic control strategies to reduce the impact of vehicles is promising in urban areas, where large variations in speed linked to the effect of congestion are known to increase noise levels. Interest in such policies has been demonstrated through pre- and post-measurement campaigns (Bendtsen and Raaberg 2006; King *et al.*, 2011; Ramis *et al.*, 2003).

As a result, greater attention is being paid to impact indicators related to safety, emissions, energy consumption, and noise that are introduced to provide a broader understanding of the impact of individual strategies, allowing for an evaluation based on multiple criteria. At the same time, they enable the definition of objective/control functions against which traffic management and control strategies can be directed.

This view is also supported by the evolution of the technological landscape (the increasingly tangible development of CCAM ecosystems). Indeed, the availability of reliable traffic measurements is generally one of the prerequisites for traffic control strategies to be effective. In the context of conventional vehicles (non-connected vehicles), such measurements are typically obtained from roadside traffic sensors placed at specific locations. This information is easily collected nowadays thanks to developments in vehicle technologies and smart infrastructures in the context of cooperative, connected, and automated mobility (CCAM).

Communication protocols between each vehicle and the infrastructure and/or between/among vehicles (Dey *et al.*, 2016) support data collection on vehicles and traffic status (Wang *et al.*, 2022) and provide users with accurate and reliable travel information and driving assistance. The automation and communication systems of vehicles provide an opportunity to expand and improve real-time measurement capabilities using information collected on board. Several studies have shown that network performance may be significantly improved by applying enhanced control strategies on networks with connected and automated vehicles rather than with human-driven ones alone (Feng *et al.*, 2015; Liebner *et al.*, 2013; Fajardo *et al.*, 2011). Using the single data provided by connected and automated vehicles, the limitations of aggregate measures may be overcome, allowing the application of disaggregated models for performance and impact estimation, leading to the development of consistent, more effective traffic control strategies. To date, several studies have shown that enhanced strategies based on connected and automated control and speed advisory can significantly reduce network performance (e.g., total time spent, travel time, delay time, etc.) compared to networks with human-driven vehicles (Fajardo *et al.*, 2011; Liebner *et al.*, 2013; Feng *et al.*, 2015).

Moreover, these strategies also affect impact indicators; for this reason, proper optimisation of traffic management strategies may lead to significant benefits.

Several studies have investigated vehicle speed optimisation strategies only through eco-driving strategies, such as variable speed algorithms for arterial traffic commonly known as the Green Light Optimized Speed Advisory (GLOSA) system (Katsaros *et al.*, 2011) or through individual dynamic speed advice aimed at optimising the speed profiles (Asadi and Vahidi 2010; Bandeira *et al.*, 2018; Kamalanathsharma *et al.*, 2015; De Nunzio *et al.*, 2016). Other studies have focused on the development of fuel economic control strategies for a single vehicle without considering the impact on other vehicles (Homchaudhuri *et al.*, 2017; Zhou *et al.*, 2017 and Ma *et al.*, 2017) while few studies have attempted to optimise the fuel consumption of platoons (Zhao *et al.*, 2015). However, very few studies have investigated for instance the use of traffic light signals for fuel consumption and emission reduction (Di Pace *et al.*, 2022; Zhao *et al.*, 2021; Zegeye *et al.*, 2009, 2013; Stevanovic *et al.*, 2009). In more detail, Stevanovic *et al.* (2009) proposed an integrated tool for offline traffic signal timing optimisation combining fuel consumption and emission reduction, while Zegeye *et al.* (2009) designed a model predictive control (MPC) strategy focusing on online control of both the total travel time and total emissions. Zegeye *et al.* (2013) proposed a framework combining macroscopic traffic flow models and microscopic emission and fuel consumption models for freeway traffic networks. At the urban level, Zhu *et al.* (2013) adopted a similar approach aimed at minimising delays and emissions. Referring to noise, one of the challenges to be considered is the implementation of a traffic control strategy that minimised noise emissions while optimising travel times. Indeed, different traffic control strategies affect the speed profile of each vehicle running on the network and can thus directly influence the noise emitted by the traffic flow (Zambon *et al.*, 2017a, 2017b). At the urban level, traffic control based on traffic lights is considered one of the most effective strategies to implement. Furthermore, according to the literature, optimising externalities such as noise is usually combined in a multi-objective network design problem with traffic management measures (Possel *et al.*, 2018), achieving several objectives, such as travel time savings and reduced impacts (e.g., noise pollution, fuel consumption, and vehicle emissions; Wang and Szeto, 2017). Only Stoilova and Stoilov (1998) and Wismans *et al.*, (2011) investigated the relationship between traffic dynamics and noise emissions at simple signalised junctions.

Furthermore, Connected and Autonomous Vehicles (CAVs) aim to eliminate human drivers, a significant contributor to many accidents. Consequently, researchers and experts believe that CAVs have the potential to improve traffic safety and efficiency (Wang *et al.*, 2020; Papadoulis *et al.*, 2019; Wu *et al.*, 2024; Sultana and Hassan, 2024; Garg and Bouroche (2023).

The objective of this chapter is to address the main issues related to externalities.

Concerning safety, two main problems must be addressed: the safety prediction and the Crash frequency and severity estimation.

The Highway Safety Manual (HSM), published by AASHTO in 2010 and 2014, serves as the main reference for safety prediction. It provides extensive guidance on how to incorporate safety analysis into highway planning, design, and operations, and outlines procedures for monitoring and reducing crash frequency and severity. The HSM includes predictive methods for assessing average crash frequency based on total crashes, severity, or collision type. The expected average crash frequency at a specific location is determined using a predictive model estimate, supplemented by observed crash frequency when available. While predictive models differ based on the type of facility and site, they all share fundamental components: Safety Performance Functions, Crash Modification Factors, and Calibration factors.

Crash frequency and severity are vital indicators for assessing the safety effectiveness of road designs and preventive measures. However, due to the infrequency and randomness of crashes, analysing them necessitates extensive historical data. This challenge is particularly evident when implementing new safety strategies, as their effects on safety may take time to appear in actual crash data.

Moreover, there are significant concerns regarding the limitations of crash data, including issues with data quality, difficulties in tracing the sequence of events leading to a crash, underreporting, and the challenges of using crash data to derive new insights or evaluate new safety improvement methods. Consequently, there is a growing need to utilise alternative safety measures to either replace or complement crash data.

The concept of Surrogate Safety Measures (SSM) is essential for providing proactive insights into transportation system safety. SSMs are metrics designed to evaluate road safety without relying exclusively on crash data. They focus on traffic conflict events as a means to observe and assess the potential for all types of crashes. Further details are provided in Chapter 15.

With respect to the noise, Chapter 16, outlines the main models and techniques for estimating the noise produced by road traffic. It particularly examines the noise emissions from individual vehicles, reviewing the characteristics of existing road traffic noise models and highlighting the input parameters necessary for assessment. This effort clarifies both the microscopic and macroscopic features of models found in the scientific literature, specifically contrasting noise emissions from single vehicles with approaches based on overall traffic flow. Noise Emission Models (NEMs), which estimate noise from individual vehicles or total road traffic, are often combined with sound propagation models to project noise levels at receivers, creating a Road Traffic Noise Model (RTNM). Additionally, these models can aggregate contributions from various types of vehicles and multiple units within the same category, moving from individual vehicle evaluations to traffic flow analyses, thereby establishing a 'micro-to-macro' modelling framework. The chapter concludes with a summary of the modelling frameworks used for predicting road traffic noise. It introduces both regression and statistical approaches, followed by an overview of the latest models and techniques available in the scientific community for assessing the noise emissions of single vehicles.

Concerning emissions and consumption, a specific chapter outlines the main models and techniques for estimating the environmental impacts produced by road traffic. Key aspects of emission estimation in road transport simulation are concisely outlined in Chapter 4.4, encompassing methodologies, functions, and tools for assessing emissions, fuel consumption, and energy utilisation in road transport. The chapter focuses on methodologies endorsed by environmental agencies and incorporates extensively documented techniques from literature, providing a robust framework for emission accounting. The models can be classified by the aggregation level but also based on the function typology. Most of the models are regression functions and results to be used in a vast variety of operational tools. They need a lot of data and some of them are not available. In other cases, data are publicly available, such as the open database on electric vehicles (EV) cycle tests by Idhao National Laboratory. A different approach is based on the physical modelling of engines and vehicles giving results formulae for fuel or energy consumption and emissions factors. Even in this case, calibration of the models requires testing activities. Aggregate models may use these detailed models, but results are grouped for an average cycle, activity, and type of vehicle. These aggregate models are the most useable. It is the case of the HBEFA guidebook.

14.1 Summary

The upcoming chapter provides a summary of the reasons for exploring externalities in the context of traffic management strategy design. It specifically highlights the key impact indicators and includes a brief overview of each chapter in Part IV, which discusses these indicators in depth.

Regarding environmental concerns, organisations such as the European Commission and various research institutions have progressively refined the concepts of alternative methodologies, including Well-to-Tank (WTT) and Well-to-Wheel (WtW) analyses. These frameworks concentrate on the upstream emissions linked to fuel production and the emissions resulting from fuel consumption in vehicles. These methodologies have been formalised and widely implemented as part of broader initiatives to assess the environmental impacts of various fuel types and vehicle technologies, particularly in the context of developing policies aimed at reducing greenhouse gas emissions and fostering sustainable transportation solutions.

In the realm of safety, verifying the claimed advantages of Connected and Autonomous Vehicles (CAVs) poses significant challenges. A key concern is that incorporating self-driving cars into current traffic may heighten the risk for road users. Moreover, there are persistent doubts about whether existing traffic management systems can effectively support these new vehicles. Additionally, road authorities must identify the minimum infrastructure needs to ensure the safe coexistence of CAVs and conventional vehicles.

Concerning the integration of the traffic management strategies a possible application may refer to the case in which the vehicle approaches the communication infrastructure (an example of the communication range and the control zone is shown in Figure 14.1).

More generally, the integration of traffic control strategies in traffic networks requires different sub-models interacting with each other in a complex multi-objective optimisation framework. One specific example is shown in Figure 14.2, concerning the case in which the objective function of the control strategy is composed of a performance indicator (e.g., the travel time) and the EV energy consumption.

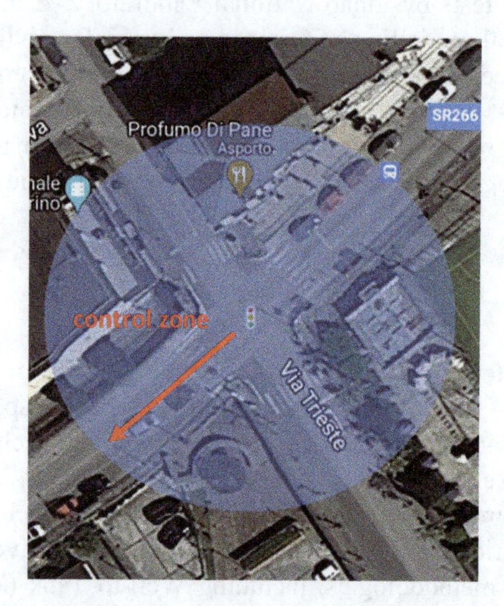

Figure 14.1 Scheme of the control zone and V-to-infrastructure communication

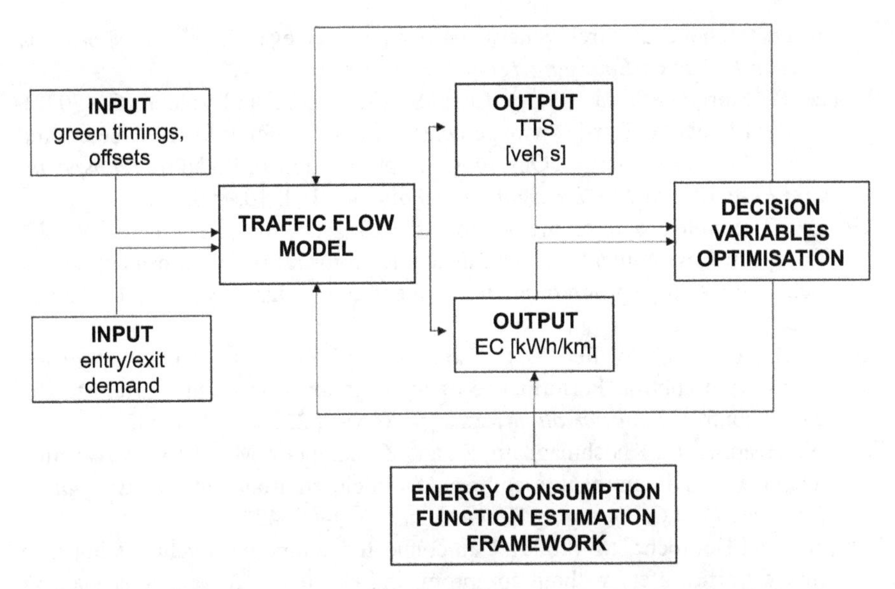

Figure 14.2 Overview of the multi-objective optimisation framework (Di Pace et al., 2022)

References

Some suggestions for further readings are reported below among the huge literature available on these topics. It is also recommended to read the documentation for the main projects mentioned in the text in detail such as the European Commission's "Road Safety 2021–2030" Strategy, the European Green Deal, the Fit for 55 Package, etc.

Asadi, B., and Vahidi, A. (2010). Predictive cruise control: Utilizing upcoming traffic signal information for improving fuel economy and reducing trip time. *IEEE Transactions on Control Systems Technology*, 19(3), 707–714.

Bandeira, J. M., Fernandes, P., Fontes, T., Pereira, S. R., Khattak, A. J., and Coelho, M. C. (2018). Exploring multiple eco-routing guidance strategies in a commuting corridor. *International Journal of Sustainable Transportation*, 12(1), 53–65.

Bendtsen, H., and Raaberg, J. (2006). French experiences on noise reducing thin layers. *Report, Denmark*, Road Directorate, Danish Road Institute.

De Nunzio, G., De Wit, C. C., Moulin, P., and Di Domenico, D. (2016). Eco-driving in urban traffic networks using traffic signals information. *International Journal of Robust and Nonlinear Control*, 26(6), 1307–1324.

Dey, K. C., Rayamajhi, A., Chowdhury, M., Bhavsar, P., and Martin, J. (2016). Vehicle-to-vehicle (V2V) and vehicle-to-infrastructure (V2I) communication in a heterogeneous wireless network–performance evaluation. *Transportation Research Part C: Emerging Technologies*, 68, 168–184.

Di Pace, R., Fiori, C., Storani, F., de Luca, S., Liberto, C., and Valenti, G. (2022). Unified network tRaffic management frAmework for fully conNected and electric vehicles energy cOnsumption optimization (URANO). *Transportation Research Part C: Emerging Technologies*, 144, 103860.

EEA. (2022). Single programming document 2023–2025, EEA Report No 16/2022. European Environment Agency, available from: https://www.eea.europa.eu/publications/singleprogramming-document-2023-2025 (accessed: 14 August 2023).

Fajardo, D., Au, T. C., Waller, S. T., Stone, P., and Yang, D. (2011). Automated intersection control: Performance of future innovation versus current traffic signal control. *Transportation Research Record*, 2259(1), 223–232.

Feng, Y., Head, K. L., Khoshmagham, S., and Zamanipour, M. (2015). A real-time adaptive signal control in a connected vehicle environment. *Transportation Research Part C: Emerging Technologies*, 55, 460–473.

Garg, M., and Bouroche, M. (2023). Can connected autonomous vehicles improve mixed traffic safety without compromising efficiency in realistic scenarios? *IEEE Transactions on Intelligent Transportation Systems*, 24(6), 6674–6689.

Homchaudhuri, B., Vahidi, A., and Pisu, P. (2017). Fast model predictive control-based fuel efficient control strategy for a group of connected vehicles in urban road conditions. *IEEE Transactions on Control Systems Technology*, 25(2), 760–767. doi:10.1109/tcst.2016.2572603

Kamalanathsharma, R. K., Rakha, H. A., and Yang, H. (2015). Networkwide impacts of vehicle ecospeed control in the vicinity of traffic signalized intersections. *Transportation Research Record: Journal of the Transportation Research Board*, 2503, 91–99.

Katsaros, K., Kernchen, R., Dianati, M., and Rieck, D. (2011). Performance study of a Green Light Optimized Speed Advisory (GLOSA) application using an integrated cooperative ITS simulation platform. *In 2011 7th International Wireless Communications and Mobile Computing Conference* (pp. 918–923). Piscataway, NJ: IEEE.

King, E. A., Murphy, E., and Rice, H. J. (2011). Implementation of the EU environmental noise directive: Lessons from the first phase of strategic noise mapping and action planning in Ireland. *Journal of Environmental Management*, 92(3), 756–764.

Liebner, M., Klanner, F., Baumann, M., Ruhhammer, C., and Stiller, C. (2013). Velocity-based driver intent inference at urban intersections in the presence of preceding vehicles. *IEEE Intelligent Transportation Systems Magazine*, 5(2), 10–21.

Ma, J., Li, X., Zhou, F., Hu, J., and Park, B. B. (2017). Parsimonious shooting heuristic for trajectory design of connected automated traffic part II: Computational issues and optimization. *Transportation Research Part B: Methodological*, 95, 421–441.

Papadoulis, A., Quddus, M., and Imprialou, M. (2019). Evaluating the safety impact of connected and autonomous vehicles on motorways. *Accident Analysis & Prevention*, 124, 12–22.

Possel, B., Wismans, L. J., Van Berkum, E. C., and Bliemer, M. C. (2018). The multi-objective network design problem using minimizing externalities as objectives: Comparison of a genetic algorithm and simulated annealing framework. *Transportation*, 45(2), 545–572.

Ramis, J., Alba, J., García, D., and Hernández, F. (2003). Noise effects of reducing traffic flow through a Spanish city. *Applied Acoustics*, 64(3), 343–364.

Stevanovic, A., Stevanovic, J., Zhang, K., and Batterman, S. (2009). Optimizing traffic control to reduce fuel consumption and vehicular emissions: Integrated approach with VISSIM, CMEM, and VISGAOST. *Transportation Research Record: Journal of the Transportation Research Board*, 105–113.

Stoilova, K., and Stoilov, T. (1998). Traffic noise and traffic light control. *Transportation Research Part D: Transport and Environment*, 3(6), 399–417.

Sultana, T., and Hassan, H. M. (2024). Does recognizability of connected and automated vehicles (CAVs) platoons affect drivers' behavior and safety? *Transportation Research Part F: Traffic Psychology and Behaviour*, 103, 368–386.

Wang, L., Zhong, H., Ma, W., Abdel-Aty, M., and Park, J. (2020). How many crashes can connected vehicle and automated vehicle technologies prevent: A meta-analysis. *Accident Analysis & Prevention*, 136, 105299.

Wang, Y., and Szeto, W. Y. (2017). Multiobjective environmentally sustainable road network design using Pareto optimisation. *Computer-Aided Civil and Infrastructure Engineering*, 32(11), 964–987.

Wang, Y., Zhao, M., Yu, X., *et al.* (2022). Real-time joint traffic state and model parameter estimation on freeways with fixed sensors and connected vehicles: State-of-the-art overview, methods, and case studies. *Transportation Research Part C: Emerging Technologies*, 134, 103444.

WHO Regional Office for Europe. (2018). What is the evidence on existing policies and linked activities and their effectiveness for improving health literacy at national, regional and organizational levels in the WHO European region?

Wismans, L. J., Van Berkum, E. C., and Bliemer, M. C. (2011). Comparison of multiobjective evolutionary algorithms for optimisation of externalities by using dynamic traffic management measures. *Transportation Research Record*, 2263(1), 163–173.

Wu, X., Postorino, M. N., and Mantecchini, L. (2024). Impacts of connected autonomous vehicle platoon breakdown on highway. *Physica A: Statistical Mechanics and its Applications*, 650, 130005.

Yang, R. J., Tseng, L., Nagy, L., and Cheng, J. (1994). Feasibility study of crash optimisation. *In International Design Engineering Technical Conferences and Computers and Information in Engineering Conference* (Vol. 97683, pp. 549–556). New York: American Society of Mechanical Engineers.

Zambon G., Roman H. E., and Benocci R. (2017). Scaling model for a speed-dependent vehicle noise spectrum. *Journal of Traffic and Transportation*

Engineering (English Edition), 4(3), 230–239. DOI: 10.1016/j.jtte.2017.05.001;

Zambon G., Roman H. E., and Benocci R. (2017). Vehicle speed recognition from noise spectral patterns. *International Journal of Environmental Research*, 11(4), 449–459. DOI:10.1007/s41742-017-0040-4

Zhao, H. X., He, R. C., and Yin, N. (2021). Modeling of vehicle CO_2 emissions and signal timing analysis at a signalized intersection considering fuel vehicles and electric vehicles. *European Transport Research Review*, 13(1), 1–15.

Zhao, J., Li, W., Wang, J., and Ban, X. (2015). Dynamic traffic signal timing optimization strategy incorporating various vehicle fuel consumption characteristics. *IEEE Transactions on Vehicular Technology*, 65(6), 3874–3887.

Zegeye, S. K. (2011). Model-based traffic control for sustainable mobility. *Thesis* Technische Universiteit Delft TUD.

Zhou, F., Li, X., and Ma, J. (2017). Parsimonious shooting heuristic for trajectory design of connected automated traffic, part I: Theoretical analysis with generalized time geography. *Transportation Research Part B: Methodological*, 95, 394–420.

Zhu, F., Lo, H. K., Lin, H.-Z. (2013). Delay and emissions modelling for signalised intersections. *Transportmetrica B: Transport Dynamics*, 1, 111–135.

Chapter 15
Safety

Maria Rella Riccardi[1] *and Antonella Scarano*[1]

Accidents, and particularly street and highway accidents, do not happen – they are caused.

Ernest Greenwood

Outline. *The following chapter provides an overview of methods that can support the impact evaluation in terms of safety. In particular, it is divided into two parts: the first part focuses on safety prediction with reference to the Highway Safety Manual (AASHTO, 2010, 2014); the second part refers to the concept of Surrogate Safety Measures (SSM) that are crucial metrics used to assess road safety without relying solely on crash data.*

The Highway Safety Manual (HSM), published by AASHTO in 2010 and 2014, is the primary reference for safety prediction. It offers comprehensive guidance on integrating safety analysis into highway planning, design, and operation, and outlines steps to monitor and mitigate crash frequency and severity. The HSM includes predictive methods to evaluate the average crash frequency by total crashes, severity, or collision type. The expected average crash frequency at a site is determined using a predictive model estimate and, when available, observed crash frequency. Although predictive models vary by facility and site type, they all share common core components: Safety Performance Functions, Crash Modification Factors, and Calibration factors. Further details are provided in Section 15.1.

Crash frequency and severity are essential indicators for evaluating the safety effectiveness of road designs or preventive measures. However, due to the rarity and randomness of crashes, studying them requires extensive historical data. This difficulty is especially pronounced when implementing new safety strategies, as their impact on safety can take time to manifest in real-world crash data.

Additionally, there are significant concerns regarding the negative aspects of crash data, such as data quality issues, challenges in tracing the sequence of events leading to a crash, underreporting, and the limitations in using crash data to gain new insights and evaluate new safety improvement methods. Therefore, the need arises to use alternative safety measures to replace or complement crash data.

[1]Department of Civil, Architectural and Environmental Engineering, University of Naples 'Federico II', Italy

The concept of Surrogate Safety Measures (SSM) is crucial in offering proactive insights into transportation system safety. SSMs are metrics used to assess road safety without relying solely on crash data. They are based on the idea of considering traffic conflict events to observe and assess the potential for all types of crashes. Further details are provided in Section 15.2.

15.1 Prediction models

Mobility plays a pivotal role in our daily lives with the road transport system being the connector from our homes to our school or work. However, mobility is not without its challenges. Among them, there are road crashes. Road crashes cause nearly 1.19 million preventable deaths and an estimated 50 million non-fatal injuries, with many incurring a disability as a result of their injury (WHO, 2023). The road fatality is now the 12th leading cause of death for all age groups surpassing HIV/AIDS and tuberculosis. Dramatically, road crashes become the leading cause of death for children and young adults aged 5–29 years. The global burden of road deaths signals the urgent need to make safe mobility a human right. In addition to the human suffering caused by road traffic injuries, road crashes also incur a heavy economic burden on victims and their families, through treatment costs for the injured as well as loss of productivity of those killed or disabled. Other multiple collateral effects of road crashes further affect the transport system with delays and congestion. Given the above, through the UN General Assembly Resolution 74/299 (United Nations, 2020), a Second Decade of Action for Road Safety 2021–2030 has been launched with the explicit aim to stabilise and then half road deaths and injuries pursuing the targets 3.6 and 11.2 presented in the 2030 Agenda for Sustainable Development Goals (United Nations, 2015). The inclusion of specific road safety-related targets reflects the universal recognition that death and injury from road crashes are now among the most serious threats to countries' sustainable development and shall no longer be neglected. Understanding the factors that influence crash occurrence is of particular concern to decision-makers and researchers to meet the global goals for a safer and sustainable transport system.

To date, due to the enormous losses to society resulting from road crashes, researchers have continually sought ways to measure road safety. In the absence of detailed driver data (i.e., acceleration, steering, and braking information), most researchers have addressed this problem by framing it in terms of understanding the factors affecting crash frequency by adopting analytical approaches and using safety prediction methods. The crash frequency is defined as the number of crashes occurring in some geographical space over a specified time period (Lord and Mannering, 2010). Since in a predictive method, the roadway is divided into homogenous roadway elements, the geographical space is usually a roadway segment or intersection, named individual sites. A contiguous set of individual intersections and roadway segments forms a facility. A number of contiguous facilities form a roadway network. Thus, first, it is possible to estimate the expected average crash frequency of an individual site over a given time period, handling geometric design characteristics of the roadway, traffic control features, and traffic volumes (also known as Annual Average Daily Traffic, AADT). Then, the cumulative sum of all sites is used as the estimate for an entire facility or network. Furthermore, the estimate can be made for multiple purposes such as to evaluate road safety for existing conditions, to predict future performance for an existing facility under alternative conditions (i.e., to select countermeasures and carry out economic evaluation for future improvements), or to predict future performance for new roadways during the preliminary phase of the design process. A powerful advantage of the predictive models is that the reliance on the availability of limited crash data for one site is reduced by incorporating predictive relationships based on data from similar sites.

The most important publication in safety prediction research is probably the Highway Safety Manual (AASHTO, 2010, 2014). The HSM provides guidance for integrating safety analysis into highway planning, design, and operation while suggesting steps to monitor and reduce crash frequency and severity. It also includes predictive methods to evaluate the average crash frequency by total crashes, crash severity, or collision type.

The expected average crash frequency of a site, $N_{expected}$, is estimated using a predictive model estimate of crash frequency, $N_{predicted}$ (also referred to as the predicted average crash frequency) and, where available, observed crash frequency (also referred to as the site's past crash frequency for a given time period), $N_{observed}$. During this time period, the site's geometric design and traffic control features are unchanged and traffic volumes are known or forecasted.

The basic elements of the predictive method are:

- A predictive model estimate of the average crash frequency for a specific site type. This is done using a statistical model developed from data for several similar sites. The model is corrected to consider specific site conditions and local conditions. Then it can be used to determine the predicted or expected average crash frequency for the conditions being analysed.
- The use of the empirical Bayes (EB) method combines the estimation from the statistical model with observed crash frequency at the specific site. The EB method uses a weighting factor that is applied to the two estimates and reflects the model's statistical reliability.

Despite a predictive model varies by facility and site type, all predictive models share the same central elements:

- Safety Performance Functions (SPFs): statistical 'base' models used to estimate the average crash frequency per year at a location as a function of traffic volume (and length for road segments) for a facility type with specified base conditions. Base conditions are specified for each SPF and may include additional conditions such as horizontal and vertical alignments, deflection angles, lane width, presence or absence of lighting, presence of turn lanes, presence of tunnels or bridges, and so on. Importantly, SPFs account for the non-linear relationships between crashes and exposure, while also capturing and quantifying the random and uncertain nature of crash occurrences (Lord and Washington, 2018).

- Crash Modification Factors (CMFs): CMFs are the ratio of the effectiveness of one condition in comparison to another condition. Multiplied by the crash frequency predicted by the SPF, CMFs account for the difference between site conditions and specified base conditions.
- Calibration factor (C): it is used to account for differences between the jurisdiction and time period for which the predictive models were developed and the jurisdiction and time period to which they are applied by the analyst.

While the functional form of the SPFs varies, the model to estimate the predicted average crash frequency $N_{predicted}$, is always formed by a multiplication of the predicted average crash frequency determined for the base conditions for a given site type (N_{spf}), Crash Modification Factors (CMFs) specific to the given site type, and the calibration factor (C) to adjust the SPF to local and temporal conditions of the given site type. The predictive model format can be shown as (AASHTO, 2010, 2014):

$$N_{predicted} = N_{spf} \times \left(CMF^1 \times CM^2 \times \ \ldots \ \times CMF_i\right) \times C \quad (15.1)$$

where:

$N_{predicted}$: *a* predictive model estimate of crash frequency for a specific year on the given site type (crashes/year),
N_{spf}: predicted average crash frequency determined for base conditions with the SPF (crashes/year),
CMF_i: Crash Modification Factors,
C: Calibration factor.

15.1.1 Safety Performance Functions, SPFs

The SPFs are multiple regression equations that estimate the average crash frequency for a specific site type (with specified base conditions) as a function of AADT and, in the case of roadway segments, the segment length (*L*). The following equation is the general form of the SPFs developed for segments:

$$N_{spf} = L \times e^{a_0 + a_1 \times \ \mathrm{ln(AADT)}} \times e^{\sum_{i=1}{}^{n} b_i \times x_i} \quad (15.2)$$

where:

a_0, a_1, b_i: model parameters,
AADT = Annual Average Daily Traffic (vehicles per day),
x_i: explanatory variables (i.e., geometric design and traffic control features).

Differently, the SPFs form for intersections consider the AADT on major roads and minor roads separately. The general form is as follows:

$$N_{spf} = L \times e^{a_0 + a_1 \times \ \mathrm{ln(AADTmaj)} + a_2 \times \ \mathrm{ln(AADTmin)}} \times e^{\sum_{i=1}{}^{n} b_i \times x_i} \quad (15.3)$$

where:

a_0, a_1, b_i: model parameters,
$AADT_{maj}$ = Annual Average Daily Traffic on major roads (vehicles per day),
$AADT_{min}$ = Annual Average Daily Traffic on minor roads (vehicles per day),
x_i: explanatory variables (i.e., geometric design and traffic control features).

Modelling SPFs requires assumptions about the distribution of crash counts. Crash counts are properly modelled with several methods that account for the non-negative and integer nature of crash data. The most popular models for analysing crash-frequency data are the Poisson and the Negative Binomial regression models. The Poisson regression approximates crash occurrence as rare-event count data has served as a starting point for crash-frequency analysis for several decades. However, researchers have often found that crash data exhibit features that make the use of the Poisson model problematic. Specifically, the Poisson model cannot handle data over- and under-dispersion. Indeed, one requirement of the Poisson distribution is that the mean of the crash count equals its variance. However, referring to crash data, the variance typically exceeds the mean. If the equality between mean and variance does not hold, the data are said to be under-dispersed (the mean exceeds the variance) or over-dispersed (the variance exceeds the mean). Hence, the regression parameters of the SPFs shall be determined by overcoming the presence of possible over-dispersion in the crash-frequency data. To account for the Poisson regression limitations, most researchers have carried out their research assuming a Negative Binomial distribution for crash frequencies. The Negative Binomial distribution is an extension of the Poisson distribution. The degree of over-dispersion in a Negative Binomial model is represented by a statistical parameter, known as the over-dispersion parameter, that is estimated along with the coefficients of the regression equation. The over-dispersion parameter also provides an indication of the statistical reliability of the SPF. The larger the value of the over-dispersion parameter, the more the crash data vary as compared to a Poisson distribution with the same mean, or in other words, the closer the overdispersion parameter is to zero, the more statistically reliable the SPF.

To sum up, in order to apply for an SPF, the following information about the site under consideration is necessary:

- Basic geometric and geographic information of the site to determine the facility type and to determine whether an SPF is available for that facility and site type.
- Detailed geometric design characteristics and traffic control features of the site to determine whether and how the site conditions vary from those used for the SPF development.
- AADT information for estimation of past periods as well as for forecast estimates of AADT for estimation of future periods.

15.1.2 Crash Modification Factors, CMFs

CMFs represent the relative change in crash frequency due to a change in one specific condition (assuming that all other conditions and site characteristics remain unchanged). CMFs are the ratio of the expected average crash frequency of a site under two different conditions. Hence, a CMF may serve as an estimate of the effect of a specific geometric design, traffic control feature, or the effectiveness of a particular treatment (i.e., illuminating an unlighted road segment, paving gravel shoulders, signalising a stop-controlled intersection, or changing a speed limit). The following equation shows the calculation of a CMF for the change in expected average crash frequency from site condition 'A' to site condition 'B':

$$\text{CMF} = \frac{(\text{Expected average crash frequency with condition } 'B')}{(\text{Expected average crash frequency with condition } 'A')} \tag{15.4}$$

Defined in this way for expected crashes, CMFs can also be applied to the comparison of predicted crashes between the two site conditions. The site condition 'A' serves as the base condition. The assumption allows us to compare treatment options against a reference condition. Under the base conditions (i.e., with no change in the conditions), the value of a CMF is 1.00. CMF values less than 1.00 suggest that the alternative treatment is effective in reducing the estimated average crash frequency in comparison to the base condition. CMF values greater than 1.00 mean that the alternative treatment is ineffective in improving site safety: the implementation of the alternative treatment increases the estimated average crash frequency in comparison to the base condition. The relationship between a CMF and the expected percent change in crash frequency is shown with the following equation:

$$\text{Percent reduction in crashes} = 100 \times (1.00 - \text{CMF}) \tag{15.5}$$

However, some treatments only impact specific collision types. Thus, their impact must be evaluated accordingly:

$$\text{CMF} = (\text{CMF}_{\text{ra}} - 1.0) \times p_{\text{ra}} + 1.0 \tag{15.6}$$

where:

CMF: Crash Modification Factor for total crashes,
CMF_{ra}: Crash Modification Factor for related accidents,
p_{ra}: proportion of related accidents affected by the countermeasure.

In Equation 15.1, the CMFs are multiplicative because the effects of the features they represent are presumed to be independent. Since the CMFs can be also used to estimate the anticipated effects of proposed future treatments or countermeasures, in the presence of multiple treatments such effects can be evaluated by multiplying each countermeasure-related CMF. Nevertheless, multiple treatments may be potentially interrelated and more than one of the treatments may affect the same type of crash. Consequently, users should exercise engineering judgment to assess the independence of different treatments being considered for implementation to avoid overestimation of the combined effects. The benefits of applying multiple CMFs shall be accurately assessed also when multiple treatments change the whole nature or character of the site; in this case, certain CMFs used in the analysis of the existing site conditions may not be compatible with the proposed treatment.

In the HSM (AASHTO 2010, 2014), SPFs are accompanied by crash severity and collision-type distribution tables and adjustment factors; similarly, some CMFs are accompanied by tables of collision-type proportions of total crashes.

15.1.3 Calibration factor, C

The use of the predictive models regardless of the geographical context requires the calibration of the HSM-SPFs, and replacement of crash severity and collision type distribution tables and adjustment factors to local and site-specific conditions. Different weather conditions, driver and animal populations, crash reporting system procedures and thresholds, and time periods (i.e., different years) are some of the reasons supporting the need for developing calibration factors for the SPFs toward more accurate local estimates of expected frequency and severity of crashes. The calibration factors will have values greater than 1.0 for those roadways that, on average, experience more crashes than the roadways used in developing the SPFs. The calibration factors for those roadways that, on average, experience fewer crashes than the roadways used in the development of the SPF, will exhibit values less than 1.0.

15.1.4 The empirical Bayes method, EB method

The outcome of SPFs provides the predicted average crash count of a homogenous element (segment or intersection), and it has been often interpreted to reflect the unobserved true safety of the site. Analysing a given observation period, at the same site, one shall conclude that the observed crash count may reflect how the site is performing with respect to safety in contrast to how it should perform, as predicted by an SPF. Nevertheless, in any period the observed crashes are likely to 'regress to the mean' so that sites that appear to perform worse than average in a given observation period, are likely to improve in the subsequent period because the number of crashes will regress to the long-term mean, even if no countermeasures have been implemented to improve the safety of such sites. Vice versa a site that experienced a low number of crashes will show an increase in the number of crashes. The number of crashes will regress toward the long-term average as well. Every site's crash history (number of crashes per year) fluctuates randomly over the years. The regression-to-the-mean bias is even stronger when observed crash averages are based on short-term periods (e.g., 1 to 3 years). To combine the predicted average crash frequency with the observed crash frequency and compensate for the potential regression to the mean bias, the empirical Bayes (EB) method is used. This fundamental notion of combining results obtained from the comparison of similar sites and the variance of crashes 'within' a specific site led to the development of the EB approach, which is the most reliable approach for accurately predicting the future safety performance of a site (Lord and Washington, 2018).

The EB method uses a weighted factor, w, which is a function of the SPFs overdispersion parameter that determines a weighting between $N_{\text{predicted}}$ and N_{observed}.

$$EB_{\text{estimate}} = w \times N_{\text{predicted}} + (1 - w) \times N_{\text{observed}} \tag{15.7}$$

The value of the weight depends on the reliability of the SPFs (how different can be the safety at a site in relation to the average predicted by the SPF) and the reliability and quality of the crash count. The weight factor w is evaluated using Equation (15.8).

$$w = 1/\left(1 + \left(E(\hat{Y})\right)/k\right) \quad (15.8)$$

The value of the weight factor decreases if the value of the overdispersion parameter increases, and thus more emphasis is placed on the crash count rather than the predicted crash frequency obtained with the SPF. Hence, less weight is placed on the SPF estimation and more weight on the observed crash frequency. This is the case of the SPF model developed using highly over-dispersed data. On the other hand, when the data used to develop a model have little or no over-dispersion, the reliability of the resulting SPF is likely to be higher and more weight is placed on the SPF estimation.

However, if a site's crash history is not available, the EB method cannot be applied and the estimate of expected average crash frequency is limited to using a predictive model (i.e., the predicted average crash frequency) that can be performed to formulate safety evaluations. A brief logical process for safety prediction is captured in Figure 15.1.

To date, many researchers have developed SPFs and studied the transferability of HSM-SPFs processes. Several studies have been performed to transfer the HSM-SPFs into different local jurisdictions, states, and even countries. Mehta and Lou (2013) conducted a study to calibrate and transfer HSM-SPFs for a rural two-lane two-way road and a four-lane divided road in Alabama, concluding that the calibrated HSM-SPFs have good crash prediction performance. A similar conclusion was also drawn by Moraldi *et al.* (2020) after performing a study to calibrate the HSM-SPF for rural two-lane two-way roads in Germany. Martinelli *et al.* (2009) and La Torre *et al.* (2014) examined the HSM-SPF transferability to Italian roads. Martinelli *et al.* (2009) applied the calibration procedure for the two-lane two-way rural road segment model suggested by HSM to a road located in Arezzo, Italy. By analysing actual versus predicted crashes and residual plots, they concluded that the best approach is the base model with CMF calculation, as defined by the HSM procedure, but with the calibration coefficient calculated not as a simple mean of each class coefficient but using a weighted average based on the total length of the sections in each class. La Torre *et al.* (2014) examined the transferability of the HSM freeway model to the Italian motorway network using four indicators (mean absolute deviation, calibrated overdispersion parameter, root mean square error, and residual plots). They concluded that the model shows good transferability to the Italian network, especially for fatal and injury crashes.

By contrast, many other studies indicated that HSM-SPFs often have low accuracy prediction performance in local jurisdictions. AlKaaf and Abdel-Aty (2015) conducted a study to calibrate and transfer the HSM-SPF for urban four-lane divided roads in Riyadh, Saudi Arabia. In this study, local CMFs were developed for the calibration process. The results indicated that employing the local CMFs

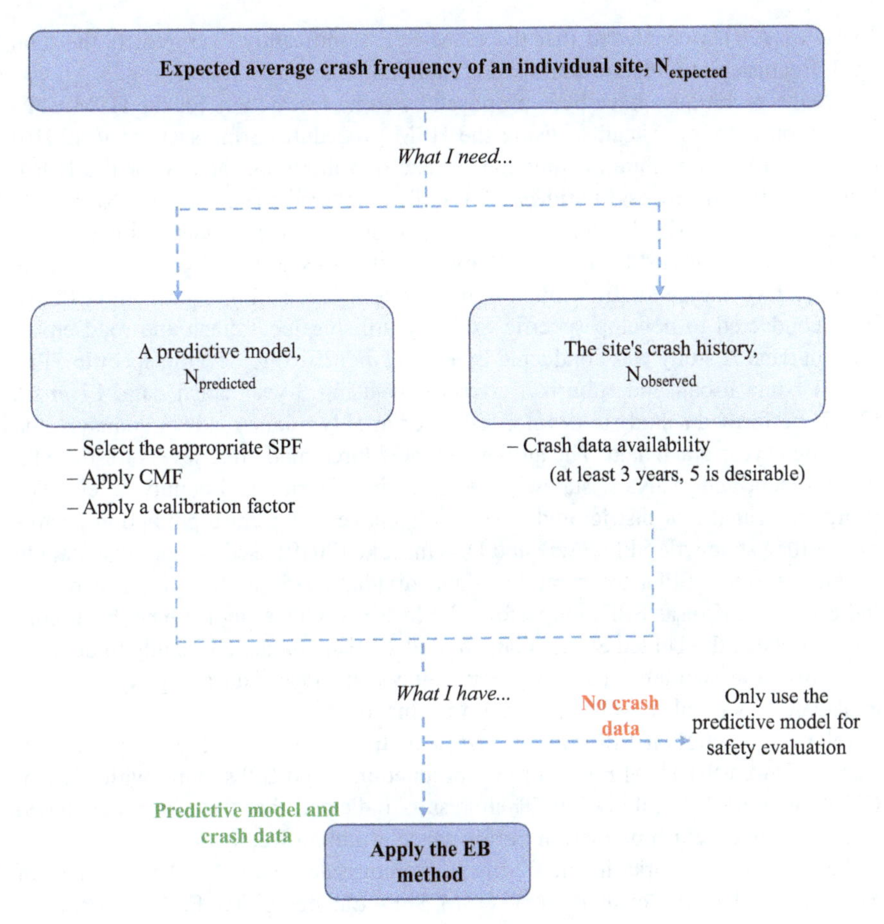

Figure 15.1 Evaluation of the expected average crash frequency of an individual site

instead of HSM-CMF values gives better prediction performance. A study by Sun *et al.* (2011) was performed to calibrate the HSM-SPF for rural multilane roads in Louisiana. The results indicated that the HSM-SPF underpredicts the crash frequency. Likewise, Cafiso *et al.* (2012) found that HSM-SPF underpredicts fatal and severe injury crash frequency in Italy dividing multilane roads by 26%. Brimley *et al.* (2012) performed a study to calibrate and transfer the HSM-SPF of rural two-lane two-way roads in Utah. They found that the HSM-SPF underpredicts the crash frequency by 16%. On the other hand, Srinivasan and Carter (2011) conducted a study to calibrate the HSM-SPF for North Carolina rural divided multilane roads. They found that the HSM-SPF slightly overpredicts (less than 5%) the crash frequency. A similar conclusion has been drawn by Sun *et al.* (2014) regarding using the HSM-SPF to predict the total crash frequency on Missouri rural divided multilane roads.

Xie *et al.* (2011) concluded that the HSM-SPF significantly overpredicts the total crash frequency at Oregon rural divided multilane roads by 22%.

Novel techniques have been proposed by some researchers for the HSM-SPFs calibration process instead of using the HSM procedure. Srinivasan *et al.* (2016) proposed using a calibration function instead of calibration factors for the HSM-SPFs calibration process. Farid *et al.* (2018) employed the K-Nearest-Neighbors regression for the HSM-SPFs calibration process. Both techniques had better performance than the HSM procedure. However, the K-Nearest-Neighbors technique outperformed the calibration function technique. Meanwhile, several studies have been conducted to develop specific SPFs by utilising local crash and road environment data. A study was conducted by Kim *et al.* (2015) to develop specific SPFs for Alabama urban and suburban arterials by using 3-year crash data. Li *et al.* (2017) performed a study to develop SPFs for rural two-lane roads in Pennsylvania by using 8-year crash data. The authors adopted three modelling levels for the SPF development and analysis: statewide, engineering district, and county levels. The results indicated that district and county SPFs have better crash prediction performance than statewide SPFs. Aziz and Dissanayake (2019) used 3-year crash data to develop specific SPFs for rural four-lane divided roads in Kansas. The results indicated that Kansas SPFs outperform HSM-SPFs. Other studies have been conducted outside the United States. Garach *et al.* (2016) conducted a study to develop SPFs for rural two-lane roads in Spain. Five-year crash data along with several explanatory variables were gathered for this purpose. La Torre *et al.* (2019) developed jurisdiction SPFs for freeways in Italy examining 5-year crash data. They followed the HSM procedure by production-based SPFs along with a set of CMFs for the SPFs calibration. Their results indicated that the newly developed SPFs have good crash prediction performance at Italian freeways.

Several other works in the pertinent literature deal with the development of SPFs rather than the evaluation of HSM-SPFs transferability. Early attempts to correlate crash frequency to traffic and road geometric characteristics were made by Abdel-Aty and Radwan (2000). In New Zealand, Turner *et al.* (2012) developed SPFs for two-lane rural roads using the GLM approach (Generalized Linear Modelling technique) for key crash types. Using a stepwise forward procedure based on the generalised likelihood ratio test, Caliendo *et al.* (2007) developed a prediction model for Italian four-lane median-divided motorways. Montella *et al.* (2008, 2015) developed SPFs for Italian rural motorways, also using GLM techniques and assuming a negative binomial distribution error structure. Separate models were calibrated for the following crash types: total, nighttime, daytime, rainy, non-rainy, Run-Off-Road, non-Run-Off-Road, injury, and Property Damage Only. Cafiso *et al.* (2010) defined SPFs for two-lane rural road sections based on a combination of exposure, geometry, consistency, and context variables directly related to the safety performance, also based on the GLM approach and assuming a negative binomial distribution error structure. Using data from interchange influence areas on urban freeways in the state of Florida, USA, Haleem *et al.* (2013) developed an SPF regarding the effect of changes in median width and inside and outside shoulder widths, applying a promising data-mining method known as multivariate adaptive regression splines. Other researchers (Cansiz, 2010; Çodur and Tortum, 2015) proposed artificial intelligence approaches (artificial neural networks and genetic algorithms) to crash prediction modelling.

Several methodologies have been used in the literature to estimate CMFs. A review of the methodologies employed for CMF estimation with a larger focus on practical implementation can also be found in A Guide to Developing Quality Crash Modification Factors (Gross *et al.*, 2010). The most basic method for estimating a CMF, the so-called naive before–after approach, involves a simple comparison of accident rates before and after the implementation of a treatment (Allaire *et al.*, 1996; Graham and Harwood, 1982; Outcalt, 2001; Pitale *et al.*, 2009). Despite the advantage that the approach is simple to apply, it has several limitations: (1) it does not take into account changes in traffic volumes that can affect crash rates and, even if crash rates are normalised by some measure of traffic volume (as was done, for instance, by Graham and Harwood, 1982; Outcalt, 2001; and Pitale *et al.*, 2009), (2) the approach still does not account for other factors that could potentially affect crash rates, such as general time trends. Some authors have implemented a full Bayesian approach to estimate CMFs (Pawlovich *et al.*, 2006; Persaud *et al.*, 2010). This approach also uses a group of reference sites but, instead of point estimates of the expected number and variance of crashes, it estimates a probability distribution for the expected crash rates. This is then used to estimate the expected number of crashes at the treatment site in the after period had the treatment not been implemented. Furthermore, CMFs can be estimated with multivariate regression models. In this case, crashes are modelled as a function of a set of explanatory variables. Typical explanatory variables are traffic volume and segment length, but other variables (e.g., geometric design, driving density, friction) are commonly also included. Typically, negative binomial (Cafiso *et al.*, 2010, 2021; Persaud *et al.*, 2012; Turner *et al.*, 2012) or Poisson (Dinu and Veeraragavan, 2011; Wichert and Cardoso, 2007) models are used, although other modelling forms have also been developed (e.g., log-linear (Zegeer *et al.*, 1988), zero-inflated Poisson (Qin *et al.*, 2004). Multivariate regression models can be useful when only cross-sectional data are available. However, simple multivariate regression models also do not consider the fact that treatment implementation is not random. The treatment variable will therefore be endogenous in the model (correlated with the error term) and more advanced modelling techniques (e.g., instrumental variables) are needed to obtain unbiased estimates of the effect of the treatment.

The empirical Bayes (EB) methodology has been applied for over 30 years now in conducting statistically defendable before–after studies of the safety effect of treatments applied to roadway sites and is considered the state-of-art approach (Hauer, 1997; Hauer *et al.*, 2002; Khan *et al.*, 2015; Montella, 2009, 2010; Persaud and Lyon, 2007). The methodology is highly appealing because it corrects for regression to the mean and traffic volume. It uses reference sites to estimate the expected number of crashes in the treatment site that would have occurred in the after period in the absence of the treatment by estimating SPFs for crash prediction using data from reference sites. These estimates are combined with observed crashes in the before period to estimate the expected number of crashes in the treatment site in the after period in the absence of the treatment.

15.2 Surrogate measures of safety

Crash frequency and severity are critical metrics for assessing the safety effectiveness of road designs or preventative measures. However, crashes are rare and random events, making them challenging to study without a substantial amount of historical data (Elvik, 2009). This challenge becomes evident when implementing new safety strategies, as their safety impact takes time to become apparent in terms of real-world crash data.

Relying on past crash data to evaluate safety strategies may not be the most optimal approach. There is also significant concern regarding the negative aspects of the data, such as data quality, the difficulty in tracing the event sequence leading to a crash, underreporting of data, and limitations in crash data for acquiring new knowledge and evaluating new safety improvement methods (Elvik and Mysen, 1999; Lord and Washington, 2018). Therefore, from the need to replace or complement crash data arises the necessity to conduct analyses using alternative safety measures.

The concept of Surrogate Safety Measures (SSM) plays a pivotal role in providing proactive insights into transportation system safety. Research into SSM dates to the early 1970s, as documented by Hayward (1971). Since then, this field has made significant advancements, growing in relevance over time. In recent years, SSMs have been widely adopted by researchers and industry. They utilise recorded videos and traffic simulation output to carry out alternative or complementary analyses compared to traditional crash data analysis, offering a good perspective on traffic safety.

SSMs are metrics used to assess road safety without necessarily relying on crash data. They are founded on the idea of considering traffic conflict events for observing and assessing all types of crash potential (Johnsson *et al.*, 2021). A road crash occurs when more vehicles converge at the same point in both space and time. Depending on various factors such as reaction times, braking efficiency, visual capabilities, attentiveness, and speed, a collision can be avoided through evasive actions.

Traffic conflicts represent an event where the interaction among multiple road users, both spatially and temporally, creates potentially hazardous scenarios that may result in a crash if no successful evasion manoeuvre is performed (Amundsen and Hyden, 1977; Chin and Quek, 1997; Tarko, 2020a) (Figure 15.2). The frequency of such conflicts and their associated crash probabilities offer valuable insights into road safety.

Notably, traffic conflicts occur more often than actual crashes and require shorter data collection periods. These characteristics make the traffic conflict technique a proactive analysis approach based on direct observation, rather than a reactive strategy relying on historical crash data.

Several forms and definitions are associated with SSMs, among which traffic conflict analysis stands out as both widely utilised and highly consistent. These measures serve as tools for quickly evaluating innovative road designs and traffic control strategies,

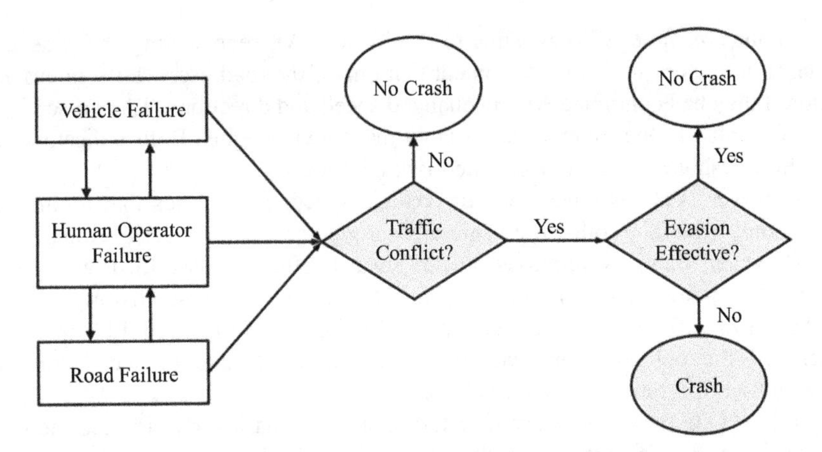

Figure 15.2 Causality model of traffic conflicts and crashes (Tarko, 2020b)

bypassing the necessity for crash history. Furthermore, SSMs facilitate the identification of safety issues and proactive countermeasures (Lord *et al.*, 2021).

SSMs are commonly employed to estimate the road crash likelihood. Fewer researchers have used surrogate measures to assess crash severity, regarding it as a consequence of the crashes that have already occurred. However, recent studies are showing a growing trend towards utilising SSMs for estimating crash severity as well (Arun *et al.*, 2021; Laureshyn *et al.*, 2010; 2017; Zhou *et al.*, 2011).

Three primary sub-groups exist within SSMs:

1. the temporal-based SSMs,
2. the deceleration and distance-based SSMs, and
3. the energy-based SSMs.

15.2.1 The temporal-proximity-based surrogate safety measures

Temporal-proximity-based SSMs are relevant for assessing road safety by considering both spatial distance and vehicle speed simultaneously.

Among these measures, Time-to-Collision (TTC), introduced by Hayward (1971), stands out as the most widely used (Figure 15.3). It estimates the remaining time before a crash occurs between two vehicles if no evasive action is taken. It is calculated as:

$$TTC_i = D/(V_i - V_{i-1}) \tag{15.9}$$

where:

i: the following vehicle,
$i-1$: leading vehicle,
D: space between vehicle,
V: vehicle speed.

Another notable SSM is Time-to-Accident (TA), representing the time that remains until a crash, from the moment that one of the road users starts an evasive action if they had continued with unchanged speed and directions. TA is essentially the TTC value at the moment an evasive manoeuvre begins. Both indicators use specific thresholds to identify elevated risk conflicts.

However, TTC assumes constant vehicle speed. It supposes that a collision occurs only when the following vehicle has a greater speed than the leading vehicle. However, these assumptions do not always reflect real-world driving behaviour. To address these limitations, the Modified Time-to-collision (MTTC) (Ozbay *et al.*, 2008) was introduced. MTTC represents a modified TTC that considers all the potential longitudinal conflict scenarios due to acceleration or deceleration differences between vehicles.

The MTTC formulas to evaluate four distinct scenarios, considering the relative distance, speeds, and accelerations between the leading and following vehicles were reported in the equation below:

$$MTTC = \{max(t_1, t_2), if\ \Delta a > 0; min\ (t_1, t_2), if\ \Delta a < 0\ and\ \Delta V > 0; t_3, if\ \Delta a = 0\ and\ \Delta V > 0; N/A, if\ \Delta a \leq 0\ and\ \Delta V \leq 0\} \tag{15.10}$$

where:

$t_1 = (-\Delta V + \sqrt{(\Delta V^2 + 2\Delta a D)})/\Delta a$
$t_2 = (-\Delta V - \sqrt{(\Delta V^2 + 2\Delta a D)})/\Delta a$
$t_3 = D/\Delta V$
$\Delta V = V_i – V_{i-1}$ (i: following vehicle, i–1: leading vehicle)
$\Delta a = a_i – a_{i-1}$
$D = X_{i-1} – X_i$
V, a, X are the position, speed, and deceleration rate, respectively.

Several more complex SSMs such as Time-Exposed TTC (TET) and Time-Integrated TTC (TIT) were derived from TTC (Minderhoud and Bovy, 2001) (Figure 15.3). TET measures the time during a conflict when TTC falls below a

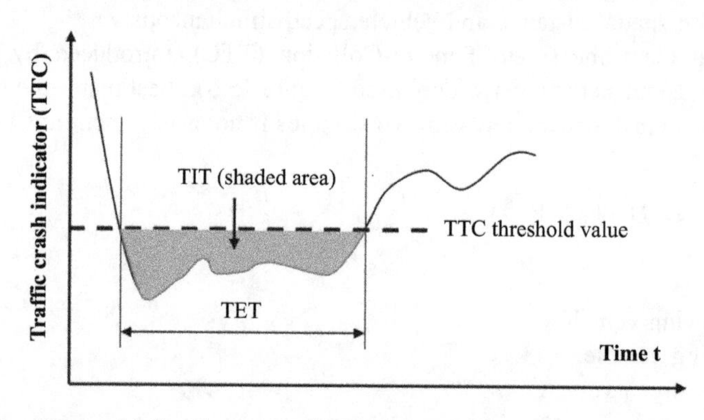

Figure 15.3 Concept of TIT and TET (Lord et al., 2021)

designated TTC threshold, while TIT represents the integral of the TTC profile during the time it is below a certain TTC threshold. Unlike TTC and TA, TET, and TIT focus on measuring the risk associated with the duration of dangerous driving conditions.

$$TIT(t) = \sum_{n=1}^{N} [1/TTC_{it} - 1/TTC^*] \cdot \Delta t,\ 0 < TTC(t) < TTC^* \quad (15.11)$$

$$TIT^* = \sum_{t=1}^{T} TIT\ (t)$$

$$TET(t) = \sum_{n=1}^{N} \delta_t \times \Delta t,\ \delta_t = \{1, 0 < TTC(t) < TTC^*;\ 0,\ \text{otherwise}\}$$

$$TET* = \sum_{t=1}^{T} TET\ (t)$$

where:

Δt: time step length,
TTC^*: TTC threshold value,
$TTC_{(it)}$: the TTC for the ith vehicle at time t,
n: the vehicle ID,
N: the total number of vehicles, and
δ: switching variable.

Post-Encroachment Time (PET) is another significant SSM, representing the time between a lead vehicle leaving a conflict point and the following vehicle arriving at it (Figure 15.4). PET quantifies instances where collisions are narrowly avoided, indicating near misses or situations with minimal safety margins. PET, introduced by Allen *et al.* (1978), has no speed and direction assumptions and

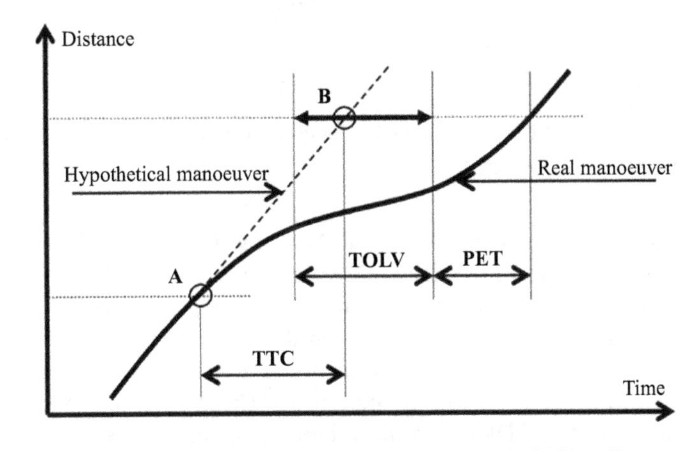

Figure 15.4 TTC and PET in the time-space domain (TOLV = Time occupied by leading vehicle at the collision point, PET = post-encroachment time, A = point of beginning of the evasive maneuver, B = point of possible collision; Gettman et al., 2008; Persaud and Bassani, 2015)

requires no assumption of the collision course like TTC.

$$PET = t_2 - t_1 \tag{15.12}$$

where:

t_2: arriving time at a conflict point of the 2nd vehicle, and
t_1: time of the 1st vehicle departing the conflict point.

Additional SSMs have been proposed based on the PET definition. For instance, Gap Time (GT) measures the time difference between the entries into the conflict point of two vehicles. Similarly, the Time Advantage (TAdv) extends the PET idea by predicting its value under the assumption that both road users maintain their initial paths and speeds (Laureshyn *et al.*, 2010). To identify conditions indicating a 'crash' or 'high-risk interactions', the threshold values of the SSMs intended for use must be established.

15.2.2 The deceleration and distance-based surrogate safety measures

Drivers commonly reduce their speeds to avoid potential collisions. Consequently, deceleration-based SSMs assess a vehicle's ability to adequately decrease its speed to evade a crash, with the Deceleration Rate to Avoid the Crash (DRAC) serving as a useful SSM. DRAC represents the minimum deceleration rate required by the following vehicle to prevent a crash with the lead vehicle. This metric assumes that one vehicle undertakes evasive manoeuvres while the other maintains its current speed and trajectory. Established thresholds are necessary for assessing crash risk within the DRAC methodology.

$$DRAC_{i,t+1} = \left(V_{i,t} - V_{i-1,t}\right)^2 / \left[\left(X_{i-1,t} - X_{i,t}\right) - L_{i-1}\right] \tag{15.13}$$

where:

t: time interval,
X: position of a vehicle (i = the following vehicle, $i - 1$ = the lead vehicle),
L: vehicle length, and
V: speed.

The collision risk increases as the DRAC value increases. When the DRAC value exceeds the Maximum Available Deceleration Rate (MADR), a crash would occur. Therefore, assessing crash risk based on DRAC requires an understanding of MADR, which is unique to each vehicle and depends on factors such as vehicle type, environmental conditions, and pavement characteristics.

To address this, Cunto (2008) introduced the Crash Potential Index (CPI) by extending the concept of DRAC to incorporate a vehicle's braking capability or MADR. The CPI represents the probability of DRAC exceeding MADR at any given moment and can be calculated as follows:

$$CPI = \sum_{t=ti}^{tfi} P\left(DRAC_{i,t} \geq MADR\right)\Delta t / (T_i) \tag{15.14}$$

where:

MADR: maximum braking rate,
Δt: time step length,
T_i: total travel time, and
t_i and tf_i: initial and final time steps, respectively.

Although distance and deceleration-based SSMs serve different purposes, both operate under assumptions of MADR. A distance-based SSM to identify hazardous conditions is the rear-end collision risk index (RCRI). RCRI is based on the 'safe stopping distance', which refers to the minimum distance required for the following vehicle to safely reduce its speed and avoid a crash with the leading vehicle, assuming the leading vehicle decelerates at the maximum rate and comes to a stop. To prevent rear-end crashes in every car-following scenario, the stopping distance of the leading vehicle must exceed that of the following vehicle (Oh *et al.*, 2006):

$$D_L > D_F$$

$$D_L = V_L h + \left(V_L^2/2a_L\right) + l_L$$

$$D_F = V_F t_R + \left(V_F^2\right)/(2a_F)$$

$$SDI = \{(0(\text{safe}),\ \text{if } D_L > D_F;\ 1\ (\text{unsafe}), \text{otherwise})\}$$

where:

SDI: stopping distance index for each car-following event,
D_L and D_F: stopping distance of the lead and following vehicles,
l_L: length of the leading vehicle,
v_F and v_L: speeds of the flowing, and lead vehicles
t_R: brak reaction time,
h: time headway, and
a_L *and* a_F: deceleration rate of the lead and following vehicles.

Thus, RCRI measures risk by comparing the stopping distances of the lead and following vehicles under the assumption that the lead vehicle begins an emergency stopping manoeuvre utilising the MADR. The RCRI proposed by Oh *et al.* (2006) is delineated as the ratio between the number of car-following events with an unsafe SDI and the maximum potential car-following instances at road detection stations. The RCRI formulation is articulated as follows:

$$RCRI = (\text{Number of rear} - \text{end conflicts})/\text{Exposure} \tag{15.15}$$

Building upon RCRI, Rahman and Aty (2018) introduced the time-exposed rear-end crash risk index (TERCRI) to assess aggregated risk over time:

$$TERCRI\ (t) = \sum_{n=1}^{N} RCRI_n(t)\ \Delta t$$

$$TERCRI = \sum_{t=1}^{Time} TERCRI\ (t)$$

Where Δt is the time step length.

15.2.3 Energy-based surrogate safety measures

Unlike the previous SSM subgroups, the energy-based SSMs are metrics used to evaluate the interaction severity. Among these metrics, Delta-V (DV) is the most commonly used, capturing the change in speed experienced by road users due to a collision event. DV is influenced by several factors such as the mass and speed of the involved vehicles, as well as the angle at which the collision occurs (Shelby, 2011).

Assuming a crash between two vehicles is an inelastic event, the law of conservation of momentum can be used to calculate the resultant speed of each road user involved in the crash. Thus, DV can be estimated as:

$$\Delta v_1 = m_2/(m_1 + m_2)\,(v_2 - v_1) \tag{15.16}$$

$$\Delta v_2 = m_1/(m_1 + m_2)\,(v_1 - v_2) \tag{15.17}$$

where:

v_1 and v_2: pre-crash velocities for the involved vehicles with potential collision course, and m_1 and m_2: masses of vehicles.

When drivers perform evasive manoeuvres involving braking, the speed decreases, potentially reducing crash severity. This reduction depends on factors such as the initial vehicle speeds, available braking time, and deceleration rate. Bagdadi (2013) proposed conflict severity (CS) SSM, integrating DV with additional parameters such as maximum average deceleration, and the time-to-accident temporal proximity-based SSM.

Laureshyn *et al.* (2017a,b) introduced an additional metric known as the extended DV indicator, integrating DV with a time indicator and a constant for deceleration. This composite measure aims to assess both the likelihood of a crash occurring and its potential severity by considering factors such as vehicle speeds, reaction time, and braking capability. Two other energy-based SSMs are the Crash Index (CAI) (Ozbay *et al.*, 2008) and the Conflict Index (CFI) (Alhajyaseen, 2015).

CAI assesses the kinetic energy associated with a car-following interaction, considering factors such as acceleration, speed, and the minimum time to collision (MTTC):

$$CAI = (V_f + a_f MTTC)^2 - (V_L + a_L\,MTTC)^2)/(2) * 1/MTTC \tag{15.18}$$

where V_L, V_f, a_L, and a_f are the speeds and accelerations of the lead and following vehicles, respectively.

CFI primarily focuses on the kinetic energy released after a crash to provide insights into the expected severity level. It combines the crash occurrence probability assessed through PET, with the speeds, masses, and angles of the involved vehicle to evaluate both the probability and severity crash.

$$CFI = (\alpha \Delta K_e)/e^{\beta PET} \tag{15.19}$$

where α represents the released energy percentage that will affect vehicle occupant (s), ΔK_e is the change in total kinetic energy before and after the crash, and $e^{\beta \mathrm{PET}}$ is used to weigh conflicts depending on the probability of a crash to occur.

15.3 Summary

This chapter provides an overview of methods for evaluating safety impacts. It is divided into two parts: the first part focuses on safety prediction, referencing the Highway Safety Manual (AASHTO, 2010, 2014); the second part introduces the concept of Surrogate Safety Measures (SSM), which are essential metrics for assessing road safety without relying solely on crash data.

Further research perspectives are referred to the context of Connected and Automated Vehicles (CAVs; see Section 14.1.3, Chapter 14) in particular for the topic of surrogate safety measures. In more detail, CAVs eliminate human drivers which is a major cause of crashes. For this reason, researchers and experts believe CAVs can improve traffic safety and efficiency. However, verifying claims regarding the benefits of CAVs presents significant challenges. One challenge arises from the increased exposure of road users to potential hazards due to the integration of self-driving cars into regular traffic. Additionally, worries persist regarding the compatibility of current traffic management plans with these new vehicles. Road authorities also require figuring out the minimum infrastructure requirements to ensure the safe coexistence of CAVs and conventional vehicles.

These factors together make a complicated situation, needing a complete approach and constant review of traffic management and road safety plans. Thus, developing strategies and frameworks solely from an empirical and experimental standpoint is challenging in terms of cost-benefit. Therefore, the impact of autonomous vehicles on traffic flow must first be thoroughly studied using traffic microsimulation. Utilising traffic microsimulation, a computational platform adept at modelling complex vehicle interactions within traffic streams can provide insights into the performance and communicative aspects of CAVs, facilitating the development of more effective traffic management strategies. Thus, traffic microsimulations play an indispensable role in understanding the implications of CAVs, making the use of SSMs essential for interpreting microsimulation output.

Among SSMs, TTC has been predominantly employed in CAV research, followed by TET (Dai *et al.*, 2022; Rahman *et al.*, 2021; Wang *et al.*, 2023; Yao *et al.*, 2020), TIT (Dai *et al.*, 2022; Yao *et al.*, 2020), and TA (Wu *et al.*, 2020). In addition to time-based SSMs, several studies have utilised deceleration and distance-based SSMs such as DRAC (Dai *et al.*, 2022), RCRI (Li *et al.*, 2018; Rahman and Aty, 2018; Rahman *et al.*, 2019a, b), TERCRI (Rahman and Aty, 2018), and Time-exposed deceleration to avoid crashing (Wang *et al.*, 2023). Other safety indicators have also been applied for evaluating CAV safety effects, such as the standard deviation of speed (Rahman and Aty, 2018; Fu *et al.*, 2019) and Maximum speed (MaxS) (Tibljas *et al.*, 2018).

The use of specialised software like VISSIM, Paramics, PreScan, and SUMO is essential for conducting traffic microsimulations and generating detailed data on vehicle behaviour (Shahdah *et al.*, 2015; Zhu and Krause, 2019; Fu and Sayed, 2021), which is then used to assess the safety of CAVs through SSMs.

In the CAV safety analysis context, microsimulation can be used to observe the implications of gradually introducing more self-driving vehicles on roads, eventually reaching a scenario where all vehicles are connected and automated. Furthermore, it helps us understand the interaction dynamics between CAVs and traditional vehicles driven by humans. To ensure an accurate simulation, it is crucial to properly configure variables affecting vehicle behaviour, such as inter-vehicle distances and driver reaction times, to reflect real-world conditions, particularly in urban environments.

References

Some suggestions for further readings are reported below among the huge literature available on these topics.

Abdel-Aty, M., and Radwan, E., 2000. Modeling traffic accident occurrence and involvement. *Accident Analysis and Prevention*, 32(5), 633–642, https://doi.org/10.1016/S0001-4575(99)00094-9.

Alhajyaseen, W.K.M., 2015. The integration of conflict probability and severity for the safety assessment of intersections. *Arabian Journal for Science and Engineering*, 40(2), 421–430, https://doi.org/10.1007/s13369-014-1553-1.

Alkaaf, K.A., and Abdel-Aty, M., 2015. Transferability and calibration of Highway Safety Manual performance functions and development of new models for urban four-lane divided roads in Riyadh, Saudi Arabia. *Transportation Research Record*, 2515, 70–77, https://doi.org/10.3141/2515-10.

Allaire, C., Ahner, D., Abarca, M., Adgar, P., and Long, S., 1996. *Relationship Between Side Slope Conditions and Collision Records in Washington State*. Washington State Department of Transportation, Olympia, WA, USA, WA-RD 425.1.

Allen, B.L., Shin, B.T., and Cooper, P., 1978. Analysis of traffic conflicts and collisions. *Transportation Research Record*, 667, 67–74.

American Association of State Highway and Transportation Officials (AASHTO), 2014. Washington, D.C. Highway Safety Manual, first edition, supplement.

American Association of State Highway and Transportation Officials (AASHTO), 2010. Washington, D.C. Highway Safety Manual, first edition.

Amundsen, F., and Hyden, C., 1977. *Proceeding of First Workshop on Traffic Conflicts*, Institute of Transport Economics, Oslo/Lund Institute of Technology, Oslo, Norway.

Arun, A., Haque, Md.M., Bhaskar, A., Washington, S., and Sayed, T., 2021. A systematic mapping review of surrogate safety assessment using traffic conflict techniques. *Accident Analysis & Prevention*, 153, 106016, https://doi.org/10.1016/j.aap.2021.106016.

Aziz, S.R., and Dissanayake, S., 2019. A comparative study of newly developed Kansas-specific safety performance functions with HSM models for rural four-lane divided highway segments. *Journal of Transportation Safety & Security*, 180–205, https://doi.org/10.1080/19439962.2019.1622614.

Bagdadi, O., 2013. Estimation of the severity of safety critical events. *Accident Analysis and Prevention*, 50, 167–174, https://doi.org/10.1016/j.aap.2012.04.007.

Brimley, B.K., Saito, M., and Schultz, G.G., 2012. Calibration of Highway Safety Manual safety performance function: Development of new models for rural two-lane two-way highways. *Transportation Research Record*, 2279(1), 82–89, https://doi.org/10.3141/2279-10.

Cafiso, S., Di Graziano, A., Di Silvestro, G., La Cava, G., and Persaud, B., 2010. Development of comprehensive accident models for two-lane rural highways using exposure, geometry, consistency and context variables. *Accident Analysis and Prevention*, 42(4), 1072–1079, https://doi.org/10.1016/j.aap.2009.12.015.

Cafiso, S., Di Silvestro, G., and Di Guardo, G., 2012. Application of Highway Safety Manual to Italian divided multilane highways. *Procedia-Social and Behavioral Sciences*, 53, 910–919, https://doi.org/10.1016/j.sbspro.2012.09.940.

Cafiso, S., Montella, A., D'Agostino, C., Mauriello, F., and Galante, F., 2021. Crash modification functions for pavement surface condition and geometric design indicators. *Accident Analysis & Prevention*, 149, 105887, https://doi.org/10.1016/j.aap.2020.105887.

Caliendo, C., Guida, M., and Parisi, A., 2007. A crash-prediction model for multilane roads. *Accident Analysis and Prevention*, 39(4), 657–670, https://doi.org/10.1016/j.aap.2006.10.012.

Cansiz, O.F., 2010. Improvements in estimating a fatal accidents model formed by an artificial neural network. *Simulation*, 87(6), 512–522, https://doi.org/10.1177/0037549710370842.

Chin, H.C., and Quek, S.T., 1997. Measurement of traffic conflicts. *Safety Science*, 26 (3), 169–185, https://doi.org/10.1016/S0925-7535(97)00041-6.

Çodur, M.Y., and Tortum, A., 2015. An artificial neural network model for highway accident prediction: A case study of Erzurum, Turkey. *Promet – Traffic & Transportation*, 27(3), 217–225, http://dx.doi.org/10.7307/ptt.v27i3.1551.

Cunto, F., 2008. *Assessing Safety Performance of Transportation Systems Using Microscopic Simulation*, Waterloo, Ontario, Canada, http://onlinepubs.trb.org/Onlinepubs/trr/1992/1376/1376-012.pdf.

Dai, Y., Yang, Y., Wang, Z., and Luo, Y., 2022. Exploring the impact of damping on Connected and Autonomous Vehicle platoon safety with CACC. *Physica A: Statistical Mechanics and its Applications*, 607, 128181, https://doi.org/10.1016/j.physa.2022.128181.

Dinu, R.R., and Veeraragavan, A., 2011. Random parameter models for accident prediction on two-lane undivided highways in India. *Journal of Safety Research*, 42(1), 39–42, https://doi.org/10.1016/j.jsr.2010.11.007.

Elvik, R., 2009. The non-linearity of risk and the promotion of environmentally sustainable transport. *Accident Analysis and Prevention*, 41(4), 849–855, https://doi.org/10.1016/j.aap.2009.04.009.

Elvik, R., and Mysen, A., 1999. Incomplete accident reporting: Meta-analysis of studies made in 13 countries. *Transportation Research Record: Journal of the*

Transportation Research Board, 1665, 133–140, https://doi.org/10.3141/1665-18.

Farid, A., Abdel-Aty, M., and Lee, J., 2018. A new approach for calibrating safety performance functions. *Accident Analysis and Prevention*, 119, 188–194, https://doi.org/10.1016/j.aap.2018.07.023.

Fu C., and Sayed T., 2021. Multivariate Bayesian hierarchical Gaussian copula modeling of the non-stationary traffic conflict extremes for crash estimation. *Analytic Methods in Accident Research*, 29, 100154, https://doi.org/10.1016/j.amar.2020.100154.

Fu, T., Wang, W., Li, Y., Xu, C.C., Xu, T., and Li, X., 2019. Longitudinal safety impacts of cooperative adaptive cruise control vehicle's degradation. *Journal of Safety Research*, 69, 177–192, https://doi.org/10.1016/j.jsr.2019.03.002.

Garach, L., de Oña, J., López, G., and Baena, L., 2016. Development of safety performance functions for Spanish two-lane rural highways on flat terrain. *Accident Analysis and Prevention*, 95, 250–265, https://doi.org/10.1016/j.aap.2016.07.021.

Gettman, D., Pu, L., Sayed, T., and Shelby, S.G., 2008. Surrogate Safety Assessment Model and Validation: Final Report. *Federal Highway Administration*, Report FHWA HRT 08-051, https://api.semanticscholar.org/CorpusID:106519100.

Graham, J.L., and Harwood, D.W, 1982. Effectiveness of clear recovery zones. *National Cooperative Highway Research Program, Transportation Research Board*, Washington, DC, USA, NCHRP report 247, https://api.semanticscholar.org/CorpusID:106454744.

Gross, F., Persaud, B., and Lyon, C., 2010. *A Guide to Developing Quality Crash Modification Factors*. Federal Highway Administration, Washington, DC, USA, report no. FHWA-SA-10-032, https://api.semanticscholar.org/CorpusID:108184726.

Haleem, K., Gan, A., and Lu, J., 2013. Using multivariate adaptive regression splines (MARS) to develop crash modification factors for urban freeway interchange influence areas. *Accident Analysis and Prevention*, 55, 12–21, https://doi.org/10.1016/j.aap.2013.02.018.

Hauer, E., 1997. *Observational Before–After Studies in Road Safety: Estimating the Effect of Highway and Traffic Engineering Measures on Road Safety*. Pergamon Press (Elsevier Science), Oxford.

Hauer, E., Harwood, D.W., Council, F.M., and Griffith, M.S., 2002. Estimating safety by the empirical Bayes method: a tutorial. *Transportation Research Record*, 1784, 126–131, https://doi.org/10.3141/1784-16.

Hayward, J.C., 1971. *Near misses as a measure of safety at urban intersections*. Pennsylvania Transportation and Traffic Safety Center.

Johnsson, C., Laureshyn, A., and D'Agostino, C., 2021. A relative approach to the validation of surrogate measures of safety. *Accident Analysis and Prevention*, 161, 106350, https://doi.org/10.1016/j.aap.2021.106350.

Khan, M., Abdel-Rahim, A., and Williams, C.J., 2015. Potential crash reduction benefits of shoulder rumble strips in two-lane rural highways. *Accident Analysis and Prevention*, 75, 35–42, https://doi.org/10.1016/j.aap.2014.11.007.

Kim, J., Anderson, M., and Gholston, S., 2015. Modeling safety performance functions for Alabama's urban and suburban arterials. *International Journal of Traffic and Transportation Engineering*, 4(3), 84–93, https://doi.org/10.3961/jpmph.14.047.

La Torre F., Domenichini L., Corsi F., and Fanfani F., 2014. Transferability of the Highway Safety Manual freeway model to the Italian motorway network. *In Transportation Research Record: Journal of the Transportation Research Board*, No. 2435, Transportation Research Board of the National Academies, https://doi.org/10.1016/j.aap.2022.106852.

La Torre, F., Meocci, M., Domenichini, L., Branzi, V., and Paliotto, A., 2019. Development of an accident prediction model for Italian freeways. *Accident Analysis and Prevention*, 124, 1–11, https://doi.org/10.1016/j.aap.2018.12.023.

Laureshyn, A., De ceunynck, T., Karlsson, C., Svensson, A., and Daniels, S., 2017. In search of the severity dimension of traffic events: Extended Delta-V as a traffic conflict indicator. *Accident Analysis and Prevention*, 98, 46–56, https://doi.org/10.1016/j.aap.2016.09.026.

Laureshyn, A., De Ceunynck, T., Karlsson, C., Svensson, Å., and Daniels, S., 2017a. In search of the severity dimension of traffic events: Extended Delta-V as a traffic conflict indicator. *Accident Analysis and Prevention*, 98, 46–56, https://doi.org/10.1016/j.aap.2016.09.026

Laureshyn, A., de Goede, M., Saunier, N., and Fyhri, A., 2017b. Cross-comparison of three surrogate safety methods to diagnose cyclist safety problems at intersections in Norway. *Accident Analysis and Prevention*, 105, 11–20, https://doi.org/10.1016/j.aap.2016.04.035.

Laureshyn, A., Svensson, A., and Hyden, C., 2010. Evaluation of traffic safety, based on micro-level behavioural data: theoretical framework and first implementation. *Accident Analysis and Prevetion*, 42, 1637–1646, https://doi.org/10.1016/j.aap.2010.03.021.

Li, L., Gayah, V. V., and Donnell, E. T., 2017. Development of regionalized SPFs for two-lane rural roads in Pennsylvania. *Accident Analysis and Prevention*, 108, 343–353, https://doi.org/10.1016/j.aap.2017.08.035.

Li, P., Abdel-Aty, M., Cai, Q., and Yuan, C., 2020. The application of novel connected vehicles emulated data on real-time crash potential prediction for arterials. *Accident Analysis and Prevention*, 144, 105658, http://dx.doi.org/10.1016/j.aap.2020.105658.

Li, Y., Fu, T., Fan, Q., Dong, C., and Wang, W., 2018. Influence of cyber-attacks on longitudinal safety of connected and automated vehicles. *Accident Analysis and Prevention*, 121, 148–156, https://doi.org/10.1016/j.aap.2018.09.016.

Lord, D., and Mannering, F., 2010. The statistical analysis of crash-frequency data: A review and assessment of methodological alternatives. *Transportation Research Part A*, 44, 291–305, https://doi.org/10.1016/j.tra.2010.02.001.

Lord, D., Qin, X., and Geedipally, S.R., 2021. *Highway Safety Analytics and Modeling* Amsterdam: Elsevier.

Lord, D., and Washington, S., 2018. Safe mobility: challenges, methodology and solutions. *Transport and Sustainability*, volume 11. Emerald.

Martinelli, F., La Torre, F., and Vadi, P., 2009. Calibration of the Highway Safety Manual's accident prediction model for Italian secondary road network. *Transportation Research Record*, 2103(1), 1–9, https://doi.org/10.3141/2103-01.

Mehta, G., and Lou, Y., 2013. Calibration and development of safety performance functions for Alabama: Two-lane, two-way rural roads and four-lane divided highways. *Transportation Research Record*, 2398(1), 75–82, https://doi.org/10.3141/2398-09.

Minderhoud, M.M., and Bovy, P.H.L., 2001. Extended time-to-collision measures for road traffic safety assessment. *Accident Analysis and Prevention*, 33(1), 89–97, https://doi.org/10.1016/S0001-4575(00)00019-1.

Montella, A., 2009. Safety evaluation of curve delineation improvements: Empirical Bayes observational before-and-after study. *Transportation Research Record*, 2103(1), 69–79, https://doi.org/10.3141/2103-09.

Montella, A., 2010. A comparative analysis of hotspot identification methods. *Accident Analysis and Prevention*, 42(2), 571–581, https://doi.org/10.1016/j.aap.2009.09.025.

Montella, A., Colantuoni, L., and Lamberti, R., 2008. Crash prediction models for rural motorways. *Transportation Research Record*, 2083, 180–189, http://dx.doi.org/10.3141/2083-21.

Montella, A., and Imbriani, L.L., 2015. Safety performance functions incorporating design consistency variables. *Accident Analysis and Prevention*, 74, 133–144, https://doi.org/10.1016/j.aap.2014.10.019.

Moraldi, F., La Torre, F., and Ruhl, S., 2020. Transfer of the Highway Safety Manual predictive method to German rural two-lane, two-way roads. *Journal of Transportation Safety & Security*, 12(8), 977–996, https://doi.org/10.1080/19439962.2019.1571546.

Oh, C., Park, S., and Ritchie, S.G., 2006. A method for identifying rear-end collision risks using inductive loop detectors. *Accident Analysis and Prevention*, 38, 295–301, https://doi.org/10.1016/j.aap.2005.09.009.

Outcalt, W., 2001. *Centerline Rumble Strips*. Report No. CDOT-DTD-R-2001-8. Colorado Department of Transportation, Denver, CO, USA.

Ozbay, K., Yang, H., Bartin, B., and Mudigonda, S., 2008. Derivation and validation of new simulation-based surrogate safety measure. *Transportation Research Record*, 105–113, https://doi.org/10.3141/2083-12.

Pawlovich, M., Li, W., Carriquiry, A., and Welch, T., 2006. Iowa's experience with road diet measures: use of Bayesian approach to assess impacts on crash frequencies and crash rates. *Transportation Research Record*, 1953, 163–171, https://doi.org/10.3141/1953-19.

Persaud, B., and Bassani, M., 2015. Calibration and application of crash prediction models for safety assessment of roundabouts based on simulated conflicts. *Conference: 94th Transportation Research Board Annual Meeting At*: Washington D.C., US.

Persaud, B., Lan, B., Lyon, C., and Bhim, R., 2010. Comparison of empirical Bayes and full Bayes approaches for before–after road safety evaluations. *Accident Analysis and Prevention*, 42(1), 38–43, https://doi.org/10.1016/j.aap.2009.06.028.

Persaud, B., and Lyon, C., 2007. Empirical Bayes before–after studies: Lessons learned from two decades of experience and future directions. *Accident Analysis and Prevention*, 39, 546–555, https://doi.org/10.1016/j.aap.2006.09.009.

Persaud, B., Lyon, C., Bagdade, J. And Ceifetz, A.H., 2012. Evaluation of safety performance of passing relief lanes. *Transportation Research Record*, 2348, 58–63, https://doi.org/10.3141/2348-07.

Pitale, J.T., Shankwitz, C., Preston, H., and Barry, M., 2009. *Benefit: Cost Analysis of In-Vehicle Technologies and Infrastructure Modifications as a Means to Prevent Crashes Along Curves and Shoulders. Minnesota Department of Transportation*, St. Paul, MN, USA. https://hdl.handle.net/11299/150627.

Qin, X., Ivan, J.N., and Ravishanker, N., 2004. Selecting exposure measures in crash rate prediction for two-lane highway segments. *Accident Analysis and Prevention*, 36(2), 183–191, http://dx.doi.org/10.1016/s0001-4575(02)00148-3.

Qin, Y., and Wang, H., 2019. Influence of the feedback links of connected and automated vehicle on rear-end collision risks with vehicle-to-vehicle communication. *Traffic Injury and Prevention*, 20(1), 79–83, https://doi.org/10.1080/15389588.2018.1527469.

Rahman, M.S., and Abdel-Aty, M., 2018. Longitudinal safety evaluation of connected vehicles' platooning on expressways. *Accident Analysis and Prevention*, 117, 381–391, https://doi.org/10.1016/j.aap.2017.12.012.

Rahman, M.S., Abdel-Aty, M., and Lee, J., 2019a. Understanding the safety benefits of connected and automated vehicles on arterials' intersections and segments. *In: Presented at 98th Annual Meeting of Transportation Research Board*. Washington, DC.

Rahman, M.S., Abdel-Aty, M., Lee, J., and Rahman, M.H., 2019b. Safety benefits of arterials' crash risk under connected and automated vehicles. *Transportation Research Part C: Emerging Technologies*, 100, 354–371, https://doi.org/10.1016/j.trc.2019.01.029.

Rahman, M.S., Abdel-Aty, M., and Wu, Y.M., 2021. A multi-vehicle communication system to assess the safety and mobility of connected and automated vehicles. *Accident Analysis and Prevention*, 124, https://doi.org/10.1016/j.trc.2020.102887.

Shahdah U., Saccomanno F., and Persaud B., 2015. Application of traffic microsimulation for evaluating safety performance of urban signalized intersections. *Transportation. Research. Part C: Emerging Technologies*, 60, 96–104, https://doi.org/10.1016/j.trc.2015.06.01

Shelby, S. G., 2011. Delta-V as a measure of traffic conflict severity. *In 3rd International Conference on Road Safety and Simulation*, 14–16, September.

Srinivasan, R., and Carter, D., 2011. *Development of Safety Performance Functions for North Carolina* (No. FHWA/NC/2010-09). Research and Analysis Group, North Carolina Department of Transportation, North Carolina, USA.

Srinivasan, R., Colety, M., Bahar, G., Crowther, B., and Farmen, M., 2016. Estimation of calibration functions for predicting crashes on rural two-lane roads in Arizona. *Transportation Research Record*, 2583, 17–24, https://doi.org/10.3141/2583-03.

Sun, C., Brown, H., Edara, P., Claros, B., and Nam, K., 2014. *Calibration of the HSM SPFs for Missouri* (No. CMR14-007). Missouri Department of Transportation.

Sun, X., Magri, D., Shirazi, H. H., Gillella, S., and Li, L., 2011. Application of highway safety manual: Louisiana experience with rural multilane highways.

Tarko, A.P., 2020a. Chapter 13 - Summary and future research directions. In: Tarko, A.P. (Ed.), *Measuring Road Safety Using Surrogate Events*. Elsevier, 229–239, https://doi.org/10.1016/B978-0-12-810504-7.00013-6.

Tarko, A.P., 2020b. Chapter 3 - traffic conflicts as crash surrogates. In: Tarko, A.P. (Ed.), *Measuring Road Safety Using Surrogate Events*. Elsevier, 31–45, https://doi.org/10.1016/B978-0-12-810504-7.00003-3.

Tibljas, A.D., Giuffre, T., Surdonja, S., and Trubia, S., 2018. Introduction of autonomous vehicles: Roundabouts design and safety performance evaluation. *Sustainability*, 10, 1060, https://doi.org/10.3390/su10041060.

Turner, S., Singh, R., and Nates G., 2012. *The Next Generation of Rural Road Crash Prediction Models: Final Report*. NZ Transport Agency, Wellington, New Zealand, research report 509, https://doi.org/10.13140/RG.2.2.31365.70885.

United Nations, 2015. Transforming our world: the 2030 Agenda for Sustainable Development. Resolution 70/1. Available at: https://documents.un.org/doc/undoc/gen/n15/291/89/pdf/n1529189.pdf?token=SJi7rzsfq2TcC1zP4j&fe=true.

United Nations, 2020. Improving global road safety. Resolution 74/299. Available at: https://www.un.org/pga/76/wp-content/uploads/sites/101/2021/11/A_RES_74_299_E.pdf.

Varhelyi, A., 1998. Driver's speed behavior at a zebra crossing: a case study. *Accident Analysis and Prevention*, 30(6), 731–743, https://doi.org/10.1016/S0001-4575(98)00026-8.

Wang, X., Jiang, X., Li, H., Zhao, X., Hu, Z., and Xu, C., 2023. Traffic safety assessment with integrated communication system of connected and automated vehicles at signalized intersections. *Transportation Research Record*, 2678(6), 956–971, https://doi.org/10.1177/03611981231201107.

Wichert, S., and Cardoso, J., 2007. Accident Prediction Models for Portuguese Single Carriageway Roads. *Laboratório Nacional de Engenharia Civil*, Lisbon, Portugal, http://dx.doi.org/10.2495/UT080601.

World Health Organization (WHO), 2018. Global status report on road safety. Available at: https://www.who.int/publications/i/item/9789241565684.

World Health Organization (WHO), 2023. Global status report on road safety 2023. Available at: https://www.who.int/publications/i/item/9789240086517.

Wu, Y., Abdel-Aty, M., Wang, L., and Rahman, M.S., 2020. Combined connected vehicles and variable speed limit strategies to reduce rear-end crash risk under fog conditions. *Journal of Intelligent Transportation Systems*, 24(5), 494–513, http://dx.doi.org/10.1080/15472450.2019.1634560.

Xie, F., Gladhill, K., Dixon, K.K., and Monsere, C.M., 2011. Calibration of highway safety manual predictive models for Oregon state highways. *Transportation Research Record*, 2241(1), 19–28, https://doi.org/10.3141/2241-03.

Yao, Z., Hu, R., Jiang, Y., and Xu, T., 2020. Stability and safety evaluation of mixed traffic flow with connected automated vehicles on expressways. *Journal of Safety Research*, 75, 262–274, https://doi.org/10.1016/j.jsr.2020.09.012.

Zegeer, C.V., Reinfurt, D.W., Hunter, W.W., Hummer, J.E., Stewart, R.D., and Herf, L., 1988. Accident effects of sideslope and other roadside features on two-lane roads. *Transportation Research Record*, 1195, 33–47, https://api.semanticscholar.org/CorpusID:55558306.

Zhou, S., Sun, J., An, X., and Li, K., 2011. The development of a conflict hazardous assessment model for evaluating urban intersection safety. *Transport*, 26, 216–223, https://doi.org/10.3846/16484142.2011.589494.

Zhu J., and Krause S., 2019. Analysis of the impact of automated lane changing behavior on the capacity and safety of merge segments. *Transportation Research Procedia*, 41, 48–51, https://doi.org/10.1016/j.trpro.2019.09.009.

Chapter 16
Road traffic noise

Claudio Guarnaccia[1]

Noise pollution is a relative thing. In a city, it's a jet plane taking off. In a monastery, it's a pen that scratches.

Robert Orben

Outline. *This chapter summarises the modelling frameworks utilised for predicting road traffic noise. Regressive and statistical are introduced, furthermore, an overview of the models and techniques for assessing single-vehicle noise emission is provided.*

This chapter aims to outline the primary models and techniques for estimating the acoustical noise generated by road traffic. In particular, it focuses on the noise emission of single vehicles, reviewing the characteristics of existing road traffic noise models, and emphasising the input parameters required for assessment.

This endeavour helped to elucidate the microscopic and macroscopic characteristics of models found in scientific literature – specifically, assessing noise emission from individual vehicles versus adopting a road traffic flow-based approach. Frequently, Noise Emission Models (NEMs), which estimate noise from individual vehicles or overall road traffic, are integrated with sound propagation models to project noise at the receiver, forming a Road Traffic Noise Model (RTNM). Moreover, these models can aggregate contributions from various vehicle categories and multiple vehicles within the same category, transitioning from individual vehicle assessment to traffic flow analysis, thus establishing a 'microtomacro' modelling framework. An extended review of the evolution and development of such models can be found in Guarnaccia *et al.* (2024).

This chapter summarises the modelling frameworks utilised for predicting road traffic noise. After introducing both regressive and statistical approaches in Section 16.2, an overview of the latest models and techniques available in the scientific community for assessing single-vehicle noise emission is provided in Section 16.3.

[1]Department of Civil Engineering, University of Salerno, Italy

16.1 Statistical models and regression-based approach for road traffic noise prediction

Numerous models have been developed and utilised to evaluate road traffic noise, with the earliest Traffic Noise Models (TNM) originating from the 1950s. Despite being based on a statistical approach that simplifies real traffic conditions, these models typically yield reasonably accurate results. TNMs are characterised by their simplicity, requiring minimal input data and providing percentile L50 as output. They often rely on regressions conducted on specific datasets, making them influenced by local conditions such as road and weather characteristics, as well as vehicle types, resulting in relatively low accuracy. Key inputs usually include traffic flows for different vehicle categories (e.g., light and heavy vehicles), road surface characteristics, and distances between roads and receivers. Additional parameters, such as road and vehicle maintenance or weather conditions, may also be considered based on country-specific peculiarities.

In 1953, an early TNM was introduced in the *Handbook of Acoustic Noise Control*. This model primarily depended on two inputs: Q, representing traffic volume in vehicles per hour, and d, representing the distance from the observation point to the centre of the traffic lane. The percentile L50 for vehicles travelling at speeds between 55 km/h and 75 km/h and at distances greater than 6 m could be computed using the formula provided.

$$L50 = 68 + 8.5 \log(Q) - 20 \log(d) \tag{16.1}$$

Notably, this model did not offer specifications regarding vehicle types or road characteristics.

Subsequently, Nickson (1965) proposed an alternative model that included a constant term, denoted as C, which was crucial for aligning the model with experimental data. The formula used in this model is as follows:

$$L_{50} = C + 10 \log(Q/d) \tag{16.2}$$

Subsequently, Johnson and Saunders (1968) introduced a model that incorporated the average vehicle speed (v) as input data, alongside corrective factors related to ground attenuation and gradient, as depicted in the following equation.

$$L_{50} = 3.50 + 10 \log(Qv^3/d)$$

Galloway *et al.* (1969) made further advancements to the previous model by introducing the percentage of heavy vehicles transiting (P) as a parameter, as shown in the following equation.

$$L_{50} = 20 + 10 \log\left(Qv^2/d\right) + 0.4P \tag{16.3}$$

More recently, with the adoption of the equivalent level L_{eq} as a standard sound level indicator in several regulations, one of the widely utilised models that utilises similar inputs to precisely compute the equivalent level is the Burgess model (Burgess, 1977).

$$L_{eq} = 55.5 + 10.2 \log(Q) + 0.3P - 19.3 \log(d) \tag{16.4}$$

Another model was proposed by Griffiths and Langdon (1968), aiming to compute the L_{eq} using percentile levels such as L10, L50, and L90, as described below.

$$L_{eq} = L_{50} + 0.018(L_{10} - L_{90})^2 \tag{16.5}$$

Each statistical percentile noise indicator can be computed as follows:

$$L_{10} = 61 + 8.4 \log(Q) + 0.15P - 11.5 \log(d) \tag{16.6}$$

$$L_{50} = 44.8 + 10.8 \log(Q) + 0.12P - 9.6 \log(d) \tag{16.7}$$

$$L_{10} = 39.1 + 10.5 \log(Q) + 0.06P - 9.3 \log(d) \tag{16.8}$$

Subsequently, Fagotti and Poggi (1995). investigated the prospect of integrating additional vehicle categories into previous models to enhance their accuracy. Specifically, their TNM incorporates flows of motorcycles and buses (i.e., QM and QBUS, respectively, alongside light-duty and heavy-duty vehicles (i.e., QL and QP), using calibrated coefficients to consider their distinct contributions. The formulation is as follows:

$$L_{eq} = 10 \log(Q_L + Q_M + 8Q_P + 88Q_{BUS}) + 33.5 \tag{16.9}$$

Another model was devised by the French C.S.T.B (Centre Scientifique et Technique du Batiment), introducing a formulation for $L_{eq,A}$ to account for varying road and vehicle flow conditions (Saracco and Leandre, 1991). Beginning with:

$$L_{eq,A} = 0.65\, L_{50} + 28.8$$

they proposed several equations to assess the percentile level L50, applicable in either urban or highway scenarios with traffic flows below 1,000 vehicles/hour:

$$L_{50,A} = 11.9 \log Q_{eq} + 31.4 \tag{16.10}$$

However, in urban settings where tall buildings line the streets, considering the width L of the road itself, they utilised the following formulation:

$$L_{50,A} = 15.5 \log Q_{eq} - 10 \log L + 36 \tag{16.11}$$

In both preceding equations, the equivalent vehicular flow is considered as follows:

$$Q_{eq} = Q(1 + P/100 \cdot (n - 1)) \tag{16.12}$$

where n represents a form of homogenisation coefficient, aiding in recognising that heavy vehicles emit more noise than light ones, also known as the acoustical equivalent of heavy vehicles. This coefficient determines the number of equivalent light vehicles that produce the same acoustic energy as a heavy one. A formulation detailing its dependence on the speed of heavy vehicles is provided in Guarnaccia (2013).

Most of these statistical Road Traffic Noise Models (RTNMs) can be derived from a general equation of equivalent noise level:

$$L_{eq} = A \log Q[1 + P/100 \cdot (n - 1)] + b \log d + C \quad (16.13)$$

where:

- Q represents the traffic volume in vehicles per hour,
- d is the distance from the observation point to the centre of the traffic lane,
- P denotes the percentage of heavy vehicles, and
- n is the acoustic equivalent coefficient discussed earlier.

The parameters A, b, and C can be determined using a linear regression approach on a dataset comprising various L_{eq} values with different input parameters. As for n, similar regression methods can be employed, or it can even be established through measurements of emissions from individual vehicles. Naturally, multiple homogenisation coefficients could be defined to consider different vehicle categories besides heavy vehicles (e.g., motorcycles, buses, etc.).

These parameters are typically calibrated through regression on field data. However, an alternative approach is outlined in Rossi *et al.* (2023), where the authors demonstrate that a computed dataset illustrating various traffic scenarios, both urban and extra-urban, can substitute field measurements for calibrating a multi-linear regression model with satisfactory outcomes. The effectiveness of such a model primarily hinges on the chosen ranges of variables and the RTNM utilised to simulate equivalent levels.

The general equation of a regression model is often adjusted using several corrective parameters that account for factors such as average speed, road type (e.g., gradient, asphalt), weather and traffic conditions, and the presence of impediments to sound wave propagation like barriers or buildings.

These characteristics are exemplified in the English standard known as the CoRTN procedure (see DoT 1988) The CoRTN model incorporates factors beyond traffic flow and composition, including mean speed. Initially, other conditions are idealised, assuming moderate wind velocity and a dry road surface, for instance. One of the equations proposed in this model enables the assessment of the hourly noise level at a distance of 10 m from the nearest roadway, as shown below. Subsequently, the computed level can be adjusted using various coefficients to account for the actual conditions of the road infrastructure under study, indicated by the multiple coefficients.

$$L_{10,1h} = 42.2 + 10 \log(Q_{1h}) + \Delta L_{\text{flow}} + \Delta L_G + \Delta L_P + \Delta L_D + \Delta L_S + \Delta L_A + \Delta L \quad (16.14)$$

Several corrective coefficients are considered, addressing traffic flow adjustment, road gradient, type of road pavement, distance adjustment, presence of shielding, angle of view adjustment, and reflection adjustment.

Another more advanced model is the German standard, RLS 90 model (für Verkehr, 1990). In addition to the 'classic' inputs, this model collects information on average hourly traffic flow, including the motorcycles category, and the average speed of each category. A unique feature of this model is its capability to assess the noise emission of a parking lot. Similar to the CoRTN model, the RLS 90 model utilises ideal conditions to evaluate $L_{(m,E)}$^((25 m)), an average noise level at a distance of 25 m from the center of the carriageway, as demonstrated below.

$$L^{25m}_{m,E} = 37.3 + 10\log[Q(1 + 0.082P)] \tag{16.15}$$

Subsequent corrective factors are added to account for speed limits, different road surfaces and vehicle speeds, rises and falls along the route, presence of buildings, air absorption, ground and atmospheric conditions, topography characteristics, and presence of traffic lights.

To tailor this model to the Italian context, the Italian C.N.R. model was developed, and later enhanced by Cocchi *et al.* (1991). The precise formula is:

$$L_{eq,A} = \alpha + 10\log(Q_L + \beta Q_p) - 10\log(d/d0) + \Delta_V + \Delta_F + \Delta_B + \Delta_P + \Delta_G + \Delta_{VB} \tag{16.16}$$

In addition to the features described for the German and CoRTN models, the C. N.R. model introduces two new parameters, α and β, to assess the characteristics of the country's roads and vehicles. Furthermore, QL and QP represent light and heavy vehicle traffic flow, respectively, while $d0$ is a fixed reference distance of 25 m. Several corrective coefficients are also considered, accounting for mean flux velocity, the presence of reflective facades near or in the opposite direction of the observation point, and the presence of traffic lights or slow traffic conditions.

In Finland, the Nordtest approach was developed under the auspices of the Nordic Council of Ministers (Nielsen, 1997). The Nordtest model allows for the assessment of L_{eq} by conducting noise measurements continuously over specific time intervals. It outlines procedures for measuring noise levels at precise points using multiple microphone positions. The formula for the 24-h equivalent noise level L_{eq},24 h is provided below, demonstrating the independence of noise levels measured during the day, evening, and night periods:

$$L_{eq,A,24h} = 10\log 1/T[\Delta t_d 10(L_d/10) + \Delta t_e 10(L_e/10) + \Delta t_n 10(L_n/10) \tag{16.17}$$

where the sum of the three considered time intervals equals 24 h.

They also proposed an alternative method to assess traffic conditions during the night or evening period by considering average traffic conditions instead of actual measurements. For light vehicles, the equations used are:

$$L_{e,A,light} = 73.5 + 25\log(v/50);\text{for } v \geq 40\text{ km/h} \tag{16.18}$$

$$L_{e,A,light} = 71.1 + 25\log(v/50);\text{for } 30 \leq v \leq 40\text{ km/h} \tag{16.19}$$

while for heavy vehicles, the formula is:

$$L_{e,A,heavy} = 80.5 + 30\log(v/50) \tag{16.20}$$

All three equations pertain to a distance of 10 m.

Finally, one can compute the equivalent noise level in a 1-h span as follows:

$$L_{eq,A,1h} = 10 \log\left[1/3600\left(n_{light}10^{Le,A,light/10} + n_{heavy}10^{Le,A,heavy/10}\right)\right] \tag{16.21}$$

where nlight and nheavy represent the mean traffic flow per hour of the two considered categories.

16.2 Single vehicle noise emission models review

Numerous models focus on estimating road traffic noise by utilising Noise Emission Models (NEMs) to assess the sound power level (L_w) of vehicles, rather than relying on statistical estimations. This section will present the single-vehicle approach and discuss some of its potential applications.

16.2.1 Lelong

In 1999, Lelong (1999) devised an empirical model to estimate the sound power level (L_w) of individual vehicles, categorising them into two groups: light and heavy. They also differentiated between three driving conditions: cruise driving, acceleration, and deceleration for both light and heavy vehicles. The formula for this model is as follows:

$$L_{w,L} = \alpha_L + \beta_L \log(v) \tag{16.22}$$

$$L_{w,H} = \alpha_H + \beta_H v \tag{16.23}$$

This formula highlights the distinct dependence of sound power levels on speed, with a logarithmic relationship for light vehicles and a linear one for heavy vehicles.

16.2.2 SonRoad

In 2004, Switzerland calibrated a Noise Emission Model (NEM) known as SonRoad to estimate the sound power level (L_w) of a single vehicle. It considers vehicle type, speed, road grade, and surface type as primary inputs. Two distinct equations are used to evaluate the sound power level for light and heavy vehicles, as follows:

$$L_{w,A,L} = 28.5 + \log(10^{0.1[7.3+35\log(v)]} + 10^{0.1\left[60.5+10\log(1+v/443\hat{.}5\right]} + \Delta S) + \Delta_{BG} \tag{16.24}$$

$$L_{w,A,H} = 28.5 + \log(10^{0.1[16.3+35\log(v)]} + 10^{0.1\left[74.7+10\log(1+v/563\hat{.}5\right]} + \Delta_s) + \Delta_{BG} \tag{16.25}$$

16.2.3 Harmonoise

An influential NEM, crucial for subsequent developments and widely adopted in Europe, is the Harmonoise model. It estimates the sound power level (L_w) for five vehicle categories: light, medium-heavy, heavy, special (tractors, trucks), and motorbikes. This model separates the contribution of engine noise, termed propulsion noise ($L_{w,\text{propulsion}}$), from that of tire rolling on the road surface ($L_{w,\text{rolling}}$). It employs coefficients to compute these contributions for each third-octave band from 25 Hz to 10 kHz. Additionally, Harmonoise introduces a corrective factor for $L_{w,\text{propulsion}}$ to account for the acceleration or deceleration of the single vehicle.

$$L_{w,propulsion} = a_p(f) + b_p(f)\left[\left(v - v_{ref}\right)/v_{ref} + \Delta_{acc}\right] \tag{16.26}$$

$$L_{w,rolling} = a_r(f) + b_r(f)\left[\left(v/v_{ref}\right]\right. \tag{16.27}$$

The overall sound power level is then computed as follows:

$$L_{w,TOT} = 10\log\left(10L_{w,propulsion}/10 + 10L_{w,rolling}/10\right) \tag{16.28}$$

16.2.4 NMPB

In France, the NMPB–Route method estimates the source sound power level by considering the contributions of Rolling (L_r) and Propulsion (L_p). These components are determined through fitting experimental data obtained with pass-by tests, where the receiver is positioned at a lateral distance of 7.5 m from the road and at a height of 1.2 m above the ground. The parameters of the model are obtained by fitting the relationship:

$$L_{A,max} = L_r + L_p = 10\log\left(10^{Lp/10} + 10^{Lr/10}\right) \tag{16.29}$$

The power level for an individual vehicle is then derived from the $L_{A,max}$ level using the following back-propagation formula:

$$L_w(v) = L_{A,max}(v) + 20\log(d/d_0) + 10\log 2\pi \tag{16.30}$$

Measurements were conducted at numerous sites, involving 450 sites for Light Vehicles (LVs) and 150 sites for Heavy Vehicles (HVs).

16.2.5 CNOSSOS

The European Union introduced the ‘CNOSSOS model’ (Common Noise Assessment Methods), aimed at assessing various sources of noise, including road traffic noise. Its primary objective is to harmonise procedures across member countries to facilitate the creation of noise maps as mandated by regulation 2002/49/CE. The model evaluates the source power level (L_w) for each octave band, considering five vehicle categories: light, medium-heavy, heavy vehicles, powered two-wheelers, and alternative propulsion cars. Various correction terms are incorporated to address factors affecting noise emissions, such as studded tires, air temperature, road gradient, acceleration and deceleration phases, and road surface type.

The relationships used to calculate the sound power level due to propulsion and rolling noise for individual vehicles are specified for each octave band from 125 Hz to 4 kHz.

$$L_{w,P,i,m} = A_{P,m}(i) + B_{P,m}(i)(v_m - v_{ref})/v_{ref} \quad (16.31)$$

$$L_{w,R,i,m} = A_{R,m}(i) + B_{R,m}(i)\log v_m/v_{ref} \quad (16.32)$$

The coefficients vary depending on the octave band and the vehicle category considered. The reference speed in the CNOSSOS model is set at 70 km/h, while v_m represents the average speed of the m-th vehicle category considered. Initially published coefficients were updated to account for surface reflection effects, and coefficients for electric vehicles were provided, derived from statistical pass-bys for a reference pavement and compared with results obtained using a crumb rubber pavement.

The utilisation of mean speed for each vehicle category renders the CNOSSOS' NEM not fully microscopic, as it does not consider the speed of individual cars. Nonetheless, the single-vehicle sound power level is estimated separately before amalgamating all vehicle emissions in the traffic flow line source.

16.2.6 Vehicle Noise Specific Power (VNSP)

The Vehicle Noise Specific Power (VNSP) model, developed by Pascale *et al.* (2021), assesses the sound power level (L_w) considering vehicle motorisation. It utilises a fundamental relationship between L_w and speed for three types of car motorisation (diesel, petrol, and hybrid). Similar to CNOSSOS and Harmonoise models, the engine contribution is assumed to have a linear dependence on speed, while the rolling noise is modelled with a logarithmic function of speed. Unlike previous models, VNSP adopts the overall sound power level instead of fitting functions for each octave or third-octave band. The practical application of this model has been compared with results obtained using other NEMs in Pacale *et al.* (2021).

16.3 Summary

This chapter outlines the modelling frameworks used for forecasting road traffic noise. It introduces regressive and statistical methods and provides an overview of models and techniques for evaluating noise emissions from individual vehicles.

The applications of single-vehicle noise emission models are a relevant topic to be discussed in more detail. Indeed, after estimating the emission of a single vehicle using any available microscopic NEM, various options become feasible. In Graziuso *et al.* (2021), two indicators based on the sound power level of a single vehicle, calculated second by second, are defined: the overall and average sound power level emitted during a route. These indicators find utility in eco-routing applications, as exemplified in Graziuso *et al.* (2020), and Bandeira *et al.* (2018), or in signal setting design, as demonstrated in Di Pace *et al.* (2023). In numerous models and applications, such as those outlined in Sakamoto (2020), Quartieri *et al.* (2010), Guarnaccia (2020), and Pascale *et al.* (2023a, 2023b), the sound power level (L_w) of individual vehicles is employed to compute the Sound Exposure Level (SEL) for each transit along a designated road segment. Subsequently, by summing the SELs of the vehicles at the receiver, the overall traffic flow impact can be assessed.

References

Some suggestions for further readings are reported below among the huge literature available on these topics.

Bandeira JM, Guarnaccia C, Fernandes P, and Coelho MC (2018) Advanced impact integration platform for cooperative road use. *International Journal of Intelligent Transportation Systems Research* 16:1–15.

Bolt RH and Rosenblith WA (1953) *Handbook of Acoustic Noise Control* Aero Medical Laboratory, Ohio, USA.

Burgess MA (1977) Noise prediction for urban traffic conditions—related to measurements in the Sydney metropolitan area. *Applied Acoustics* 10(1):1–7.

Cocchi A, Farina A, Lopes G, *et al.* (1991) Modelli matematici per la previsione del rumore stradale: verifica ed affinamento del modello CNR in base a rilievi sperimentali nella città di Bologna, Atti XIX Convegno Nazionale AIA, Napoli, Aprile 1991.

Di Pace R, Storani F, Guarnaccia C, and de Luca S (2023) Signal setting design to reduce noise emissions in a connected environment. *Physica A: Statistical Mechanics and its Applications* 632:129328.

DoT U. K. (1988) Calculation of road traffic noise.

Dutilleux G, Defrance J, Ecoti'ere D, *et al.* (2010) Nmpb-routes-2008: The revision of the french method for road traffic noise prediction. *Acta Acustica united with Acustica* 96(3):452–462.

Fagotti C, and Poggi A (1995) Traffic noise abatement strategies: The analysis of real case not really effective. In: Proc. of 18th International Congress for Noise Abatement, pp 223–233.

für Verkehr B (1990) Richtlinien für den Lärmschutz an straßen RLs 90, *Allg. Rundschreiben strassenbau,* (8).

Galloway WJ, Clark WE, and Kerrick JS (1969) Highway noise measurement, simulation, and mixed reactions. *NCHRP report* (78).

Graziuso G, Mancini S, Francavilla AB, Grimaldi M, and Guarnaccia C (2021) Geo-Crowdsourced sound level data in support of the community facilities planning. A Methodological Proposal. *Sustainability* 13(10):5486.

Graziuso G, Mancini S, Guarnaccia C, *et al.* (2020) Comparison of single vehicle noise emission models in simulations and in a real case study by means of quantitative indicators. *International Journal of Mechanics* 14:198–207.

Griffiths I, and Langdon FJ (1968) Subjective response to road traffic noise. *Journal of Sound and Vibration* 8(1):16–32.

Guarnaccia C (2013) Advanced tools for traffic noise modelling and prediction. *WSEAS Transactions on Systems* 12(2):121–130.

Guarnaccia C (2020) Eagle: Equivalent acoustic level estimator proposal. *Sensors*, 20(3):701.

Guarnaccia C, Mascolo A, Aumond P, *et al.* (2024) From early to recent models: A review of the evolution of road traffic and single vehicles noise emission modelling. *Current Pollution Reports* 10:662–683. https://doi.org/10.1007/s40726-024-00319-5.

Heutschi K (2004) Sonroad: New swiss road traffic noise model. *Acta Acustica united with Acustica* 90(3):548–554.

Johnson D, and Saunders E (1968) The evaluation of noise from freely flowing road traffic. *Journal of Sound and Vibration* 7(2):287–309.

Kephalopoulos S, Paviotti M, and Anfosso-Ledee F (2012) Common noise assessment methods in europe (CNOSSOS-EU).

Kok A, and van Beek A (2019) Amendments for CNOSSOS-EU: Description of issues and proposed solutions.

Lelong J (1999) Vehicle noise emission: Evaluation of tyre/road and motor-noise contributions. In: Proceedings of Internoise 99, Fort Lauderdale, Florida, USA, 6-8 December 1999, volume 1.

Licitra G, Bernardini M, Moreno R, *et al.* (2023) Cnossos-eu coefficients for electric vehicle noise emission. *Applied Acoustics* 211:109511.

Nickson A (1965) Can community reaction to increased traffic noise be forecast?, *Proc. of 5th International Congress on Acoustics*.

Nielsen H (1997) *Road Traffic Noise: Nordic Prediction Method. Temanord Series*, Stationery Office.

Pascale A, Fernandes P, Bahmankhah B, *et al.* (2020) A vehicle noise specific power concept. In: *2020 Forum on Integrated and Sustainable Transportation Systems (FISTS)*, IEEE, pp 170–175.

Pascale A, Fernandes P, Guarnaccia C, *et al.* (2021) A study on vehicle noise emission modelling: Correlation with air pollutant emissions, impact of kinematic variables and critical hotspots. *Science of the Total Environment* 787:147647.

Pascale A, Guarnaccia C, Macedo E, *et al.* (2023a) Road traffic noise monitoring in a smart city: Sensor and model-based approach. *Transportation Research Part D*, Accepted and in press.

Pascale A, Macedo E, Guarnaccia C, *et al.* (2023b) Smart mobility procedure for road traffic noise dynamic estimation by video analysis. *Applied Acoustics* 208:109381.

Quartieri J, Iannone G, and Guarnaccia C (2010) On the improvement of statistical traffic noise prediction tools. In: Proceedings of the 11th WSEAS International Conference on "Acoustics & Music: Theory & Applications" (AMTA10), Iasi, Romania, pp 201–207.

Rossi D, Mascolo A, and Guarnaccia C (2023) Calibration and validation of a measurements-independent model for road traffic noise assessment. *Applied Sciences* 13(10):6168.

Sakamoto S (2020) Road traffic noise prediction model "ASJ RTN-Model 2018: Report of the research committee on road traffic noise. *Acoustical Science and Technology* 41(3):529–589.

Saracco G, and Leandre J (1991) Etude experimentale de la propagation acoustique a travers le Dioptre air-eau en presence d 'une houle sinusoidale (January 1991).

Watts G (2005) Harmonoise prediction model for road traffic noise.

Chapter 17

Emissions and energy consumption

Anna Laura Pala[1]

As business leaders, you must make it clear to your leaders that doing the right thing for the climate is also the smart thing for global competitiveness and long-term prosperity.

Ban Ki-moon, "Speech at the World Business Summit on Climate Change," May 24, 2009. Published by Grist*

Outline. *The following chapter provides an overview of methods and tools for emissions, fuel, and energy consumption estimation in road transport simulation models. It categorises these methods based on their detail level, with particular emphasis on those endorsed by environmental agencies. Notably, well-documented techniques from the literature are highlighted, offering a comprehensive approach to emission accounting in transport simulation models.*

Key aspects of emission estimation in road transport simulation are concisely outlined below, encompassing methodologies, functions, and tools for assessing emissions, fuel consumption, and energy utilisation in road transport. The chapter focuses on methodologies endorsed by environmental agencies and incorporates extensively documented techniques from literature, providing a robust framework for emission accounting.

The models can be classified by the aggregation level but also based on the function typology. Most of the models are regression functions and results to be used in a vast variety of software and tools. They need a lot of data and some of them are not available. In other cases, data are publicly available, such as the open database on electric vehicles (EV) cycle tests by Idaho National Laboratory. A different approach is based on the physical modelling of engines and vehicles giving results formulae for fuel or energy consumption and emissions factors. Even in this case, calibration of the models requires testing activities. Aggregate models may use these detailed models, but results are grouped for an average cycle, activity, and type of vehicle. These aggregate models are the most useable. It is the case of the HBEFA guidebook.

[1]Department of Computer Engineering, Automatic and Management, Sapienza University of Rome, Italy

*https://grist.org/article/2009-05-25-un-ban-climate-action-speech/

Moreover, the chapter reports a selection of data sources and sensors that might be very useful to consider in the CNMOST project. However, some of them are not free and it may be worth considering their acquisition.

17.1 Types and source of emissions

Emissions depend on different characteristics such as:

- vehicle type,
- vehicle load,
- road characteristics,
- environmental conditions,
- traffic conditions,
- driving habits,
- weather conditions.

The estimation of emissions requires indeed a variety of data of high quality. However, in some cases, the scope of the work could be an aggregate analysis of emissions of a transport system in predefined conditions. Moreover, in some cases, the estimation of emissions is used as an evaluation of a transport solution. In this case, a transport model is used to simulate a scenario and the result is evaluated in relation to emissions. In other cases, the emissions can be considered inside the model and not as a post-process evaluation module. In this latter case, online evaluation of emissions must be integrated with simulation models to account for user behaviour or regulatory actions that may be influenced by emission levels.

The emissions vary in nature, and different vehicle types do not generate the same level of emissions. Moreover, the accounting of emissions can be done by considering different domains. If we consider the usage phase and the tank-to-wheel model, EVs may be considered as zero-emission vehicles with the propulsion engine (not for the braking system and the wheels). However, if we consider the production of electricity, we must also assess the energy consumption of the vehicles to evaluate the emissions of these ones. In particular, if we refer to ISO 14040 and ISO 14044, we must define the domain of the study in order to correctly assess the environmental impact of a transport system overall (Life Cycle Assessment method). In general, considering the European regulations and the most used vehicle emissions accounting standards, such as the COPERT method, the emissions can be categorised into seven types:

- ozone precursors (CO, NO_x, *NMVOCs* – non-methane volatile organic compounds).
- greenhouse gases (CO_2, CH_4, N_2O); However, also emissions contributing to aerosol, which is considered a greenhouse gas, should be considered. Emissions contributing to aerosol formation are BTX, SOx, NOx, and VOC.
- acidifying substances (NH_3, SO_2, NO_x).
- particulate matter mass (*PM*) including black carbon (*BC*) and organic carbon (*OC*); We remind that actual EURO 6 and EURO 7 norms also take into account particulate number PN.
- carcinogenic species (*PAHs* – polycyclic aromatic hydrocarbons, and *POPs* – organic pollutants).

- toxic substances (dioxins and furans).
- heavy metals.

Greenhouse sources can be aggregated. Following the LCA methodology and the ISO14040 and ISO14040 standards, it is possible to convert these climate-altering emissions into equivalent carbon dioxide emissions (CO_2e). For example, 1 gram of N_2O corresponds to 298 grams of CO_2e (with 1 [g CO_2e] = 1 [g CO_2]).

17.2 The European Environment Agency guidelines for emissions factors

The primary reference for calculating road transport emissions is the document released and regularly updated by the European Environmental Agency (EEA). In particular, we refer to the chapter '1.A.3.b.i, 1.A.3.b.ii, 1.A.3.b.iii, 1.A.3.b.iv: Passenger cars, light commercial trucks, heavy-duty vehicles including buses and motorcycles"[†] in Ntziachristosband and Samaras (2023).

Considering the wide array of propulsion technologies in use, emissions estimating from road vehicles presents a complex and rigorous process that hinges on the availability of reliable activity data and emission factors. The chapter by Ntziachristos and Samaras (2023) within the Guidebook in EEA (2023) is designed to comprehensively address emissions generated by all currently prevalent technologies, facilitating the creation of precise and top-tier emission inventories. For detailed calculations is highly recommended to refer to this guidebook. Also, the guidebook equations are embedded in multiple software programs, such as the latest version of COPERT. In this chapter, we present only the aggregate calculations and tables to provide easy support for emissions evaluation. However, it is important to understand that the emissions derive not only from the driving phase of the vehicle, even if these account for the biggest part. This part is called exhaust emissions and is covered by the cited work. Other chapters of the guidebook cover the other emissions.

Another important factor to consider is that vehicle emissions vary substantially depending on whether the engine is in a cold phase or a warm phase (hot emissions). A detailed accounting of emissions takes both phases into account using different emissions equations. The total emissions calculation includes the fraction (β) of the trip during which the vehicle operates in the cold phase. Therefore, it is important to estimate the average trip length. This chapter provides an estimation based on country-specific data.

Considering the level of detail, an important concept is the different levels of detail, also referred to as tiers. The method defines three tiers, ranging from the most aggregated level (tier 1) to the most detailed level (tier 3). All emission factors are also stored in an Excel file included with the chapter[‡]. This file contains emission factors that can be used in the estimation functions of EMEA/EPA and the COPERT methodology.

[†]https://www.eea.europa.eu/publications/emep-eea-guidebook-2023/part-b-sectoral-guidance-chapters/1-energy/1-a-combustion/1-a-3-b-i/view

[‡]https://www.eea.europa.eu/publications/emep-eea-guidebook-2023/part-b-sectoral-guidance-chapters/1-energy/1-a-combustion/1-a-3-b-i-1/at_download/file.

Regarding the specific Excel file at issue, it is noteworthy that the pollutants column does not encompass CO_2. This is because CO_2 emissions produced by the combustion of biofuels (bioethanol, biodiesel, and biogas) should not be included. According to IPCC 2006 Guidelines, CO_2 emissions from biofuel production are accounted for in the Land Use, Land-Use Change, and Forestry sectors. However, other greenhouse gases emitted during biofuel combustion, such as CH_4 and N_2O, should be reported in greenhouse gas emissions from road transport. Additionally, in countries where CNG or LPG processes use coal-derived gas, there is a risk of double-counting. In such cases, CO_2 produced from coal should not be included in road transport emissions. In the following paragraph, it will be described how to calculate CO_2 emissions.

The columns of the Excel file correspond to the parameters of the COPERT equations that are documented in a few sources. One of these sources is Achour *et al.* (2011) in which the emission factors function is explained as the following regression function:

$$EF_{i,m,n} = \left(\alpha + \gamma\chi + \varepsilon\chi^2 + \zeta\chi^{-1}\right)/\left(1 + \beta\chi + \delta\chi^2\right)(1 - RF) \tag{17.1}$$

where, $EF_{i,m,n}$ is the emissions factor, in grams per kilometre travelled [g/km] for a given species i, of age m, and engine size n. χ is the average vehicle speed in kilometres per hour, and α, β, δ, ε, and ζ are related to the legislative emission factors for that car, i.e., Euro 1, 2, 3, etc. RF are coefficients specific to a given engine size n, and technology level m.

Depending on the available level of detail and the chosen approach for emissions calculations, the mentioned pollutants can be categorised into the following four groups:

Group 1: These pollutants benefit from a comprehensive methodology that relies on specific emission factors. This methodology accounts for various traffic scenarios (urban, rural, highway) and engine conditions, ensuring high-quality results.
Group 2: Emissions of Group 2 pollutants are estimated based on fuel consumption, producing results of similar quality to those in Group 1.
Group 3: Pollutants falling into this category are subject to a simplified methodology due to the lack of detailed data.
Group 4: These pollutants are calculated as a proportion of the total non-methane volatile organic compounds (NMVOC) emissions. A minor fraction of these 'residual' NMVOCs are designated as polycyclic aromatic hydrocarbons (PAHs).

The methodology to use depends on the scope of the study and the availability of the data. We report in Figure 17.1 the schema followed by the EEA guidebook and implemented into the Copert software:

17.2.1 Carbon dioxide (CO_2) emissions calculation

Emissions of ultimate CO_2 originate from three sources:

- Combustion of fuel,
- Combustion of lubricant oil,
- Addition of carbon-containing additives in the exhaust.

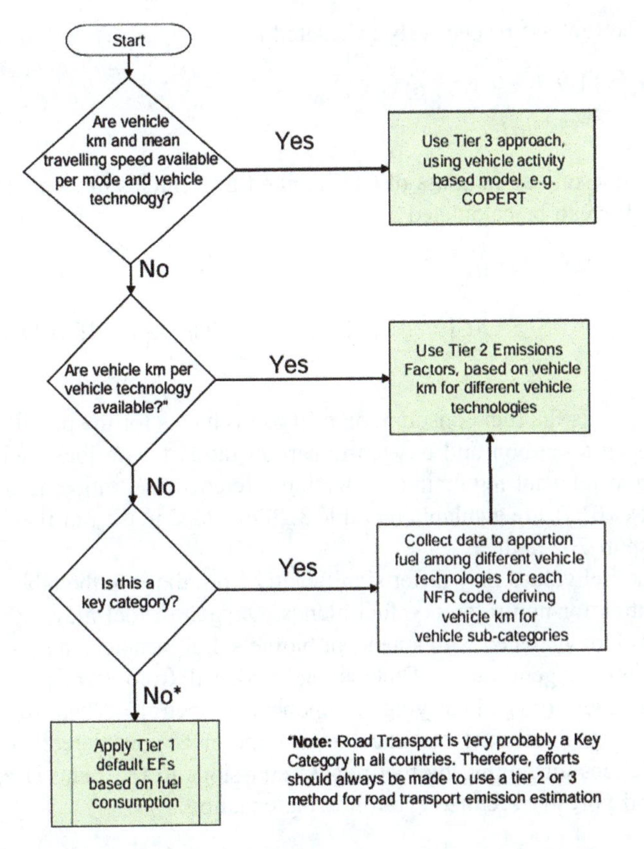

Figure 17.1 Decision tree for exhaust emissions from road transport in Ntziachristos and Samaras (2023)

Ultimate in this case means that the carbon contained in either of the three sources is fully oxidised into CO_2. The following paragraphs describe the methodology to calculate CO_2 in each case.

CO_2 due to fuel combustion

In the case of an oxygenated fuel described by the generic chemical formula CxHyOz the ratio of hydrogen to carbon atoms, and the ratio of oxygen to carbon atoms, are, respectively:

$$R_{H:C} = y/x$$

$$R_{O:C} = z/x$$

If the fuel composition is known from ultimate chemical analysis, then the mass fractions of carbon, hydrogen, and oxygen atoms in the fuel are c, h, and o, where $c + h + o = 1$. In this case, the ratios of hydrogen to carbon and oxygen to carbon in the fuel are respectively calculated as:

$$R_{H:C} = 11.916\, h/c$$

$$R_{O:C} = 0.7507\, o/c$$

With these ratios, the mass of CO_2 emitted by vehicles in technology k, combusting fuel m can be calculated as

$$E_{CO2,k,m}{}^{CALC} = 44.011 \times FC_{k,m}{}^{CALC} / \left(12.011 + 1.008 r_{H:C,m} + 16.000 r_{O:C,m}\right) \tag{17.2}$$

where FC^{CALC} is the fuel consumption of those vehicles for the period considered.

Hydrogen-to-carbon and oxygen-to-carbon ratios for various fuel types, originating from relevant regulations reflecting reference fuel ratios used for vehicle testing (UN, 2015), are available in Table 3.29 included in the pdf file "1.A.3.b.i-iv Road transport 2023.pdf"[§].

Market fuel values may differ significantly from those in the table, which also includes ratios for non-reference fuel blends. Oxygen in fuel may increase due to blending with oxygenated components or biofuels. For instance, in diesel, biodiesel is a common oxygen source. Bioethanol, produced from fermenting sugars, is another common oxygen-carrying component in petrol. When reporting CO_2 emissions, only fossil fuel statistical consumption is considered, aligning with IPCC guidelines attributing biofuel-related emissions to the Land Use, Land-Use Change, and Forestry sectors. The following equation:

$$E_{CO2,k,m}{}^{CORR} = E_{CO2,k,m}{}^{CALC} \times FC_m{}^{STAT,FOSSIL} / \left(\sum\nolimits_k FC_{k,m}{}^{CALC}\right) \tag{17.3}$$

corrects CO_2 emissions per vehicle category, excluding the oxygen content of biofuels. Notably, biodiesel production processes may yield a non-zero fossil fuel fraction, resulting in partially carbon-neutral fuel. Ofgem (2015) highlighted variations in fossil fuel carbon content across production pathways. Table 3.30 (included in the pdf file[‡] previously mentioned) lists fossil CO_2 per kg of fatty acid methyl esters (FAME) for different production methods.

In the equation above, the calculated CO_2 emission should be derived from the previous equation, without considering the oxygen content of the biofuel part.

It should be noted that, depending on the production process, biodiesel might have a non-zero fossil fuel fraction, resulting in a not entirely carbon-neutral fuel. Ofgem (2015) has, for example, shown that different production pathways may result in a fossil fuel carbon content of about 5.5%.

[§]1.A.3.b.i–iv Road transport 2023 — European Environment Agency (europa.eu)

CO_2 due to lubricant oil

New and well-maintained vehicles typically consume small amounts of lubrication oil, which forms an oil film on cylinder walls and is burned during combustion. Wear from extended engine operation can increase oil consumption, especially with older vehicles. Vehicles with 2-stroke engines consume significantly more lubricant oil, often injected into the intake or blended with fuel, and this oil is almost entirely combusted in the cylinder. While oil combustion contributes less to CO_2 emissions than fuel combustion, it should still be considered in national emission totals. Table 3.31 provides typical oil consumption rates for various vehicle types, fuels, and ages, gathered from multiple sources including internet references and interviews with maintenance experts and fleet operators. The term 'old' vehicle is subjective but generally refers to vehicles at or beyond their typical useful life, typically around 150,000 kilometres for passenger cars.

CO_2 emissions due to lube oil consumption can be calculated by means of an equation used to calculate $E^{CALC}_{CO2,k,m}$, but here fuel consumption should be replaced by the values of Table 3.31. This will lead to CO_2 emitted in kg per 10,000 km which has to be converted to tonnes/km by multiplying with 10^{-7}. Typical values for lube oil hydrogen to carbon ratio ($r_{H:C}$) is 2.08, while oxygen to carbon ratio ($r_{O:C}$) is 0.

CO_2 due to exhaust additive

Aftertreatment systems in Euro V and Euro VI heavy-duty vehicles, and increasingly in Euro 6 diesel light commercial vehicles, use urea as a reducing agent to reduce NOx emissions. When urea is injected upstream of a hydrolysis catalyst in the exhaust line, it reacts to form ammonia and carbon dioxide. While ammonia primarily reacts with nitrogen oxides to reduce them, the hydrolysis reaction also produces carbon dioxide, contributing to total CO_2 emissions from these vehicles. Commercial urea solution specifications for mobile use are regulated by DIN 70070, with a typical concentration of 32.5% wt and a density of 1.09 g/cm^3. The ultimate CO_2 emissions from using the additive can be calculated based on the total commercial urea solution sales (UC in litres), using the following equation:

$$E_{CO2,urea} = 0.26 \cdot UC$$

which considers factors such as solution density and urea content. The coefficient 0.26 (kg CO_2/litre urea solution) takes into account the density of urea solution, the molecular masses of CO_2 and urea, and the content of urea in the solution. If total urea consumption is known, a different coefficient is applied. If urea solution consumption is unknown, it can be estimated as a percentage of fuel consumption, typically around 5%–7% for Euro V and 3%–4% for Euro VI vehicles. Calculating selective catalytic reduction (SCR)-equipped vehicle shares and fuel consumption allows for the estimation of urea solution consumption and subsequent CO_2 emissions using the equation above.

17.2.2 Aggregated emissions evaluation: Tier 1 method

The Tier 1 approach for exhaust emissions uses the following general equation:

$$E_i = \sum_j \left(\sum_m \left(FC_{jm} EF_{ijm} \right) \right) \tag{17.4}$$

where:

E_i = emission of pollutant i [g/km],
$FC_{j,m}$ = fuel consumption of vehicle category j using fuel m [kg/km],
$EF_{i,j,m}$ = fuel consumption-specific emission factor of pollutant i for vehicle category j and fuel m [g/kg].

The vehicles are grouped in the following categories:

- Passenger Cars PC
- Light Commercial Vehicles – LCV
- Heavy Duty Trucks – HDC (>3.5 tons)
- Buses
- L-Category (Moped, Motorcycles, mini cars, all terrains – ATVs)

The emission factors are presented in Figures 17.2–17.10, which correspond to Tables 3.5–3.12 from Ntziachristos and Samaras (2023).

Table 3.5: Tier 1 emission factors for CO and NMVOCs

Category	Fuel	CO (g/kg fuel)			NMVOC (g/kg fuel)		
		Mean	Min	Max	Mean	Min	Max
PC	Petrol	84.7	49.0	269.5	10.05	5.55	34.42
	Diesel	3.33	2.05	8.19	0.70	0.41	1.88
	LPG	84.7	38.7	117.0	13.64	6.10	25.66
LCV	Petrol	152.3	68.7	238.3	14.59	3.91	26.08
	Diesel	7.40	6.37	11.71	1.54	1.29	1.96
HDV	Diesel	7.58	5.73	10.57	1.92	1.33	3.77
	CNG (Buses)	5.70	2.20	15.00	0.26	0.10	0.67
L-category	Petrol	497.7	331.2	664.5	131.4	30.0	364.8

Figure 17.2 Table 3.5 by Ntziachristos and Samaras (2023) for emission factors Tier 1

Table 3.6: Tier 1 emission factors for NOx and PM

Category	Fuel	NOx (g/kg fuel)			PM (g/kg fuel)		
		Mean	Min	Max	Mean	Min	Max
PC	Petrol	8.73	4.48	29.89	0.03	0.02	0.04
	Diesel	12.96	11.20	13.88	1.10	0.80	2.64
	LPG	15.20	4.18	34.30	0.00	0.00	0.00
LCV	Petrol	13.22	3.24	25.46	0.02	0.02	0.03
	Diesel	14.91	13.36	18.43	1.52	1.10	2.99
HDV	Diesel	33.37	28.34	38.29	0.94	0.61	1.57
	CNG (Buses)	13.00	5.50	30.00	0.02	0.01	0.04
L-category	Petrol	6.64	1.99	10.73	2.20	0.55	6.02

Figure 17.3 Table 3.6 by Ntziachristos and Samaras (2023) on emission factors Tier 1

Table 3.7: Tier 1 emission factors for N_2O and NH_3

Category	Fuel	N_2O (g/kg fuel)			NH_3 (g/kg fuel)		
		Mean	**Min**	**Max**	**Mean**	**Min**	**Max**
PC	Petrol	0.206	0.133	0.320	1.106	0.330	1.444
	Diesel	0.087	0.044	0.107	0.065	0.024	0.082
	LPG	0.089	0.024	0.202	0.080	0.022	0.108
LCV	Petrol	0.186	0.103	0.316	0.667	0.324	1.114
	Diesel	0.056	0.025	0.072	0.038	0.018	0.056
HDV	Diesel	0.051	0.030	0.089	0.013	0.010	0.018
	CNG (Buses)	0.24	0.15	0.4	n.a.	0.000	0.000
L-category	Petrol	0.059	0.048	0.067	0.059	0.048	0.067

Figure 17.4 Table 3.7 by Ntziachristos and Samaras (2023) on emission factors Tier 1

Table 3.8: Tier 1 emission factors for ID(1,2,3–cd)P and B(k)F

Category	Fuel	ID(1,2,3–cd)P (g/kg fuel)			B(k)F (g/kg fuel)		
		Mean	**Min**	**Max**	**Mean**	**Min**	**Max**
PC	Petrol	$8.90E^{-06}$	$5.90E^{-06}$	$1.33E^{-05}$	$3.90E^{-06}$	$3.90E^{-06}$	$3.90E^{-06}$
	Diesel	$2.12E^{-05}$	$1.11E^{-05}$	$4.05E^{-05}$	$1.18E^{-05}$	$3.00E^{-06}$	$4.58E^{-05}$
	LPG	$2.00E^{-07}$	$2.00E^{-07}$	$2.00E^{-07}$	$2.00E^{-07}$	$2.00E^{-07}$	$2.00E^{-07}$
LCV	Petrol	$6.90E^{-06}$	$3.90E^{-06}$	$1.21E^{-05}$	$3.00E^{-06}$	$2.60E^{-06}$	$3.50E^{-06}$
	Diesel	$1.58E^{-05}$	$8.70E^{-06}$	$2.84E^{-05}$	$8.70E^{-06}$	$2.40E^{-06}$	$3.21E^{-05}$
HDV	Diesel	$7.90E^{-06}$	$7.30E^{-06}$	$8.6E^{-06}$	$3.44E^{-05}$	$3.18E^{-05}$	$3.72E^{-05}$
	CNG (Buses)	n.a			n.a		
L–category	Petrol	$1.02E^{-05}$	$1.00E^{-05}$	$1.04E^{-05}$	$6.80E^{-06}$	$6.70E^{-06}$	$7.00E^{-06}$

Figure 17.5 Table 3.8 by Ntziachristos and Samaras (2023) on emission factors Tier 1

Table 3.9: Tier 1 emission factors for B(b)F and B(a)P

Category	Fuel	B(b)F (g/kg fuel)			B(a)P (g/kg fuel)		
		Mean	**Min**	**Max**	**Mean**	**Min**	**Max**
PC	Petrol	$7.90E^{-06}$	$5.40E^{-06}$	$1.14E^{-05}$	$5.50E^{-06}$	$4.80E^{-06}$	$6.20E^{-06}$
	Diesel	$2.24E^{-05}$	$9.60E^{-06}$	$5.26E^{-05}$	$2.14E^{-05}$	$1.00E^{-05}$	$4.55E^{-05}$
	LPG				$2.00E^{-07}$	$2.00E^{-07}$	$2.00E^{-07}$
LCV	Petrol	$6.10E^{-06}$	$3.60E^{-06}$	$1.03E^{-05}$	$4.20E^{-06}$	$3.20E^{-06}$	$5.60E^{-06}$
	Diesel	$1.66E^{-05}$	$7.50E^{-06}$	$3.69E^{-05}$	$1.58E^{-05}$	$7.90E^{-06}$	$3.19E^{-05}$
HDV	Diesel	$3.08E^{-05}$	$2.84E^{-05}$	$3.33E^{-05}$	$5.10E^{-06}$	$4.70E^{-06}$	$5.50E^{-06}$
	CNG (Buses)	n.a			n.a		
L-category	Petrol	$9.40E^{-06}$	$9.20E^{-06}$	$9.60E^{-06}$	$8.40E^{-06}$	$8.20E^{-06}$	$8.60E^{-06}$

Figure 17.6 Table 3.9 by Ntziachristos and Samaras (2023) on emission factors Tier 1

Table 3.10: Tier 1 emission factors for lead (Pb)

Category	Fuel	Pb (g/kg fuel)		
		Mean	Min	Max
PC	Petrol	$3.30E^{-05}$	$1.70E^{-05}$	$2.00E^{-04}$
	Diesel	$5.20E^{-05}$	$1.60E^{-05}$	$1.94E^{-04}$
	LPG	n.a		
LCV	Petrol	$3.30E^{-05}$	$1.70E^{-05}$	$2.00E^{-04}$
	Diesel	$5.20E^{-05}$	$1.60E^{-05}$	$1.94E^{-04}$
HDV	Diesel	$5.20E^{-05}$	$1.60E^{-05}$	$1.94E^{-04}$
	CNG (Buses)	n.a		
L-category	Petrol	$3.30E^{-05}$	$1.70E^{-05}$	$2.00E^{-04}$

Figure 17.7 Table 3.10 by Ntziachristos and Samaras (2023) on emission factors Tier 1

Table 3.11: Tier 1 BC fractions of PM

Vehicle category	f-BC
Petrol passenger cars	0.12
Petrol light-duty vehicles	0.05
Diesel passenger cars	0.57
Diesel light-duty vehicles	0.55
Diesel heavy-duty vehicles	0.53
Petrol L-category	0.11

Figure 17.8 Table 3.11 by Ntziachristos and Samaras (2023) on emission factors Tier 1

Table 3.12: Tier 1 CO_2 emission factors for different road transport fossil fuels

Subsector units	Fuel	kg CO_2 per kg of fuel[1]
All vehicle types	Petrol	3.169
All vehicle types	Diesel	3.169
All vehicle types	LPG[2]	3.024
All vehicle types	CNG[3](or LNG)	2.743
All vehicle types	E5[4]	3.063
All vehicle types	E10[4]	2.964
All vehicle types	E85[4]	2.026
All vehicle types	ETBE11[5]	3.094
All vehicle types	ETBE22[5]	3.021

Notes:
[1] CO_2 emission factors are based on an assumed 100% oxidation of the fuel carbon (ultimate CO^2).
[2] LPG assumed to be 50% propane + 50% butane.
[3] CNG and LNG are assumed to be 100% methane.
[4] E5, E10 and E85 blends are assumed to consist of 5, 10 and 85% vol. respectively ethanol (bio-ethanol or synthetic ethanol) and 95, 90 and 15% respectively petrol.
[5] ETBE11 and ETBE22 blend assumed to consist of 11 and 22% vol. respectively ETBE and 89 and 78% respectively petrol.

Figure 17.9 Table 3.12 by Ntziachristos and Samaras (2023) on emission factors Tier 1

Table 3.13: Tier 1 CO_2 emission factors from the combustion of lubricant oil[1]

Category	Fuel	CO_2 from lubricant (g/kg fuel)		
		Mean	Min	Max
PC	Petrol	8.84	7.83	9.89
	Diesel	8.74	8.01	11.3
	LPG	8.84	7.83	9.89
LCV	Petrol	6.07	4.76	7.28
	Diesel	6.41	5.41	7.72
HDV	Diesel	2.54	1.99	3.32
	CNG (Buses)	3.31	3.09	3.50
L-category	Petrol	53.8	33.3	110

Note:[1] These emission factors assume typical consumption values for lubricant oil used in automotive applications.

Figure 17.10 Table 3.13 by Ntziachristos and Samaras (2023) on emission factors Tier 1

Table 3.14: Tier 1 - Typical sulphur content of the fuel (1 ppm = 10^{-6} g/g fuel)

Fuel	1996 Base fuel (Market average)	Fuel 2000	Fuel 2005	Fuel 2009 and later
Petrol	165 ppm	130 ppm	40 ppm	5 ppm
Diesel	400 ppm	300 ppm	40 ppm	3 ppm

Figure 17.11 Table 3.14 by Ntziachristos and Samaras (2023) on emission factors Tier 1

The SO_2 emissions per fuel type m must consider the type of fuel and its sulphur content. It is assumed that all sulphur in the fuel is transformed completely into SO_2, using the formula $E(SO_{2,m}) = 2\, k_{s,m}\, FC_m$ where $E(SO_{2,m})$ is the emission in SO_2 for the fuel m in [g/km], $k_{s,m}$ is the weight-related sulphur content in the fuel of type m [g/g fuel], and FC_m is the fuel consumption of fuel m [g/km].

Ntziachristos and Samaras (2023) report the following values (Figure 17.11) for sulphur content.

Up until now, we have focused on the Tier 1 method, which is the most aggregated approach. However, this is not the only available method. The emissions models can be classified into three tiers based on their level of detail:

The Tier 1 method require an aggregate evaluation of fuel consumption for vehicle categories. This is reported in the following table (Figure 17.12).

Tier 2 method covers the specific technology of each vehicle type. The technology is regulated by the European standard, where the last standard is the Euro 6d. Subsequently, it is crucial to estimate the annual mileage for each vehicle technology, along with fuel consumption, in addition to emissions per unit of fuel. These data are reported in the chapter in Tables 3.16–3.27 (included in the pdf file, refer to footnote on page 362).

Tier 3 is the more detailed method. It considers emissions summing up the phases of cold start and hot engine. Moreover, it distinguishes the type of trip as Urban, Rural or Highway. In addition, velocity, and other detailed considerations are considered. A detailed explanation of the model is presented in Ntziachristos and Samaras (2023), where they provide an overview of the method's schema (Figure 17.13).

Table 3.15: Tier 1 – Typical fuel consumption figures, per km, by category of vehicle

Vehicle category (j)	Fuel	Typical fuel consumption (g/km)
Passenger cars	Petrol	70
	Diesel	60
	LPG	57.5
	E85	86.5
	CNG	34.2
LCV	Petrol	100
	Diesel	80
HDV	Diesel	240
	CNG (buses)	500
L-category	Petrol	35

Figure 17.12 Table 3.15 by Ntziachristos and Samaras (2023) on emission factors Tier 1

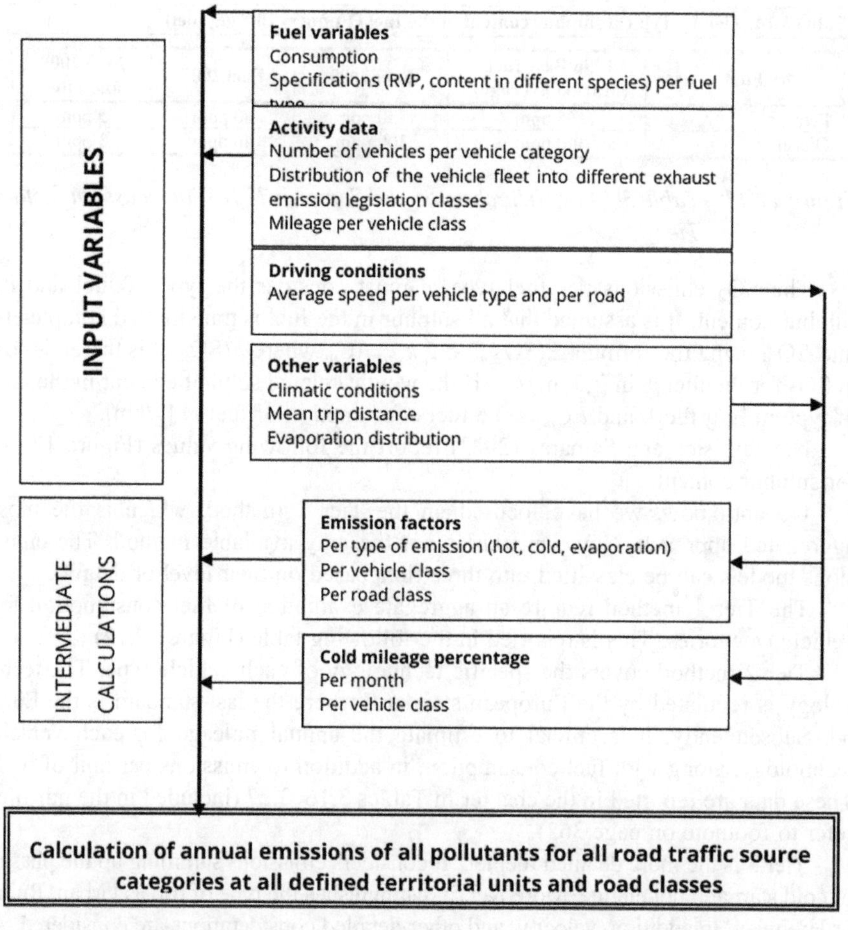

Figure 17.13 Tier 3 schema as in Figure 3.2 by Ntziachristos and Samaras (2023)

17.3 Methods and tools for calculating energy consumption

The methods for accounting emissions can be expressed based on functions that specify the mechanical characteristics of the engine and the vehicle, with parameters tuned with data resulting from validation tests (Tier 3 methods). These functions can be modelled starting from the external conditions, going *backward* to the efficiency of the engine, computing the emissions through the fuel consumption, or going *forward* from the physical and mechanical characteristics of the engine and the vehicle. In general, the simpler approach is the backward one.

A different approach can be applied, using data analytics and machine learning methods.

However, all these aspects may be difficult to consider in detail in a transportation model. Furthermore, aggregate accounting can benefit from extensive parameter testing conducted over years of data collection and simulation runs spanning entire regions. Thus, emissions aggregate modelling is very important and appears to be among the most practical way to estimate traffic emissions.

An aggregate emissions model accounts for the emissions e for each emission type k, of a vehicle category i over space or time (e^k_{si} or e^k_{ti}) as a function of different kinematic and exogenous variables. Moreover, it is common to specify different functions corresponding to different conditions and leaving in the function only the velocity v and eventually the acceleration a. For internal combustion vehicles, the emission function can be considered inversely proportional to the vehicle's velocity, thus we can write, following the work by Cascetta *et al.* (1988):

$$e_{si}{}^k(\bar{V}) = a_i{}^k + (b_i{}^k)/\bar{V}c_i{}^k)[\mathrm{g/km}] \tag{17.5}$$

Where $a^k{}_i$, $b^k{}_i$, $c^k{}_i$, are respectively pollutant, vehicle, and velocity range-specific parameters.

The total quantity of emissions $E^k{}_l$ of a pollutant k over an arc of length L_l, with respect to a vehicular flow f_l, can be indeed computed considering the fraction w_i of vehicles i of the flow f_l with the following formula:

$$E_l^k = f_l \sum_i w_i e_{si}^k \left[\bar{V}(f_l) L_l\right] [\mathrm{g/h}] \tag{17.6}$$

These methods can be used when mean flows are available and, in this regard, for example, we find in Cascetta *et al.* (1988) models computed for USA and UK being equal to

$$e_s 4 + 742\bar{V}^{-1} [\mathrm{g/m}] \tag{17.7}$$

and to

$$E_s = 1.031\bar{\mathrm{V}}^{-0.795} 10^{-4} f [\mathrm{g/(m \cdot s)}] \tag{17.8}$$

respectively.

If we look at the purpose of estimating emissions considering coherently with simulation models, we can consider two main purposes:

- Macro simulation
- Microsimulation

In this regard, emissions models can be classified according to their detail level. Greater details require data such as instantaneous velocity and acceleration, and cold start times, which are not addressed in aggregated simulation models. Thus, it is important to classify the model according to simulation one. Following the EMEA guidelines, also specified in the COPERT model (EMEP/EEA, 2023), the model detail level can be classified into three levels based on its detail and data requirements. The level is named as 'tier'.

- **Tier 1**. Methods designed for aggregate modes. They rely on default or average emission factors, which are applied to vehicle fleets and traffic activity. Tier 1 is suitable for aggregate analysis, macrosimulation, and situations where refined data are not available.
- **Tier 2**. Tier 2 models consider more refined data and allow the accounting of emissions based on different characteristics of the transport system, vehicle type, and traffic conditions. Tier 2 models consider also driving cycles and emission standards. They can be used mostly on macro simulations but also microsimulations.
- **Tier 3**. Tier 3 models emissions consider all the parameters affecting them. For instance, real-time driving conditions, road characteristics, vehicle type and age are considered in these models. Tier 3 can be used in microsimulation and dynamic models when refined and real-time data is available. Tier 3 is used in COPERT model.

In the following sections, the most used methods and tools used to account for emissions are presented.

17.3.1 The Handbook of Emission Factors for Road Transport

The Handbook of Emission Factors for Road Transport model (HBEFA)[**] is a database application for road transport emission factors containing:

- Energy/fuel consumption, GHG, regulated and unregulated pollutant emission factors
- For all relevant road transport vehicles (cars, vans, trucks, buses, coaches, motorcycles)
- Emission factors available from very detailed levels (vehicle type, technology, emission standard, traffic situation) to aggregated levels (e.g., average car in country X and year Y)

[**]https://www.hbefa.net

HBEFA is widely used for various purposes, including environmental impact assessments, air pollution inventories from city to national levels, climate/GHG reporting, and energy and climate impact scenarios. Furthermore, it serves as an input for many third-party applications such as COPERT, EcoTransIT, LCA tools (e.g., eco-invent), and the PTV transport/logistics suite.

Therefore, it allows the computation of emissions at the link level or the system level. The inputs are the traffic situation, the traffic volumes, and the fleet composition.

The calculation is different for warm and cold start emissions. Warm emissions calculations require as input the type of road (urban, rural), speed limits, static traffic situation, country, type of vehicle, and year. For cold start emission, the accounting must consider the origin/destination points of each route. Thus, the accounting can be done considering the convex hull of polygons associated with zones where demand originated, or considering each route, accounting for the cold start to the starting link with a decay function. For the sake of completeness, in the following, we list all the emissions (not only pollutants) accounted for by the HBEFA model:

- Noise calculation RLS-19: Calculation of the emissions of the length-based acoustic power level of a source line per direction of travel according to the guidelines of the Forschungsgesellschaft für Straßen-und Verkehrswesen (FGSV).
- Noise-Emis-Nordic: Calculation of noise emission levels in accordance with the Nordic Council of Ministers (1996).
- Pollution-Emis: Calculation of air pollution emissions in accordance with emission factors of the Swiss Federal Office for the Environment (BAFU).

The HBEFA model is actually a specification book – The Handbook of Emission Factors for Road Transport and can be bought at the following site: https://www.hbefa.net. Currently, it does not consider Italy as a country. However, is one of the most used methods and is embedded in different tools such as for example the PTV Visum simulation software. Moreover, it is used in a vast variety of studies and tools also outside of Europe. Thus, the Handbook of Emission Factors for Road Transport (HBEFA) is a widely recognised resource for estimating emissions from road vehicles. The HBEFA provides emission factors that consider various vehicle categories, fuel types, and driving conditions. To use HBEFA data to estimate emissions from a specific vehicle, these steps can be followed:

1. *Identify vehicle characteristics*: Gather information about the vehicle you want to estimate emissions for, including the vehicle category, engine type, fuel type, and any relevant parameters such as vehicle weight.
2. *Select the appropriate emission factor*: Refer to the HBEFA database to find the relevant emission factor for your specific vehicle category and conditions. HBEFA provides emission factors for various pollutants, including CO_2, CO, NOx, and PM, based on vehicle technology and fuel type.

3. *Calculate emissions*: Use the selected emission factor to calculate emissions based on the distance travelled (D) and the emission factor (EF) for the specific pollutant of interest. The general equation for emissions calculation in this case is:

   $E = D \cdot EF$ with E being the emissions (e.g., in grams or kilograms of the specific pollutant), D, the distance travelled (e.g., in kilometres or miles), EF, the emission factor for the specific vehicle, pollutant, and driving conditions.
4. *Repeat for different pollutants*: repeat the process using the corresponding emission factors for each pollutant of interest.
5. *Consider driving cycle*: In some cases, the emissions may vary based on driving conditions and cycles. HBEFA may provide different emission factors for urban, suburban, and highway driving cycles. The most appropriate driving cycle that matches the vehicle's operating conditions must be selected.

It is important to note that using HBEFA emission factors ensures a more accurate estimation of emissions specific to the vehicle category and fuel type. HBEFA provides detailed emissions data for a wide range of vehicles, making it a valuable resource for governments, researchers, and organisations aiming to assess and manage emissions from road transport.

The HBEFA emission factors are not freely available currently and must be bought to be used.

17.3.2 Emissions functions based on regression analysis

We have already covered an important regression function giving the base for EMEA and COPERT methods, documented in Achour *et al.* (2011). In this section, we recall other related methods. The emissions equation for vehicles can be computed based on regression analysis of test cycles. In this case, good estimations give emissions as polynomial functions of velocity and or acceleration. Good estimations in the literature consider 3-grade polynomials. However, it is important to differentiate among vehicle technologies. An important work considering ICE vehicles gives a 3-grade polynomial for emissions and can be found in Ahn *et al.* (2002). The study is based on the Oak Ridge National Laboratory – ORNL database. A slightly different and more detailed approach is used in Bakhit *et al.* (2015), in which emissions are computed through regression considering the instantaneous power and square of power that are in turn depending linearly by the velocity and the square of the velocity. We report for sake of clarity these functions (Bakhit *et al.*, 2015):

$$FC = \alpha_0 + \alpha_1 \cdot P(t) + \alpha_2 \cdot P(t)^2 \text{ if } P(t) \geq 0 \qquad (17.9)$$

$$FC = \alpha_0 \text{ if } P(t) < 0$$

where $\alpha_0, \alpha_1, \alpha_2$ are vehicle specific constants and *P(t)* Instantaneous power time

$$P(t) = ((R(t) + 1.04ma(t))/(3600\eta_d)v(t)) \qquad (17.10)$$

where m corresponds to vehicle mass (kg), $a(t)$ is the acceleration of the vehicle at time t (m/s^2), η_d represents the driveline efficiency, which denotes the effectiveness of the vehicle in transferring power from the motor to the tires (%), and $v(t)$ is the speed at time t (m/s). $R(t)$ is a resistance function determined by drag, rolling resistance (friction), and grade resistance and is defined as (Bakhit *et al.*, 2015):

$$R(t) = \rho/25.92\ C_D C_h A_f v(t)^2 + 9.80066m\ c_r/1000\ (C_1 v(t) + C_2) + 9.8066mG(t) \tag{17.11}$$

where, ρ is the density of air at sea level at a temperature of 15°C (59°F) (equal to 1.226 kg/m^3), C_D is the vehicle drag coefficient, C_h is the correction factor for altitude (equal to 1-0.085H, where H is the altitude in kilometers), A_f is the frontal area of the vehicle (m^2), C_r, C_1, and C_2 are coefficients associated with rolling, and $G(t)$ is the grade at time t. For this model, the altitude will be 48 m assumed sea level ($H = 0$ km, $C_h = 1.0$) and the grade will be assumed level causing the last term to drop. After the instantaneous power is determined, the α parameter associated with vehicle parameters can be found and the article provides insights on this.

It is important to note that this function is widely used in tier 3 models. A regression approach is also provided for modern EV vehicles. One of the most important references is Galvin (2017). He defines the following equation for estimating power at wheels (P_w) and power demand at battery respectively (P_{db})

$$P_w = \alpha v + \beta v^2 + \gamma v^3 + mva \tag{17.12}$$

$$P_{db} = rv + sv^2 + tv^3 + uva$$

where v and a are velocity and acceleration respectively, m is the mass, and the others are parameters to be found with regressions.

Just as an example, we report the power demanded at the battery (P_{db}) and the energy required per unit of distance (E/S) for Nissan SV as

$$P_{db} = 479.1\text{V} - 18.93\text{V}^2 + 0.7876\text{V}^3 + 1507\text{A} \tag{17.13}$$

$$E/S = 479.1 - 18.93\text{V} + 0.7876\text{V}^2 + 1507\text{A}$$

17.3.3 Emissions and fuel consumption models for internal combustion engine (ICE) vehicles

Tier 3 emission standards refer to the emission standards established by various regulatory agencies, such as the US Environmental Protection Agency (EPA) and the European Union, to limit the emissions of pollutants from internal combustion engine (ICE) vehicles. These standards primarily focus on reducing emissions of pollutants like nitrogen oxides (NOx), particulate matter (PM), and non-methane hydrocarbons (NMHC).

To estimate emissions or fuel consumption for ICE vehicles, emission factors and emission equations are used. Emission factors are specific to different vehicle categories, engine types, and driving conditions. In relation to Tier 3 models, the US EPA provides the MOVES (Motor Vehicle Emission Simulator) model, which can estimate emissions for various pollutants and fuel consumption based on a wide range of parameters, including vehicle type, speed, and driving cycles. In Europe, the EMEA/EPA implemented the emissions factor guidebook Tier 3 model into the COPERT software.

The fuel consumption of an internal combustion engine (ICE) vehicle based on velocity and acceleration can be estimated using a variety of equations, depending on the specific characteristics of the vehicle, engine, and driving conditions. One common approach is to use a simplified power or energy consumption model.

Detailed emission calculation can be done by accounting for different variables such as instantaneous velocity and acceleration, slope, engine efficiency, and so on. Several papers and tools proposed well-studied models that are also embedded in official computation tools such as COPERT and MOVE or into detailed simulation tools such as CO2MPAS. We report the works of Fontaras *et al.* (2014), Fiori *et al.* (2016), Tsiakmaki *et al.* (2017), and Tsiakmakis *et al.* (2019). Specifically, Tsiakmakis *et al.* (2019) presented an equation to enable a detailed computation of internal combustion engine (ICE) energy consumption. In this context, the drive-train energy

P_{dtr} is defined as:

$$P_{dtr} = \left(F_0 \cdot \cos(\phi) + F_1 \cdot v + F_2 \cdot v^2 + m \cdot a + m \cdot g \cdot \sin(\phi)\right) \cdot v/\eta_{dtr} \tag{17.14}$$

F_0, F_1, and F_2 are parameters, namely the road load values, and can be found in the research literature. A very useful work is from Moskalik (2020) where F_i (i = 0,1,2) is expressed as functions $F_i = L_i + D_i$ where L_i is the vehicle loss and D_i is the dynamometric loss. In turn, D_i depends on aerodynamic and rolling resistance while L_i depends on transmission coefficient and rolling resistance (for public data see https://www.epa.gov/compliance-and-fuel-economy-data/data-cars-used-testing-fuel-economy). Therefore, v and a are the velocity and acceleration of the vehicle respectively. The numerator represents the main motive power required to move the engine at the defined velocity profile and surpass the aerodynamic and rolling resistances, while the denominator represents the efficiency of the complete transmission system, final drive, gearbox, and clutch/torque converter. Variables m and g are the mass and acceleration of gravity respectively while ϕ is the road gradient. The transmission efficiency, η_{dtr}, is assumed to be a function of the torque, the rotational speed at the powered shaft, and empirical constant factors, which are derived from detailed losses, and the part that is proportional to the square of the vehicle's velocity (aerodynamic component). Given the importance of vehicle mass and road loads for an accurate estimate of the engine's power demand, and thus, ultimately, the engine's fuel consumption and CO_2 emissions, a dedicated module for the calculation of the vehicle's road loads must developed.

Another clear treatment of the fuel consumption model considering engine properties can be found in Treiber *et al.* (2008), in which physics models are introduced in contraposition to regression methods. The derived function is:

$$\tilde{P}(v, \dot{v}) = P\_0 + v\left[m\left\{\dot{v} + (\mu 0 + \mu 1\, v + \beta)g\right\} + 1/2\, cw\, \rho A v^2\right] \tag{17.15}$$

In Figure 17.14 we report Treiber *et al.*'s (2008) explanation table for the function stated above (16.15).

Given the critical importance of vehicle mass and road loads in achieving precise estimations of the engine's power demand, and subsequently, the engine's fuel consumption and CO_2 emissions, a dedicated module for calculating the vehicle's road loads must be developed. This is encompassed in the model described by Tsiakmatis *et al.* (2017). Recent studies by Makridis *et al.* (2020) use these types of functions to estimate behavior and the impact of connectivity and automation on flow and emissions. Standard computation software embeds these functions for tier 3 calculations. In this regard, the CO2MPAS model is explained on the website of the Model CO2MPAS platform, maintained by the Joint Research Centre (JRC) of the European Commission.[††]

17.3.4 Electric vehicles consumption and the Virginia-Tech Comprehensive Power-based EV Energy consumption Model VT-CPEM for EV

Electric vehicle emissions estimation differs from the ICE vehicle in some respects. The energy consumption of an electric vehicle is not directly connected to a local emission caused by the power unit. Conversely, it is needed to consider the emissions of the source of energy. For this purpose, it can be assumed the emission of the country's energy mix or different scenarios of energy generation. The energy consumption of electric vehicles can be estimated considering different parameters and specifics of the power unit. For instance, it is often important to consider that EVs are equipped with efficient energy recovery from the braking system. Regenerative braking allows the EV to recover some of the energy typically lost as heat during traditional friction braking. Energy recovery can be accounted for by adjusting the energy consumption component of the emission equation. The energy saved through regenerative braking can be subtracted from the total energy consumption, thus reducing emissions. As explained in the paper of Fiori *et al.* (2016), energy recovery is computed when the power at wheels is negative. Its efficiency is computed based on the results of Gao *et al.* (2007).

Parameter	VW Passat Synchro	VW Polo Diesel
Effective cylinder volume C_{cyl}	1.8 l	1.4 l
Basic power consumption P_0	3 kW	2 kW
Vehicle mass m	1600 kg	1050 kg
Friction coefficient μ_0	0.015	0.015
Friction coefficient μ_1	0.0003 s/m	0.0003 s/m
Cross-sectional area A	2.03 m^2	1.70 m^2
Air-drag coefficient (cd-value) c_w	0.32	0.36

Table 1: Car data for typical passenger cars like a VW Passat and a VW Polo.

Figure 17.14 Table 1 by Treiber et al. (2008) with parameter explanation

[††]https://web.jrc.ec.europa.eu/policy-model-inventory/explore/models/model-co2mpas/

The VT-Comprehensive Power-based EV Energy consumption Model (VT-CPEM) of Fiori *et al.* (2016) is a sophisticated, backward power-centric model. To clarify, this model takes in the vehicle's current speed and its characteristics as inputs and generates several outputs, including energy consumption (EC) measured in kilowatt-hours per kilometre (kWh/km) for a given driving cycle, the instantaneous power consumption in kilowatts (kW), and the state of charge (SOC) of the electric battery as a percentage at the conclusion of the simulation.

The EV energy model computes an EV's instantaneous energy consumption using second-by-second vehicle speed, acceleration, and roadway grade data as input variables. In doing so, the model estimates the instantaneous braking energy regeneration.

In particular, the power at wheels at time *t*, defined as $P_{Wheels}(t)$, is computed as:

$$P_{Wheels}(t) = \left(ma(t) + (mg\cos\vartheta C_{\neg r})/1000\,(c_1 v(t) + c_2) + 1/2 p_{Air} A_f C_D v^2(t) + mg\sin\vartheta\right)v(t) \tag{17.16}$$

The proposed model is general and is applied to the Nissan Leaf for illustration purposes. However, data for a vast variety of electric and hybrid vehicles can be found on the website of the Advanced Vehicle Testing program of the National Renewable Energy Laboratory (NREL)[‡‡].

In the formula above, the parameters are set with exemplificative values for a specific electric vehicle (Nissan Leaf is reported):

- *m* vehicle mass [kg],
- $a(t) = d\,v(t)/dt$ is the acceleration of the vehicle in [m/s^2] ($a(t)$ takes negative values when the vehicle decelerates),
- $g = 9.8066$ [m/s^2] is the gravitational acceleration,
- *h* is the road slope,
- $C_r = 1.75$,
- $c_1 = 0.0328$ and $c_2 = 4.575$ are the rolling resistance parameters that vary as a function of the road surface type, road condition, and vehicle tire type.
- The typical values of vehicle coefficients are reported in Rakha *et al.* (2001):
  - $\rho_{Air} = 1.22563$ [kg/m^3] is the air mass density, $A_f = 2.3316$ [m^2] is the frontal area of the vehicle, and $C_D = 0.28$ is the aerodynamic drag coefficient of the vehicle and $v(t)$ is the vehicle speed in [m/s].

The power at the electric motor $P_{Electric\ motor}$ (t) is computed, given the power at the wheels, considering the driveline efficiency $\eta_{Driveline} = 92\%$ and assuming that the efficiency of the electric motor is $\eta_{\text{Electric Motor}} = 91\%$ in the studied case.

When the power is negative (negative acceleration), a recovery of the battery due to regenerative braking can be computed. The efficiency of the regenerative system is often assumed (as cited in Fiori *et al.* 2016) to be a negative exponential with the acceleration. The authors synthetise several literature experiments through

[‡‡]https://avt.inl.gov/

the minimum squares method, and the efficiency is computed as

$$\eta_{rb}(t) = \left\{ ([e^{\wedge}0.0411/|a(t)|))]^{\wedge(-1)}, \forall a(t) < 0;\ 0, \forall a(t) \geq 0 \right\} \tag{17.17}$$

Thus, the Power is multiplied by $\eta_{rb}(t)$ when it is negative.

17.3.5 Tools for energy consumption estimation

CO2MPAS vehicle simulator

CO2MPAS is a vehicle simulation model for the calculation of CO_2 emissions, energy, and fuel consumption from passenger cars and light commercial vehicles. It uses experimental data, retrieved from chassis dyno or on-road tests, to simulate the operation of vehicles under different operating conditions. It is used also to estimate and type-approve CO_2 emissions of vehicles by simulating NEDC conditions based on the emissions measured during WLTP tests, according to the EU legislation.

The CO2MPAS model is used in several research papers and is well documented[§§].

It is also available as an open-source package hosted on git hub and the references are in the following web locations:

- Webpage: https://co2mpas.readthedocs.io/en/stable/
- GIT-HUB repository: https://github.com/JRCSTU/CO2MPAS-TA

The balance of power for the simulation of the engine is computed in a detailed way as reported in Figure 17.15 taken from the CO2MPAS documentation.

CO2MPAS has a more detailed approach if compared to the approaches used as standard compliant COWLTP-GTR test. Considering the documentation[***], the approach can be synthesised in the following equations:

$$\begin{aligned} P_{engine} &= P_{wheel}/\,\eta_{dt} + P_{elec} + P_{mech} \\ P_{wheel} &= (F_0 + F_1 \cdot v + F_2 \cdot v^2 + m \cdot a)v \\ P_{elec.} &= P_{elec.dem.}/\,(\eta_{alt/or} \cdot \eta batt) \\ P_{mech.} &= T_{const} \cdot RPM \end{aligned} \tag{17.18}$$

The primary use foreseen for the model is to correlate type approval CO_2 emissions determined for light-duty vehicles in accordance with the new WLTP test procedure set out in Commission Regulation 2017/1151 with the CO_2 emissions determined pursuant to the old NEDC test procedure set out in Commission Regulation (EC) No. 692/2008. The CO2MPAS model calculates CO_2 emissions of light-duty vehicles on the NEDC test procedure using input data retrieved during the vehicle type approval carried in accordance with the WLTP test procedure (Commission Implementing Regulations (EU) 2017/1152 and 2017/1153, hereinafter 'Correlation Regulations'). The correlation procedure and the model allow the transition from the NEDC to the WLTP without affecting the CO_2 targets defined

[§§]https://jrcstu.github.io/co2mpas/Presentations/2016_12_12_CO2MPAS_workshop_public.pdf
[***]https://jrcstu.github.io/co2mpas/Presentations/20181114_co2mpas_presentation.pdf

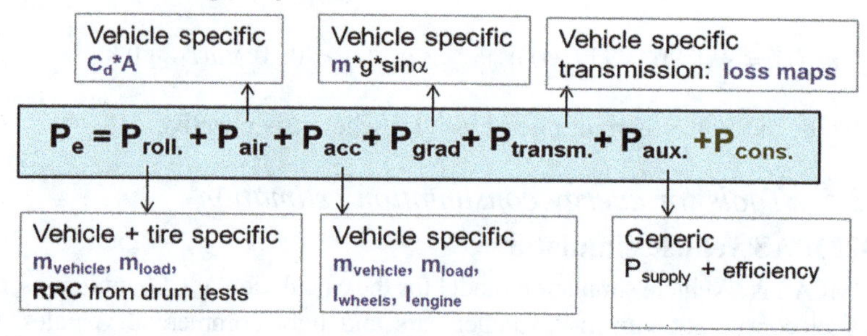

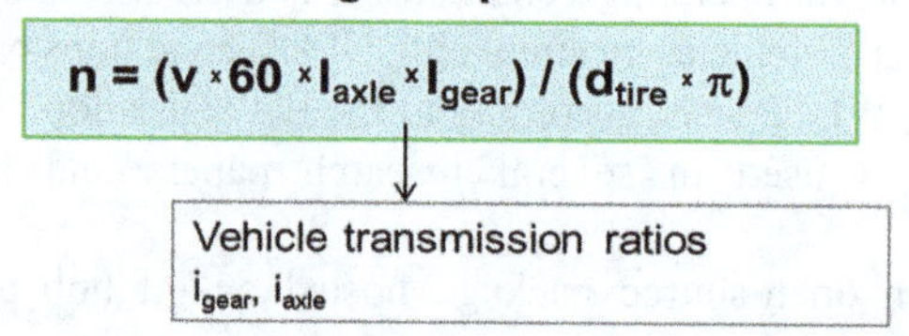

Figure 17.15 CO2MPAS model. Source: https://jrcstu.github.io/co2mpas/Presentations/20181114_co2mpas_presentation.pdf

in Regulations 443/2009 and 510/2011 (replaced since 1 January 2020 by Regulation (EU) 2019/631). The model requires an extensive list of inputs, as set out in Table 1 of Annex I (source: https://eur-lex.europa.eu/eli/reg_impl/2017/1152/oj/eng; i.e., Commission Implementing Regulations (EU) 2017/1152 and 2017/1153), to ensure the required level of accuracy in the correlation of the CO_2 emission values. The inputs include in brief:

- type of fuel,
- engine/powertrain characteristics,
- gear-box characteristics,
- electric components,
- vehicle characteristics,
- specific technologies of the vehicle.

The results of the vehicle tested under WLTP conditions.

Several default input values are defined for vehicle classes for the model to be used for other purposes. For example, it might be used to carry out scenario analyses on the effect on global CO_2 emissions due to the introduction of specific vehicle technologies in the vehicle fleet or to assess the effect of modifying some aspects of the vehicle type-approval test. In this case, achieving high accuracy on the single estimation is not important and the default values can allow its use to derive aggregated statistics.

Main output. CO2MPAS is able to provide fuel consumption, energy consumption, and CO_2 emissions for a vehicle following a predefined trajectory. Default simulation conditions are the NEDC and WLTP test procedures, for which it has been designed.

Transport European Simulation Tool

TRUST is a European-scale transportation network model. It is developed and maintained by TRT. It simulates various transportation modes, including road, rail, inland waterways, and maritime activities, covering the entirety of Europe and its neighbouring countries. TRUST enables the allocation of passenger and freight origin-destination matrices at a detailed NUTS3 level, comprising approximately 1,600 zones, across the multimodal transport network. Utilising data from Eurostat, national statistics, and the ETISPLUS database (CORDIS RCN: 92896), TRUST is calibrated to accurately replicate the tons-kilometres and passenger-kilometres by country, consistent with the statistics presented in the DG MOVE Transport in Figures pocketbook.

TRUST serves as a valuable tool for conducting impact assessments and supporting policy development and evaluation in the transportation sector. This transport network model operates within the PTV-VISUM software environment and incorporates fuel consumption and emissions factors for road modes based on COPERT IV functions. Documentation on the tool can be found at the following links:

https://web.jrc.ec.europa.eu/policy-model-inventory/explore/models/model-trust/
http://www.trt.it/en/tools/trust/

Vehicle Energy Consumption Calculation Tool – VECTO

VECTO is a simulation tool that has been developed by the European Commission and is used for CO_2 emissions determination and fuel consumption from Heavy Duty Vehicles (trucks, buses, and coaches) with a Gross Vehicle Weight of over 3,500 kg. Scientific specifications of the calculation tool can be found in Fontaras *et al.* (2013) and Fontaras *et al.* (2016). Further documentation can be found on the dedicated website provided by the European Commission[‡‡‡].

COPERT (Computer Program to calculate Emissions from Road Transport)

COPERT is the standard program for emissions estimating with the emissions factors specified in the EMEP/EEA air pollutant emissions inventory guidebook. COPERT is widely recognised for its ability to estimate emissions from road transport, and even though the exact equations and algorithms used within COPERT are not publicly disclosed, the model employs complex and detailed methodologies to estimate emissions.

The guidebook is described in detail in previous sections of the present chapter. For COPERT documentation and installation instructions the reader can refer to the following link: https://www.eea.europa.eu/publications/copert-4-2014-estimating-emissions.

[‡‡‡]https://climate.ec.europa.eu/eu-action/transport/road-transport-reducing-co2-emissions-vehicles/vehicle-energy-consumption-calculation-tool-vecto_en

The EMEP/EEA air pollutant emission inventory guidebook can be downloaded from: https://www.eea.europa.eu/publications/emep-eea-guidebook-2023.

COPERT architecture and method are presented by Ntziachristos *et al.* (2009). The key components and parameters that COPERT takes into consideration to estimate emissions from ICE vehicles include:

1. Vehicle characteristics: COPERT considers information about the vehicle, such as its type, age, engine technology, and emission control systems. Different types of vehicles (e.g., passenger cars, trucks, buses) and engine technologies (e.g., gasoline, diesel, natural gas) have distinct emission characteristics.
2. Fuel consumption: Estimating fuel consumption is a crucial part of emissions estimation. COPERT uses data related to fuel consumption, taking into account driving conditions and specific technology characteristics.
3. Driving patterns: COPERT considers driving patterns, such as speed, acceleration, and deceleration, which affect emissions. The model also takes into account factors like urban, suburban, or highway driving.
4. Emission standards: COPERT considers the emission standards in place and how they impact emissions. It incorporates emission factors that represent how well vehicles conform to these standards.
5. Cold start and warm-up emissions: The model takes into account the higher emissions during the cold start and warm-up phases of the vehicle.
6. Road type: The type of road (urban, suburban, rural, highway) can affect emissions, and COPERT considers this variable.
7. Emission factors: COPERT uses a comprehensive database of emission factors that are specific to different vehicle types, engine technologies, and driving conditions. These emission factors are based on laboratory testing and real-world measurements.
8. Temporal factors: The model also accounts for temporal factors, including the effect of congestion and traffic flow.

It is important to note that the specific equations and algorithms used within COPERT are not publicly disclosed and are part of the proprietary model. Access to and use of COPERT typically requires a license and permission from the EEA. Users are encouraged to refer to the official COPERT documentation and resources provided by the EEA for detailed guidance on how to use the model effectively and accurately estimate emissions from ICE vehicles in the European Union.

COPERT data are provided, not freely by EMESA, and can be very useful for aggregate analysis and what-if scenario analysis.

However, the COPERT software can be freely downloaded and used with its GUI interface. The information needed to run the computation may differ considerably based on the different Tier of detail required. The data required are grouped in the following categories and tabs:

- Country,
- Required Tier level,

- Type of dataset (time series or event),
- Year of the study (to account for the vehicle's technology in the studied country),
- Environmental information (monthly temperature, humidity, etc.),
- Fuel specifications,
- Lubricant specifications,
- Statistical fuel consumption,
- Stock configuration (percentage of vehicle's type and technology, e.g. EURO 4, EURO5, etc.),
- Stock and activity data for vehicle such as circulation data, mileage on rural, urban, highway, mean speed, etc.,
- Fuel balance,
- Parameters – cold and hot emissions parameters, etc.

Results consider all types of emissions and can be summarised in different tables. An example is shown in Figure 17.16.

The methodology for the COPERT calculation of exhaust emissions may follow Tier 2 or Tier 3. In Tier 2 methodology, the calculations consider the amount of fleet and the vehicle kilometres per technology whereas in Tier 3, emission factors are mode detailed, and they depend also on temperature profile, driving pattern, parking pattern, and other activity parameters.

We recall in the following list the different Tiers managed by the method:

1. Tier 1: This is the simplest and the most basic level of emissions estimation. Tier 1 methods rely on default or average emission factors, which are applied

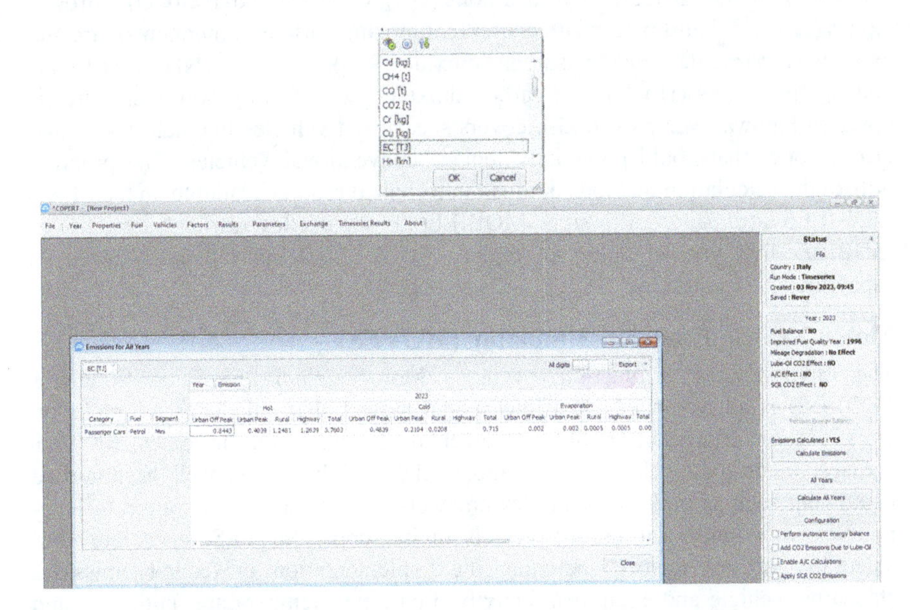

Figure 17.16 Example of COPERT emission results

to aggregated data on vehicle fleets and traffic activities. These methods are relatively straightforward and can provide a rough estimate of emissions. Tier 1 is suitable for situations where detailed data are limited.

2. Tier 2: Tier 2 methods are more advanced and accurate than Tier 1. They involve more refined and region-specific data and emission factors. These methods consider various parameters such as vehicle types, driving cycles, and emission standards to provide a more accurate estimate of emissions. Tier 2 methods are recommended when more detailed data is available, and a higher level of accuracy is required.
3. Tier 3: Tier 3 represents the most sophisticated and accurate level of emissions estimation in the COPERT system. These methods require the most detailed and specific data on vehicle fleets, activity, and emissions. Tier 3 methods take into account a wide range of variables, including vehicle age, technology, emission control systems, and real-world driving conditions. This level of detail makes Tier 3 the most accurate but also the most resource-intensive method.

The choice of tier to use depends on the availability of data and the specific analysis needs. Tier 1 is often used when data are limited, or a quick estimate is required. Tier 2 provides a more accurate estimate and is suitable for most regular assessments. Tier 3 is reserved for situations where very detailed data and the highest level of accuracy are essential.

MOBILE6

MOBILE6 is a software application designed to calculate emission factors for various pollutants, including hydrocarbons (HC), carbon monoxide (CO), nitrogen oxides (NOx), exhaust particulate matter (comprising various components), tire and brake wear particulate matter, sulphur dioxide (SO_2), ammonia (NH_3), six hazardous air pollutants (HAPs), and carbon dioxide (CO_2) from gasoline and diesel-powered highway vehicles. It also covers specialised vehicles like natural gas and electric ones that could potentially replace conventional vehicles. The program utilises the calculation methods outlined in technical reports available on the EPA's MOBILE6 webpage. Although MOBILE6.0 supersedes earlier versions of MOBILE, this version does not replace MOBILE6.0 but rather enhances its capabilities.

Motor Vehicle Emission Simulator (MOVES)

The Environmental Protection Agency's Motor Vehicle Emission Simulator (MOVES) is an advanced emissions model system that calculates air pollution emissions for criteria air pollutants, greenhouse gases, and air toxics. MOVES focuses on on-road vehicles like cars, trucks, and buses, as well as non-road equipment such as bulldozers and lawnmowers. It does not include aircraft, locomotives, or commercial marine vessels in its scope. MOVES takes into consideration various factors, including the implementation of federal emissions standards, vehicle and equipment activity, fuel types, temperature, humidity, and emission control measures like inspection and maintenance (I/M) programs.

MOVES covers modelling from the year 1990 through 2060 and allows for emissions analysis at both national and county levels using default parameters or user-provided inputs. For on-road sources, it also provides a more detailed 'project' scale option when users furnish specific input data describing project characteristics. The on-road module employs emission rates based on operating modes to ensure consistency across all scales. MOVES is a bottom-up emissions model, meaning it estimates emissions from distinct physical emission processes based on the source. It focuses on calculating emissions as fleet averages rather than for individual vehicles or equipment types. Additionally, MOVES adjusts emission rates to reflect real-world conditions accurately.

MOVES can be thought of as the American counterpart of the European COPERT model. Different from COPERT, MOVES is provided with a GNU General Public License (GPL) and is available on GIT HUB.

Documentation and repository for MOVES are available at the following links:

https://www.epa.gov/moves
https://github.com/USEPA/EPA_MOVES_Model
https://www.epa.gov/moves/moves-onroad-technical-reports#moves4
https://climate.ec.europa.eu/eu-action/transport/road-transport-reducing-co2-emissions-vehicles/co2-emission-performance-standards-cars-and-vans_en

17.3.6 Data sources and sensors

- EMEP/EEA emission factors: Free information on emissions factors already discussed in Section 17.2.
- HBEFA data: HBEFA is a standard data source for emission calculations in numerous studies and other applications. The book can be bought for around 250€ (2023 price) and contains useful information for emissions estimation.
- Advanced Vehicle Testing Activity (inl.gov): Free information on test data and test cycles on the high variety of EV vehicles sold in the USA.
- COPERT data. Results of a wide variety of computations with the COPERT model which can be bought in https://www.emisia.com/utilities/copert-data/

In order to calibrate emissions and fuel consumption models, it can be useful to gather information from local sensors. One such device is AirQuino[§§§], provided by a company and a CNR institute. It can track emissions within a 1-min time frame and there are already available studies that correlate emissions to traffic flow.

17.4 Summary

This chapter covers the most popular methods functions and tools for estimating emissions, fuel, or energy consumption of road transport. Methods are classified by detail level and higher attention is given to methods used by environmental agencies.

[§§§]https://www.airqino.it/

Vehicle consumption simulation is a widely explored area of research. Many studies have focused on the simulation of conventional vehicles (Brooker *et al.*, 2015; Kim *et al.*, 2012; Lee *et al.*, 2013; Newman *et al.*, 2016; Wipke *et al.*, 1999), while simulation models on electrified vehicles have been investigated only in the recent years (Hayes *et al.*, 2011).

Vehicle consumption (and emission) models can be classified for example in microscopic and macroscopic models (Othman *et al.*, 2019). In general, the introduction of electric vehicles will change the well-known relationship between average speed and energy consumption and, in turn, the relationship of the latter to traffic conditions. As a result, in the future with most electric vehicles, policymakers will have to manage an additional level of complexity in transportation planning. At that point, in fact, local pollution will no longer be due to vehicles, so policymakers will have to choose between the efficiency of the transportation sector (which is directly related to the average speed of vehicles) and the overall consumption of electricity (the electricity generation system is now responsible for pollution and greenhouse gas emissions from traffic). In particular, this circumstance will be amplified by the spread of vehicle connectivity and automation, which can increase the capacity of the transportation system. See also Fiori *et al.* (2018, 2019; 2021); Di Pace *et al.* (2022).

References

Some suggestions for further readings are reported below among the huge literature available on these topics.

Achour, H., Carton, J., and Olabi, A. G. (2011). Estimating vehicle emissions from road transport, case study: Dublin city. *Applied Energy*, 88(5):1957–1964.

Ahn, K., Rakha, H., Trani, A., and Van Aerde, M. (2002). Estimating vehicle fuel consumption and emissions based on instantaneous speed and acceleration levels. *Journal of Transportation Engineering*, 128(2):182–190.

Bakhit, P., Said, D., and Radwan, L. (2015). Impact of acceleration aggressiveness on fuel consumption using comprehensive power based fuel consumption model. *Civil and Environmental Research*, 7(3):148–156.

Brooker, A., Gonder, J., Wang, L., Wood, E., Lopp, S., and Ramroth, L. (2015). FASTSim: A model to estimate vehicle efficiency, cost and performance. SAE Technical Paper 2015-01-0973.

Cascetta, E. (1998). Teoria e metodi dell'ingegneria dei sistemi di trasporto. Utet.

Di Pace, R., Fiori, C., Storani, F., de Luca, S., Liberto, C., and Valenti, G. (2022). Unified network traffic management framework for fully connected and electric vehicles energy consumption optimization (URANO). *Transportation Research Part C: Emerging Technologies*, 144:103860.

European Environment Agency. (2023). *EMEP/EEA air pollutant emission inventory guidebook 2023. EEA Report 06*, European Environment Agency. Available from: https://www.eea.europa.eu/en/analysis/publications/emep-eea-guidebook-2023.

Fiori, C., Ahn, K., and Rakha, H. A. (2016). Power-based electric vehicle energy consumption model: Model development and validation. *Applied Energy*, 168:257–268.

Fiori, C., Arcidiacono, V., Fontaras, G., *et al.* (2019). The effect of electrified mobility on the relationship between traffic conditions and energy consumption. *Transportation Research Part D: Transport and Environment*, 67:275–290.

Fiori, C., and Marzano, V. (2018). Modelling energy consumption of electric freight vehicles in urban pickup/delivery operations: Analysis and estimation on a real-world dataset. *Transportation Research Part D: Transport and Environment*, 65:658–673.

Fiori, C., Montanino, M., Nielsen, S., Seredynski, M., and Viti, F. (2021). Microscopic energy consumption modelling of electric buses: Model development, calibration, and validation. *Transportation Research Part D: Transport and Environment*, 98, 102978.

Fontaras, G., Franco, V., Dilara, P., Martini, G., and Manfredi, U. (2014). Development and review of euro 5 passenger car emission factors based on experimental results over various driving cycles. *Science of the Total Environment*, 468:1034–1042.

Fontaras, G., Grigoratos, T., Savvidis, D., *et al.* (2016). An experimental evaluation of the methodology proposed for the monitoring and certification of CO_2 emissions from heavy-duty vehicles in Europe. *Energy*, 102:354–364.

Fontaras, G., Rexeis, M., Dilara, P., Hausberger, S., and Anagnostopoulos, K. (2013). The development of a simulation tool for monitoring heavy-duty vehicle CO_2 emissions and fuel consumption in Europe. *Technical report, SAE Technical Paper*.

Galvin, R. (2017). Energy consumption effects of speed and acceleration in electric vehicles: Laboratory case studies and implications for drivers and policymakers. *Transportation Research Part D: Transport and Environment*, 53:234–248.

Gao, Y., Chu, L., and Ehsani, M. (2007, September). Design and control principles of hybrid braking system for EV, HEV and FCV. In 2007 *IEEE Vehicle Power and Propulsion Conference* (pp. 384–391). IEEE.

Hayes, J. G., De Oliveira, R. P. R., Vaughan, S., and Egan, M. G. (2011). Simplified electric vehicle power train models and range estimation. *2011 IEEE Vehicle Power and Propulsion Conference*, pp. 1–5.

Kim, N., Rousseau, A., and Rask, E. (2012). Autonomie model validation with test data for 2010 Toyota Prius. SAE Technical Paper 2012-01-1040.

Lee, B., Lee, S., Cherry, J., Neam, A., Sanchez, J., & Nam, E. (2013). Development of advanced light-duty powertrain and hybrid analysis tool (No. 2013-01-0808). SAE Technical Paper.

Makridis, M., Mattas, K., Mogno, C., Ciuffo, B., and Fontaras, G. (2020). The impact of automation and connectivity on traffic flow and CO_2 emissions. A detailed microsimulation study. *Atmospheric Environment*, 226:117399.

Moskalik, A. (2020). Using transmission data to isolate individual losses in coast-down road load coefficients. *SAE International Journal of Advances and Current Practices in Mobility*, 2(4):2156.

Newman, K. A., Doorlag, M., and Barba, D. (2016). Modeling of a Conventional Mid-Size Car with CVT Using ALPHA and Comparable Powertrain Technologies. SAE Technical Paper 2016-01-1141.

Ntziachristos, L., Gkatzoflias, D., Kouridis, C., and Samaras, Z. (2009). Copert: a european road transport emission inventory model. In *Information Technologies in Environmental Engineering: Proceedings of the 4th International ICSC Symposium Thessaloniki, Greece, May 28-29, 2009*, pages 491–504. Springer.

Ntziachristos, L. and Samaras, Z. (2023). *1.a.3.b.i, 1.a.3.b.ii, 1.a.3.b.iii, 1.a.3.b.iv passenger cars, light commercial trucks, heavy-duty vehicles including buses and motorcycles*. EEA Report 06, European Environment Agency.

Ofgem. (2015). Renewables obligation: Biodiesel and fossil-derived bioliquids guidance. Available from: https://www.ofgem.gov.uk/publications/renewables-obligation-biodiesel-and-fossil-derived-bioliquids-guidance-2015.

Othman, B., De Nunzio, G., Di Domenico, D., and Canudas-de-Wit, C. (2019). Ecological traffic management: A review of the modeling and control strategies for improving environmental sustainability of road transportation. *Annual Reviews in Control*, 48:292–311.

Rakha, H., Lucic, I., Demarchi, S. H., Setti, J. R., and Aerde, M. V. (2001). Vehicle dynamics model for predicting maximum truck acceleration levels. *Journal of Transportation Engineering*, 127(5):418–425.

Treiber, M., Kesting, A., and Thiemann, C. (2008). How much does traffic congestion increase fuel consumption and emissions? Applying a fuel consumption model to the NGSIM trajectory data. *In 87th Annual Meeting of the Transportation Research Board*, Washington, DC, volume 71, pages 1–18.

Tsiakmakis, S., Fontaras, G., Cubito, C., *et al.* (2017). *From NEDC to WLTP: effect on the type-approval CO2 emissions of light-duty vehicles*. Publications Office of the European Union: Luxembourg, 50.

Tsiakmakis, S., Fontaras, G., Dornoff, J., *et al.* (2019). From lab-to-road and vice-versa: Using a simulation-based approach for predicting real-world CO_2 emissions. *Energy*, 169:1153–1165

UN. (2015). Uniform provisions concerning the approval of vehicles with regard to the emission of pollutants according to engine fuel requirements, Addendum 82: Regulation No. 83, Geneve, Switzerland.

Wipke, K. B., Cuddy, M. R., and Burch, S. D. (1999). ADVISOR 2.1: A user-friendly advanced powertrain simulation using a combined backward/forward approach. *IEEE Transactions on Vehicular Technology*, 48(6):1751–1761.

Appendix A
Control theory

Angelo Coppola[1] and Roberta Di Pace[2]

The goal of the scientist is to comprehend the phenomena of the universe he observes around him.

Richard Bellman

Outline. *This appendix delves into the fundamental aspects of automation control, with a particular emphasis on its application in traffic control. As the complexities of modern transportation systems continue to grow, the integration of automation control becomes increasingly critical to enhance efficiency, safety, and sustainability. Accordingly, this section aims to provide readers with a solid foundation in automation control principles, tailored to the unique demands and challenges of traffic systems.*

A.1 Introduction to automatic control

Automatic control, often referred to simply as control, refers to the field of engineering and applied science concerned with the design and implementation of systems capable of autonomously regulating processes to achieve desired performance. It encompasses both theoretical methodologies and practical techniques that enable technical systems to execute predefined tasks with minimal human intervention. In other words, its primary objective is to regulate processes to achieve desired behaviors or performance outcomes.

Automatic control systems are typically composed of different components, as depicted in Figure A.1.

The process/system is the dynamical system that needs to be controlled (e.g., traffic flow in an extra-urban network). Note that a dynamical system is a system that evolves over time; at any given time, a dynamical system has a state representing a point in an appropriate state space. This state is often given by a tuple of real numbers or by a vector in a geometrical manifold. The evolution rule of the dynamical system, known as the dynamics of the system, is a function that describes what future states follow from the current state. The dynamic of the system depends upon:

[1]Department of Civil, Architectural and Environmental Engineering, University of Naples 'Federico II', Italy
[2]Department of Civil Engineering, University of Salerno, Italy

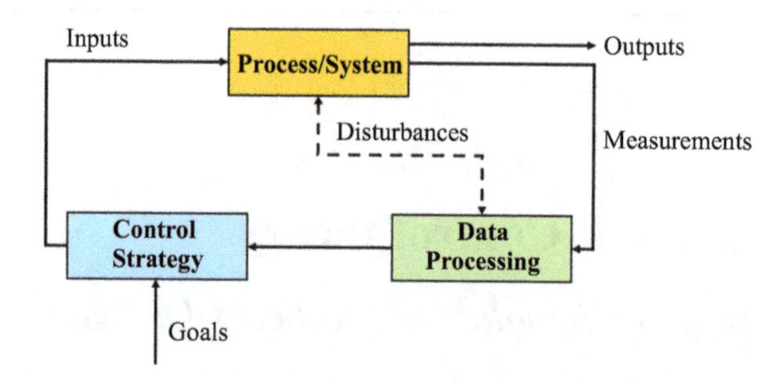

Figure A.1 Overview of the fundamental components of an automatic control system

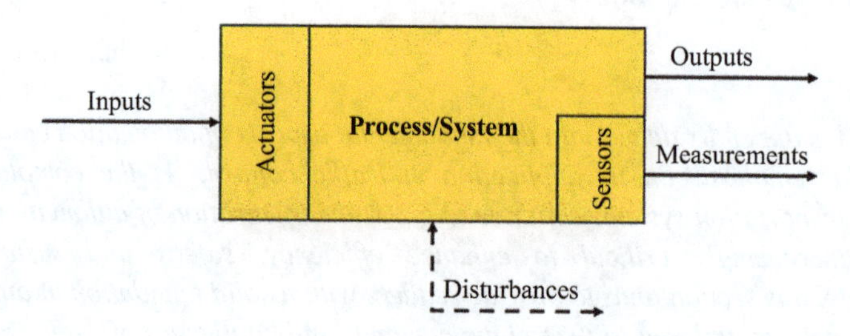

Figure A.2 Detail of the Process/System block

- External excitations (or quantities), presumed to remain unaffected by the dynamic progression of the process. In the case of extra-urban traffic, the external quantities can be exemplified by traffic demand, origin-destination patterns, and environmental conditions.
- The inherent behaviour of the process is due to its technical and physical attributes, illustrated by factors such as travel times or the accumulation of vehicles in a queue on an extra-urban link, as well as the flow capacity.

The process/system requires and includes two sub-components, as depicted in Figure A.2.

- Actuators: a set of instruments that allow acting on the process, affecting its behaviour or output. With specific regard to the traffic field, examples are panels that allow for variable messages.
- Sensors: instruments that measure the output or state of the process and convert it into a form that can be used by the control system. The information obtained from the sensor is called feedback.

The process/system is characterised by variables that enter the process/system itself or exit from it.

The variables entering the process are external excitations that can be categorised as follows:

- Input variables, come from an acceptable control region to regulate/control the process/system; an input variable can be exemplified by components such as variable speed limits.
- Disturbances are exogenous variables that may be estimated or predicted through appropriate algorithms but cannot be controlled.

The variables that come out of the process are usually referred to as output variables. The process outputs are the quantities chosen to represent the behavioural aspects of interest (e.g., the outputs of extra-urban traffic may be the average speed or density). However, not all the output variables of the process can be measured. As depicted in Figure A.1, it is possible to distinguish between the generic set of process outputs and measurements, indicating the sub-set of measurable outputs.

Real-time measurements of the process outputs and, eventually, disturbances are exploited by the data processing block to perform estimation and/or prediction tasks. The data processing block has a crucial role since it processes the received data in a way that makes it usable/useful for the control strategy.

Finally, the control strategy block performs the following task:

> *To specify in real time the process inputs, based on available measurements/ estimations/predictions, so as to achieve the prespecified goals regarding the process outputs despite the influence of various disturbances.*
>
> Papageorgiou (1998)

The control strategy block is composed of an algorithm that takes the feedback from the data processing block and compares it with a reference or desired value. Based on this comparison, the algorithm generates a control signal (input signal) that is sent to the process. The control algorithm can range from a very simple control law to highly sophisticated control strategies and must follow given control requirements related to the goals and the characteristics of the controller itself. Examples of control requirements are the time to compute the control signal, a specific desired behaviour, or a given performance level of the process.

A.2 Process and system models

The process/system block depicted in Figure A.3 can be either real (physical) or a mathematical model describing the functional behaviour of a real entity. In this case, a set of equations describes, with varying degrees of precision, the intrinsic (internal) behaviour of the process/system within the specified context. The level of complexity, resolution, and accuracy of a mathematical model can vary, and different models may be employed based on the specific requirements of the analysis or simulation. Consequently, a mathematical model, when provided with input and disturbance parameters, can be utilised to compute, within a defined level of precision, the corresponding output values or other pertinent internal variables. To illustrate, an extra-urban traffic flow model, furnished

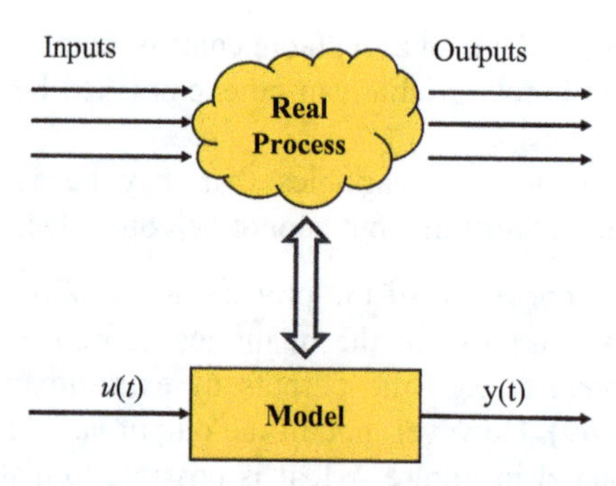

Figure A.3 Schematic representation of a real process and its corresponding model

with data pertaining to traffic demand, origin-destination patterns, and vehicle speed can be employed to ascertain queue formation/lengths and travel times across all network links.

In system theory, the functional relationship between inputs, models, and outputs is represented graphically by means of a flow chart, as Figure A.3 shows.

In the following, $u(t) = [u_1(t), \ldots, u_m(t)] \in \mathbb{R}^m$ will indicate the input vectors, $y(t) = [y_1(t), \ldots, y_p(t)] \in \mathbb{R}^p$ the outputs vector, $t \in \mathbb{R}$ the time variable.

Mathematical models can be classified into several categories based on their characteristics, purposes, and applications:

- Nature of the model
  - Physical models: Derived from physical laws that govern the system's behavior.
  - Empirical models: Developed from experimental data and observations, without relying on the underlying physical laws.
- Model behaviour
  - A model is static or algebraic if its output $y(t)$ depends only on the actual value of the input $u(t)$. These models do not take into account inputs at previous time instants, so do not give any information about the regime transient, i.e., on the time trend of the outputs during the transition from one steady state to another. The functional behaviour is described by an algebraic equation.
  - A model is dynamic if its output $y(t)$ depends on actual and previous values of the inputs.
- Signal types
  - Continuous-Time models (CT): Represent systems where the state changes continuously over time. These models are described by differential equations.
  - Discrete-Time models (DT): Represent systems where the state changes at discrete time intervals. They are represented by a set of finite-difference equations.
- Input-Output relationship
  - Linear models: Assume a linear relationship between input and output. Superposition and homogeneity principles apply.
  - Nonlinear models: Involve nonlinear relationships between inputs and outputs.

- Time dependency
  - Time-Invariant models: The system's behaviour and parameters do not change over time, and the relationship between input and output remains constant.
  - Time-Variant models: The system's dynamics or parameters change over time.
- Determinism
  - Deterministic models: The signals of a deterministic model can be described by analytical relationships.
  - Stochastic models: The signals can be given by probabilistic variables and/or contain randomness and uncertainties in the system's dynamics, inputs, or parameters.
- Number of input and output variables
  - Single Input Single Output (SISO), if $u(t) \in \mathbb{R}$ and $y(t) \in \mathbb{R}$ are scalar
  - Multi-Input Single Output (MISO), if $y(t) \in \mathbb{R}$ is scalar, but $u(t) \in \mathbb{R}^{m \times 1}$ is not
  - Single Input Multi Output (SIMO) if $u(t) \in \mathbb{R}$ is scalar, but $y(t) \in \mathbb{R}^{p \times 1}$ is not
  - Otherwise it is called Multiple Inputs Multiple Outputs (MIMO)

A.3 Dynamical models

A *dynamical system* is a system that evolves in time according to a predefined behaviour. In other words, the system interacts with the surrounding environment and, accordingly, evolves over time by means of some causes (inputs) operating on it which determines some effects (outputs), that are the response of the system to such stimulations. The knowledge of the input at time t is not enough to univocally determine the value of the output at time t. Indeed, the output $y(t)$ depends on the actual and previous values of the inputs. The functional behaviour is described by one or more differential equations that express static links between the input and output variables and their derivatives compared to time. In so doing, the dynamic model allows us to determine the trend of the output signal corresponding to a given input signal, i.e., to determine the response of the system to a given excitation.

To characterise a dynamical system, it is mandatory to define the 'specific object' (which part of the system) that evolves over time and the 'rules' that specify the evolution behaviour of the object. In so doing, a dynamical system becomes simply a model describing the temporal evolution of a system.

The first step is to define the set of variables that give a complete description of the system at any specific time, i.e., the variables must completely summarise the past history of the mathematical system. These variables are called *state variables*, while the set of all the possible values of the state variables is the *state space*. The number of state variables n is the *order* of the system/process. The state space can be discrete, consisting of isolated points, such as if the state variables could only take on integer values. It could be continuous, consisting of a smooth set of points, such as if the state variables could take on any real value.

Definition. The *state* $x(t_0) = [x_1(t_0), \ldots, x_n(t_0)] \in \mathbb{R}^n$ of a process/system in a generic time instant $t_0 \in \mathbb{R}$, assumed as the *initial instant,* is the minimum information necessary to uniquely determine the output $y(t)$, $\forall\, t \geq t_0$, once assigned the input $u(t)$.

The second step is to specify the rule for the time evolution of the dynamical system. This rule must be such that the value of the state variables at a particular time must completely determine the evolution of all future states. The time evolution rule will be based on a function f that takes as its input the state of the system at one time and gives as its output the state of the system at the next time. The time evolution rule could involve discrete or continuous time. If the time is discrete, then the system evolves in time steps, and usually integer time points are used $t = 0,1,2,\ldots$. In this case, starting from the initial conditions $x0$ at time $t = 0$, it is possible to apply the function f once to determine the state $x1 = f(x0)$ at time $t = 1$. If one applies the function repeatedly, it is possible to determine all future states, obtaining a trajectory of the points $x0, x1, x2, x3, \ldots$. If the time is continuous, the state $x(t)$ at time t can be thought of as a point that moves smoothly through the state space. In this case, starting with an initial state $x(0)$ at time $t = 0$, the trajectory of all future times $x(t)$ will be a curve through state space.

Putting together what was said above, it is possible to describe a generic dynamical system using the state-space representation, which is a system of two vectorial equations that fully characterise the system's configuration at any point in time:

$$\dot{x}(t) = f(x(t), u(t), t),\; x(t_0) = x_0 \tag{A.1}$$

$$y(t) = \eta(x(t), u(t), t) \tag{A.2}$$

if the time is continuous;

$$x(k+1) = f(x(k), u(k), k),\; x(k_0) = x_0 \tag{A.3}$$

$$y(k) = \eta(x(k),\, u(k), k) \tag{A.4}$$

if the time is discrete.

The first equation, named *state equation*, characterises the evolution of the state $x(t)$, $\forall\, t \geq t_0$, given the initial state $x(t_0)$ and the input $u(t)$. The second equation, named *output transformation*, defines the output $y(t)$, at the generic time instant t, given the state $x(t)$ and the input $u(t)$. The terms $\dot{x}(t)$ and $x(k+1)$ are the derivatives of the state vector with respect to time, in continuous and discrete time respectively, indicating how the state changes over time. The function f, named *transition* or *state function*, maps the system's condition from its initial state and time to subsequent states, capturing how the system evolves. It is represented as a vector-valued function in the case of continuous dynamical systems or as a sequence in discrete systems. The function g, named *output transformation*, maps the current state and input to an output.

A.3.1 Continuous time systems

Here the properties of continuous-time systems are summarised and discussed.

Let us start considering a system of n first-order linear ordinary differential equations (ODEs) with inputs:

$$\begin{cases} \dot{x}_1(t) & = & a_{11}x_1(t) + \ldots + a_{1n}x_n(t) + b_1u(t) \\ \dot{x}_2(t) & = & a_{21}x_1(t) + \ldots + a_{2n}x_n(t) + b_1u(t) \\ \vdots & \vdots & \vdots \\ \dot{x}_n(t) & = & a_{n1}x_1(t) + \ldots + a_{nn}x_n(t) + b_nu(t) \\ x_1(0) & = x_{10}, & \ldots \quad x_n(0) = x_{n0} \end{cases} \qquad \left[\dot{x} = \frac{dx}{dt}\right]$$

where $x = [x_1 \ \ldots \ x_n]' \in \mathbb{R}^n$ is the state of the system of order n. The equivalent matrix form of the linear ODE system is the so-called linear system:

$$\dot{x}(t) = Ax(t) + Bu(t) \tag{A.5}$$

with initial condition $x(0) = x_0$, with vector $x_0 = [x_{1,0} \ \cdots \ x_{n,0}]' \in \mathbb{R}^n$.

The nth-order ODE can be recast in a more compact notation as a 1st-order linear system of ODEs:

$$\begin{aligned} \dot{x}(t) &= Ax(t) + Bu(t) \\ y(t) &= C_x(t) + Du(t) \end{aligned}$$

$$A = \begin{bmatrix} 0 & 1 & 0 & \ldots & 0 \\ 0 & 0 & 1 & \ldots & 0 \\ \vdots & \vdots & & \ddots & \vdots \\ 0 & 0 & 0 & \ldots & 1 \\ -a_0 & -a_1 & -a_2 & \ldots & -a_{n-1} \end{bmatrix}, B = \begin{bmatrix} 0 \\ 0 \\ \vdots \\ 0 \\ 1 \end{bmatrix}$$

$$C = [b_0 \ b_1 \ b_2 \ldots b_{n-1}], \ D = 0$$

The linear system of 1st-order ODEs is called the state-space realisation of the nth-order ODE. Starting from the initial condition $x(0) = x_0$, the continuous-time linear system $x' = Ax + Bu$ has the unique solution $x(t)$ obtained employing the Lagrange's formula:

$$x(t) = e^{At}x_0 + \int e^{A(t-\tau)}Bu(\tau)d\tau \tag{A.6}$$

where the first term is the natural response of the system (i.e., if no external input is applied), while the second term is the forced response. Given $x(0)$ and $u(t)$, $\forall\, t \in [0, T]$, Lagrange's formula allows computing $x(t)$ and $y(t)$, $\forall\, t \in [0, T]$.

Let us introduce now different classes of dynamical systems.

A system is said Linear time-varying (LTV) if A, B, C, and D are constant matrices.

$$\begin{cases} \dot{x}(t) = A(t)x(t) + B(t)u(t) \\ y(t) = C(t)x(t) + D(t)u(t) \end{cases}$$

A generalisation of LTV systems is linear parameter-varying (LPV) systems:

$$\begin{cases} \dot{x}(t) = A(p(t))x(t) + B(p(t))u(t) \\ y(t) = C(p(t))x(t) + D(p(t))u(t) \end{cases}$$

A system is multivariable if it can have m inputs ($u(t) \in \mathbb{R}^m$) and p outputs ($y(t) \in \mathbb{R}^p$).

$$\begin{cases} \dot{x}(t) = Ax(t) + Bu(t) \\ y(t) = Cx(t) + Du(t) \end{cases}$$

For linear systems, it still holds that $A \in \mathbb{R}^{n \times n}$, $B \in \mathbb{R}^{n \times m}$, $C \in \mathbb{R}^{p \times n}$ *and* $D \in \mathbb{R}^{p \times m}$.

In nonlinear systems, the evolution of the system and the output are characterised via nonlinear functions.

More precisely, f: $\mathbb{R}^{n+m} \rightarrow \mathbb{R}^n$, g: $\mathbb{R}^{n+m} \rightarrow \mathbb{R}^p$ are (arbitrary) nonlinear functions.

$$\begin{cases} \dot{x}(t) = f(x(t),\, u(t)) \\ y(t) = g(x(t),\, u(t)) \end{cases}$$

Time-varying nonlinear systems:

$$\begin{cases} \dot{x}(t) = f(t, x\,(t),\, u(t)) \\ y(t) = g(t, x(t),\, u(t)) \end{cases}$$

A.3.1.1 Stability

Let us consider the following continuous-time nonlinear system:

$$\begin{cases} \dot{x}(t) = f(x(t),\, u(t)) \\ y(t) = g(x(t),\, u(t)) \end{cases}$$

Definition: A state $x_r \in \mathbb{R}^n$ and an input $u_r \in \mathbb{R}^m$ are an equilibrium pair if for initial condition $x(0) = x_r$ and constant input $u(t)$ u_r the state remains constant: $x(t) = x_r$, $\forall t \geq 0$.

x_r is called the equilibrium state, while u_r equilibrium input.

Now let us consider the nonlinear system:

$$\begin{cases} \dot{x}(t) = f(x\,(t),\, u_r) \\ y(t) = g(x(t),\, u_r) \end{cases}$$

Being x_r an equilibrium state, $f(x_r, u_r) = 0$.

Definition: The equilibrium state x_r is stable if $\forall \epsilon > 0$ $\delta > 0$: $\|x(0)\text{-}x_r\| < \delta \Rightarrow \|x(0)\text{-}x_r\| < \epsilon$, $\forall t \geq 0$.

In other words, if for each initial condition $x(0)$ 'close enough' to x_r, the corresponding trajectory $x(t)$ remains near x_r, $\forall t \geq 0$.

The equilibrium point x_r is called asymptotically stable if it is stable and $x(t) \rightarrow x_r$ for $t \rightarrow$. Otherwise, the equilibrium point x_r is called unstable.

A.3.1.2 Linearisation of nonlinear systems

Consider the nonlinear system:

$$\begin{cases} \dot{x}(t) = f(x(t),\ u(t)) \\ y(t) = g(x(t),\ u(t)) \end{cases}$$

Let (x_r, u_r) be an equilibrium, $f(x_r, u_r) = 0$. The linearisation aims to investigate the dynamic behaviour of the system under small perturbations $\Delta u(t) \triangleq u(t) - u_r$ and $\Delta x(0) \triangleq x(0) - x_r$. The evolution of $\Delta x(t) \triangleq x(t) - x_r$ is given by

$$\begin{aligned} \dot{\Delta x}(t) &= \dot{x}(t) - \dot{x}_r = f(x(t), u(t)) \\ &= f(\Delta x(t) + x_r, \Delta u(t) + u_r) \\ &\approx \underbrace{\frac{\partial f}{\partial x}(x_r, u_r)}_{A} \Delta x(t) + \underbrace{\frac{\partial f}{\partial u}(x_r, u_r)}_{B} \Delta u(t) \end{aligned}$$

$$\Delta y(t) \approx \underbrace{\frac{\partial g}{\partial x}(x_r, u_r)}_{C} \Delta x(t) + \underbrace{\frac{\partial g}{\partial u}(x_r, u_r)}_{D} \Delta u(t)$$

Similarly, where $\Delta y(t) \triangleq y(t) - g(x_r, u_r)$ is the perturbation of the output from its equilibrium.

The perturbations $\Delta x(t)$, $\Delta y(t)$, and $\Delta u(t)$ are (approximately) ruled by the linearised system:

$$\begin{cases} \dot{\Delta x}(t) = A\Delta x(t) + B\Delta u(t) \\ \Delta y(t) = A\Delta x(t) + D\Delta u(t) \end{cases}$$

A.3.1.3 Discrete-time systems

Discrete-time models describe relationships between sampled variables $x(kT_s)$, $u(kT_s)$, $y(kT_s)$, $k = 0, 1, \ldots$. The value $u(kT_s)$ is kept constant during the sampling interval $[kT_s, (k+1)T_s)$. A discrete-time signal can either represent the sampling of a continuous-time signal or be an intrinsically discrete signal.

Let us consider the set of n first-order linear difference equations forced by the input $u(k) \in \mathbb{R}$.

$$\begin{cases} x_1(k+1) = a_{11}x_1(k) + \ldots + a_{1n}x_n(k) + b_1u(k) \\ x_2(k+1) = a_{21}x_1(k) + \ldots + a_{2n}x_n(k) + b_2u(k) \\ \quad \vdots \qquad\qquad \vdots \qquad\qquad \vdots \\ x_n(k+1) = a_{n1}x_1(k) + \ldots + a_{n\,n}x_n(k) \quad + b_nu(k) \\ \qquad x_1(0) = x_{10}, \qquad \cdots \qquad x_n(0) = x_{n0} \end{cases}$$

The system can be rewritten using a compact matrix form:

$$\begin{cases} x(k+1) = Ax(k) + Bu(k) \\ x(0) = x_0 \end{cases}$$

where x = [x1, …, xn] ∈ ℝ.

The solution is

$$x(k) = \underbrace{A^k x_0}_{\text{natural response}} + \underbrace{\sum_{i=0}^{k-1} A^i Bu(k-1-i)}_{\text{forced response}}$$

where the first term is the natural response, and the second term is the force response.

Let us consider the following discrete-time nonlinear system:

$$\begin{cases} x(k+1) = f(x(k),\ u(k)) \\ y(k) = f(x(k),\ u(k)) \end{cases}$$

Definition: A state $x_r \in \mathbb{R}n$ and an input $u_r \in \mathbb{R}m$ are an equilibrium pair if for initial condition $x(0) = x_r$ and constant input $u(k)\ \ u_r$, $\forall\ k \in \mathbb{N}$, the state remains constant: $x(k)\ \ x_r$, $\forall\ k \in \mathbb{N}$.

x_r is called equilibrium state, u_r equilibrium input.

Definition: The equilibrium state x_r is stable if $\forall\ \epsilon > 0\ \ \delta > 0$: $\|x(0)\text{-}x_r\| < \delta \Rightarrow \|x(k)\text{-}x_r\| < \epsilon$, $\forall\ k \in \mathbb{N}$.

In other words, if for each initial condition $x(0)$ 'close enough' to x_r, the corresponding trajectory $x(k)$ remains close to x_r, $\forall\ k \in \mathbb{N}$.

The equilibrium point x_r is called asymptotically stable if it is stable and $x(t) \longrightarrow x_r$ for $t \longrightarrow$. Otherwise, the equilibrium point x_r is called unstable.

A.4 Control strategy design

A control strategy is an algorithm that takes decisions regarding the control actions to apply at each instant in time, i.e., it calculates the input values. Accordingly, the control strategy design involves developing a set of systematic approaches and methodologies to regulate the behaviour of a process.

The design task of a control strategy is an iterative process that may involve revisiting and refining the strategy based on performance evaluations and changes in the system or requirements. It requires a combination of theoretical knowledge, practical experience, and a thorough understanding of the specific process being controlled. For instance, a potential approach involves endeavouring to replicate, or model, the conduct of a human operator, whether real or hypothetical, following the expert system methodology. An alternative option is to strive for comprehension, or modelling, of the process behaviour, subsequently applying systematic methods within the realm of automatic control. These methods aim to formulate an appropriate control strategy.

Some common steps and considerations in control strategy design are the following:

- System analysis: analysis and understanding of the behaviour of the process; identification of key variables, parameters, and interactions that could affect the process.
- Definition of objectives: define the goals and objectives of the control system, so as to have a precise understanding of the desired outcomes.
- Modelling: a control strategy can be formulated based on a mathematical model, or it may explicitly incorporate a mathematical model as a means of dynamically assessing the effectiveness of specific control actions in real time. Hence, it is crucial to develop a mathematical model that represents the dynamics of the process, as accurately as possible.
- Controller selection: identification of the most appropriate control strategy and controller type based on the characteristics and objectives of the process.
- Tuning: fine-tune the controller parameters to achieve the desired process response. The task may involve simulation, analysis of stability criteria, and iterative adjustments to optimise the controller's performance.
- Robustness/Sensitivity analysis: understand how the performance of the process is affected by variations in parameters or external disturbances.
- Perform sensitivity analysis to understand how variations in parameters or disturbances impact the system's response.
- Testing and validation: test and validate the control strategy under various operating conditions and scenarios. This step ensures that the control strategy performs as intended and meets the specified objectives.

A.4.1 Open-loop and closed-loop control

Open loop control is a type of control system where the control action is applied to the process independently of the process output. In other words, the controller does not use feedback from the output to determine its actions but operates only based on the initial input or predetermined settings. The control action is based solely on the input or predefined settings.

Let us assume a process model of the type $y = f(u,d)$, being y, u, d the outputs, the inputs, and the disturbances vectors, while f is a generalised operator. Moreover, assume the availability of desired output values y_d. If the relation is invertible, it is possible to obtain $u_d = f^1(d,y_d)$, which corresponds to an open-loop control strategy as depicted in Figure A.4.

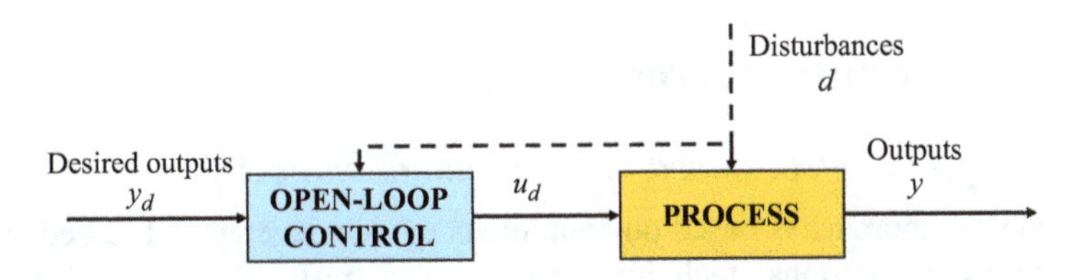

Figure A.4 Schematic representation of an open-loop control approach

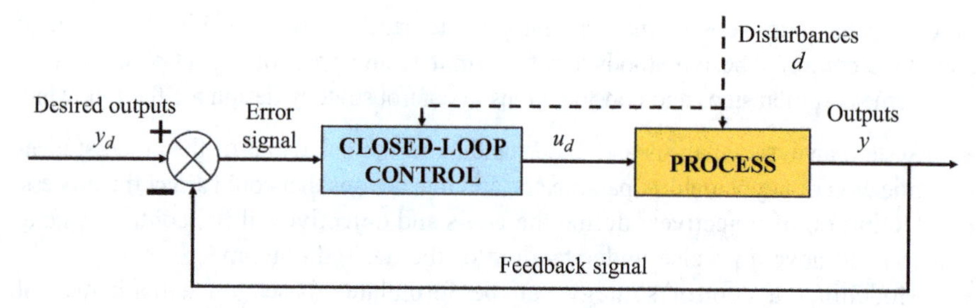

Figure A.5 Schematic representation of an closed-loop control approach

This control approach is generally simple to design, and if the process itself is stable then there is no risk of obtaining an unstable system. For these reasons, such a control scheme is suitable for situations where the relationship between input and output is well-known and does not change. However, the approach has some major disadvantages:

- Accuracy: since there is no feedback, open loop systems are not so accurate and can be affected by disturbances or changes in the system parameters;
- Inefficiency: May overdo or underdo the task;
- No error correction: Cannot correct any deviations from the desired output.

Because of these drawbacks, any automatic control system needs to include a closed-loop structure.

Closed loop control is a type of control system where the control action is dependent on the output of the system. The system uses feedback to compare the actual output with the desired output and make adjustments to minimise the error. This feedback loop allows the system to be more accurate and responsive to changes and disturbances. The control scheme can be represented in Figure A.5. The main characteristics of closed-loop control are:

- Feedback mechanism: the system continuously monitors the output and compares it to the desired setpoint. Any deviation from the setpoint is corrected by adjusting the input.
- Self-correcting: closed-loop systems can automatically correct errors by compensating for disturbances and changes in the system dynamics.
- Not simple design: require sensors, controllers, and actuators to implement the feedback mechanism.
- High accuracy: the feedback mechanism ensures accuracy and reliability

A.5 Traffic control approaches

A.5.1 Traffic control based on continuous models

Traffic control approaches based on continuous models leverage advanced mathematical and computational techniques to optimise traffic flow in real time by treating traffic as a fluid-like entity. These models, such as the Lighthill–Whitham–Richards (LWR) model, describe traffic dynamics through partial differential equations, capturing variations in traffic density, flow rate, and speed. Continuous traffic control strategies, including adaptive traffic signal control, ramp metering, and variable speed limits, rely on real-time data from sensors and predictive algorithms to dynamically adjust control measures. This real-time optimisation helps mitigate congestion, reduce travel times, and improve overall traffic efficiency. The related traffic control schemes rely on a robust infrastructure of sensors and data processing capabilities, enabling a real-time, proactive approach to traffic management that significantly reduces congestion and improves overall traffic efficiency. Despite the complexity and high data requirements, continuous models offer high scalability and predictive capabilities, making them crucial for modern, intelligent traffic management systems.

As stated in Siri *et al.* (2021), traffic control schemes falling into this category can be further divided into in-domain and boundary control. The first category refers to control inputs that are continuous both in time and space over the whole road stretch or network; the latter, instead, refers to control inputs acting at the boundaries of the considered road area.

A.5.1.1 Feedback control

Continuous time feedback control is a control strategy where the system continuously monitors its output in real time and adjusts its input based on the feedback to achieve the desired performance. The main idea is to minimise the difference between the actual output and the desired setpoint by using a feedback loop that operates continuously over time. This approach enhances the efficiency, safety, and reliability of transportation systems by dynamically responding to traffic conditions.

Most control laws depend on measurements of the state of a portion of the system and integrate the Lighthill–Whitham–Richards (LWR) model. For instance, let us consider a first-order model of the following form:

$$\partial Q(x,t)/\partial t + \partial/\partial x(u(x,t)f(Q(x,t))) = 0 \tag{A.7}$$

where $u(x,t)$ is the control input.

If one employs a variable speed limit on a road stretch as a control action, then the model becomes:

$$\partial Q(x,t)/\partial t + \partial/\partial x(F(Q(x,t)l(x,t))) = 0 \tag{A.8}$$

Being $l(x,t) \in (0, 1]$ the ratio between the imposed speed limit and the maximum speed; $F(Q,l)$ is a nonlinear function defining the Fundamental Diagram in which the variable speed limit is integrated.

A.5.1.2 Backstepping control

Backstepping control is an advanced control strategy used to deal with nonlinear systems. It systematically constructs a stabilizing controller by breaking down a complex nonlinear system into smaller subsystems and designing control laws for each step, ensuring stability at every stage. In traffic engineering, backstepping control can be applied to various scenarios, such as adaptive traffic signal control, ramp metering, and vehicle platooning.

An example of backstepping control is in Yu and Krstic (2018), where a downstream ramp metering (named DORM) and an upstream ramp metering (named UORM) control strategies were proposed. The DORM strategy consists of a boundary feedback law, while the UORM consists of a full-state feedback boundary control. The control law is obtained via a backstepping transformation.

A.5.1.3 Optimal PDE control

Optimal control of Partial Differential Equations (PDEs) involves the use of mathematical techniques to derive control strategies that optimise the performance of traffic systems modelled by PDEs. These models typically describe the evolution of traffic density, flow, and speed over time and space. The goal is to minimise congestion, travel time, or other performance criteria while ensuring smooth and safe traffic flow. The key components are: the macroscopic traffic flow model, objective function, and control input.

Then, to formulate an optimal control problem, the following steps are required:

- Identification of the state equation. This allows us to define the main variables of the macroscopic traffic flow model (e.g., LWR model) employed to emulate the behaviour of a road stretch or network. It is important to correctly define the state variables,
- Definition of control constraints, i.e., the physical and operational constraints on the control inputs and input variables (e.g., $umin \leq u(x,t) \leq umax$),
- Optimisation problem. This step requires the definition of the objective function to minimise (or to maximise) subject to the state equation and constraints.

As an example, Delle Monache *et al.* (2017) proposed an optimal controller for flow regulation via a variable speed limit. The considered single road is represented with the LWR model, and the optimal problem aims to minimise the quadratic difference between the achieved outflow and the given target flow, considering that the state of the system only depends on free-flow conditions.

A.5.2 Traffic control based on discrete models

Discrete-time control strategies involve making decisions at specific time intervals based on traffic data collected at those intervals. These strategies can be applied to various aspects of traffic management, such as signal control, ramp metering, and dynamic speed limits. The control decisions are updated periodically rather than continuously, which can simplify implementation and reduce computational requirements.

The key common components of discrete-time control strategies are:

- Data collection: Traffic data is collected at regular intervals (e.g., every minute, every 5 min).
- Control intervals: The system updates control actions at discrete time steps (e.g., $t = 0$, T, 2T, ...)

- Control laws: specific algorithms that determine control actions based on the current state of the system and desired objectives.

A.5.2.1 Feedback control

Discrete-time feedback control involves updating traffic control strategies at specific time intervals based on the real-time feedback of the traffic state. This approach is particularly useful for adaptive traffic signal control, ramp metering, and variable speed limits, where timely adjustments can significantly enhance traffic flow and reduce congestion.

A general formulation for this type of controller is presented as follows:

$$g_{k+1} = g_k + K_p(e_k) + K_i \sum_{i=0}^{k} e_i + K_d(e_k - e_{k-1})$$

where k is the discrete time step, e_k is the error at time step k, K_p K_i, and K_d are proportional, integral, and derivative gains, respectively. The error e_k is the difference between the actual value of a state variable and the desired one. Most common controllers employ solely proportional or proportional-integral actions.

This kind of controller can be extended to take into account the state variables measured in a larger portion of the road network. This is useful since in some cases the control action does not depend only on the traffic state of a portion of the road network but on a wider area.

A.5.2.2 Optimal control

Optimal control methods aim to optimise a certain performance criterion, taking into account the evolution of the dynamical system over a given time horizon. The choice of inputs u is usually limited due to physical or technical constraints that define an admissible control region via a set of inequalities This leads to optimal (or suboptimal) solutions according to a predefined objective function.

A general formulation for optimal controller is presented in the following:

Given the initial conditions $x(0) = x0$ of the system and the estimated sequence of disturbance inputs $d(k)$, $k = 0, \ldots, K\text{-}1$, find the optimal control sequence $u(k)$, $k = 0, \ldots, K\text{-}1$ that minimises the criterion J.

$$J = \vartheta[\underline{x}(K)] + \sum_{k=0}^{K-1} \varphi[\underline{x}(k), \underline{u}(k), \underline{d}(k)]$$

subject to the system dynamics and constraints (e.g., control variable constraints $u_{min} \leq u(k) \leq u_{max}$, $k = 0, \ldots, K\text{-}1$.

The solution to the problem is the optimal state trajectory $x^{\circ}(k)$, $k = 1, \ldots, K$ and the optimal control inputs $u^{\circ}(k)$, $k = 0, \ldots, K\text{-}1$ over the considered time horizon.

In the case of urban traffic, this criterion typically corresponds to the minimisation of total travel time in the considered network.

A.5.2.3 Model predictive control

Model Predictive Control (MPC) approaches allow to control a dynamic system in real-time on the basis of the prediction of its evolution over a finite horizon, considering an objective function to optimise. This is achieved by iteratively solving a Finite-Horizon Optimal Control Problem that is updated on the basis of real system measurements and can account for different types of constraints.

Considering a generic MPC scheme in a discrete-time framework, the problem to be solved at each time step k, $k = 0, \ldots, K$, over a given prediction horizon of K_p time steps can be stated as:

Given the initial conditions on the system state $x(k)$ and the estimated sequence of exogenous inputs $d(h)$, $h = k, \ldots, k+K_p$-1, find the optimal control sequence $u(h)$, $h = k, \ldots, k+K_p$ -1 that minimizes:

$$J(k) = \vartheta[\underline{x}(k + K_p)] + \sum_{h=k}^{k+k_{p-1}} \varphi[\underline{x}(h),\ \underline{u}(h), \underline{d}(h)]$$

subject to

$$x(h+1) = \underline{f}[\underline{x}(h),\ \underline{u}(h),\ \underline{d}(h)]\ h = k, \ldots, k + K_p - 1$$

$$u^{\min} \leq \underline{u}(h) \leq u^{\max}\ h = k, \ldots, k + k_p - 1$$

And further possible additional constraints on the state variables $x(h)$, $h = k+1, \ldots, k+K_p$, and on the control variables $u(h)$, $h = k+1, \ldots, k+K_p$.

The solution to the problem is the optimal state trajectory $x^{\circ}(h|k)$, $h = k+1, \ldots, k+K_p$ and the optimal control input sequence $u^{\circ}(h|k)$, $h = k, \ldots, k + K_p$-1 over the considered prediction horizon.

According to the receding-horizon framework, only the first z elements (with $z < h$) of the optimal control sequence are actually implemented in the system.

A.6 Summary

This appendix explores the core elements of automation control, with a special focus on its use in traffic management. As the complexity of contemporary transportation networks increases, the incorporation of automation control is becoming essential for improving efficiency, safety, and sustainability.

This section aims to equip readers with a comprehensive understanding of automation control principles, specifically adapted to address the unique requirements and challenges of traffic systems.

References

Some suggestions for further readings are reported below among the huge literature available on these topics.

Delle Monache, M. L., Piccoli, B., and Rossi, F. (2017). Traffic regulation via controlled speed limit. *SIAM Journal on Control and Optimization*, 55(5), 2936–2958.

Papageorgiou, M. (1998). Automatic control methods in traffic and transportation. In *Operations Research and Decision Aid Methodologies in Traffic and Transportation Management* (pp. 46–83). Berlin, Heidelberg: Springer.

Siri, S., Pasquale, C., Sacone, S., and Ferrara, A. (2021). Freeway traffic control: A survey. *Automatica*, 130, 109655.

Yu, H., and Krstic, M. (2018). Traffic congestion control on Aw-Rascle-Zhang model: Full-state feedback. In *2018 Annual American Control Conference (ACC)* (pp. 943–948). IEEE.

Postface

Roberta Di Pace[1]

Steering a big ship requires patience and gradual adjustments.

Joan Walker – Closing Ceremony of *17th International Conference on Travel Behavior Research (IATBR)*, July 14–18, 2024, Vienna, Austria.

Outline. *This chapter presents some considerations about the content of this book.*

The book is designed to equip readers with all the necessary tools for comprehension, making it accessible even to those without prior knowledge.

It is divided into four parts and addresses three main areas: traffic flow theory (Part I), traffic control (Part II), Intelligent Transportation Systems (Part III), and the evaluation of impacts and externalities (Part IV).

Each chapter is outlined below:

Part I: Traffic Flow Theory

Chapter 1: It provides a review of common basic definitions and notations on the part topic

Chapter 2: It provides a review of traffic flow steady-state models

Chapter 3: It provides a review of traffic flow dynamic models

Part II: Traffic Control

Chapter 4: It provides a review of common basic definitions and variables on the part topic

Chapter 5: It provides a review of priority junction analysis

Chapter 6: It provides a review of roundabout analysis

Chapter 7: It provides a review of signalised junction analysis

Chapter 8: It provides a review of signalised junction steady-state control (static control) [including Appendix A about the definition of some basic elements]

Chapter 9: It provides a review of signalised junction dynamic control (dynamic control)

Part III: Intelligent Transportation Systems

Chapter 10: It provides a review of advanced traffic management systems

Chapter 11: It provides a review of advanced traveller information systems

[1]Department of Civil Engineering, University of Salerno, Italy

Chapter 12: It provides a review of advanced driving assistance systems
Chapter 13: It provides a review of cooperative connected and automated mobility ecosystem
Part IV: Impacts
Chapter 14: It examines externalities in traffic management strategy design
Chapter 15: It provides a review of safety models
Chapter 16: It provides a review of noise models
Chapter 17: It provides a review of emissions and energy consumptions models

The book examines the principles of fixed and adaptive urban signal design at both individual junctions and the network level. Additionally, it discusses the latest developments in Intelligent Transportation Systems, particularly the influence of connected and cooperative vehicles on traffic analysis and control. Finally, the it presents essential strategies aimed at optimising factors such as safety, fuel consumption and emissions, and noise. It primarily emphasises theoretical concepts and does not extensively cover practical or numerical aspects. The authors plan to publish a second volume in this series, which will concentrate specifically on operational topics.

P.1 Brief history of this book

I started working in the field of Intelligent Transportation Systems (ITS) development 20 years ago with my master's thesis, which focused on traffic flow modelling and control, particularly studying and implementing ramp-metering strategies. I was in Naples, where I began studying and working with Cino Bifulco, whom I always credit for sparking my passion for the Theory of Transportation Systems. Later, I deepened my exploration of modelling choice behaviours in the presence of Advanced Traveler Information Systems (ATIS) during my PhD thesis, which was partially developed at TU Delft University in the Netherlands.

Subsequently, around 2010, I had the opportunity to continue my academic career at the Department of Civil Engineering at the University of Salerno, where I began working with Professors Giulio E. Cantarella and Stefano de Luca. Collaborating with them presented a new challenge and a tremendous opportunity for growth, as their expertise and perspectives were completely different from what I was accustomed to in the past. Moreover, working in Salerno fulfilled my desire to serve the territory from which I came and where I grew up.

In Salerno, at the beginning, it was just the three of us, united by a single desire to grow together. During these years of 'shoveling snow', we never explicitly stated it, but I am sure that this shared thought has always been on our minds. It has been and continues to be, a challenging yet rewarding personal and professional journey, during which we have also been able to think about the realisation of this work. This book contains various contributions that reflect how much we have grown as a group and in our collaborative relationships over these 15 years together.

Additionally, I was a temporary researcher for several years, and at the end of 2018, I became a tenure-track researcher and began teaching some of the topics covered in this book in the master's program. In 2021, I finally became an associate professor. It has been a 'long journey' during which we have collectively developed a broader cultural project: the master's degree in transportation engineering and sustainable mobility. This experience highlighted the need to share and convey to our students the topics of traffic flow modelling, management, and control. Thus, this book is conceived as a means of transferring skills and knowledge primarily to our students.

P.2 Final memories and comment

Referring to Joan Walker's phrase, in relation to my big ship, below, I share two photos that are memorable to me and represent my personal growth journey: the first is from my childhood school in my hometown—what it was and what I would have liked it to be, with my eyes today—to remind me that education is everything, from the first day of kindergarten to the last day of university; the second photo is with Professor Markos Papageorgiou from the course on Dynamic Traffic Flow Modeling and Control, which I had the opportunity to attend in Chania in October 2012. Several colleagues are also in the photo.

My childhood school in my hometown, Agnone Cilento, a small town in southern Italy (the photo was taken by Bruno Di Pace)

Prof. M. Papageorgiou – Course on Dynamic Traffic Flow Modeling and Control, Chania in October 2012

Index

www.ingramcontent.com/pod-product-compliance
Lightning Source LLC
LaVergne TN
LVHW020245020825
817679LV00003B/163

* 9 7 8 1 8 3 9 5 3 7 1 6 5 *